IMAGES
of America

SOUTHWEST MISSOURI MINING

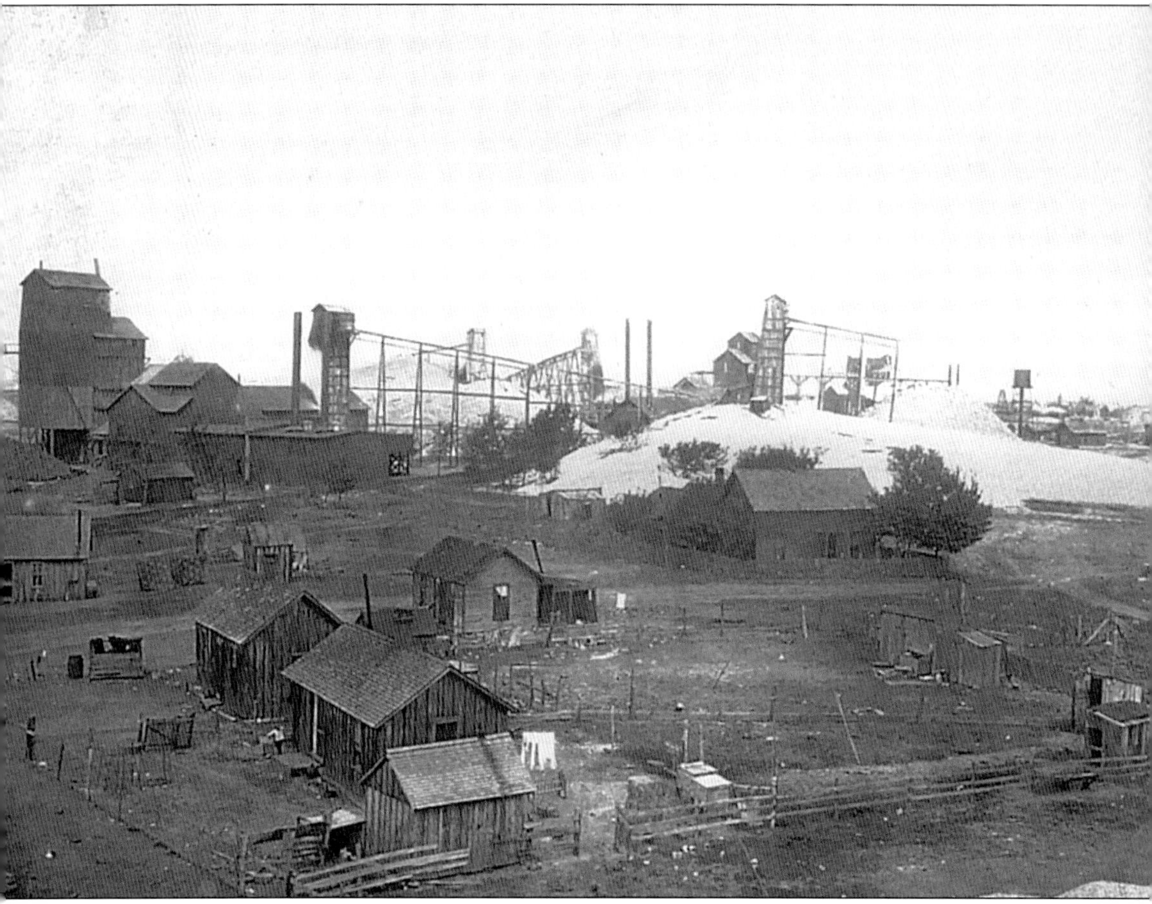

Sometimes little more than shanties, the early miners' homes were built adjacent to the mines where they worked. This photograph, shot from a tailing pile, may have been taken on a Monday, which was the traditional wash day of the era. At one corner of the house, in the foreground, a young woman bends over a washboard. At another corner, a beautiful rocking cradle and a wagon or goat cart sit beneath clothes hung to dry.

The lives of the women were little easier than those of the men who labored underground. It was said that the only diamonds a miner's wife ever had glistened in her eyes when the hardships and disappointments broke her heart. (Unless otherwise indicated, all photos Courtesy Ron Mosbaugh.)

IMAGES
of America

SOUTHWEST MISSOURI MINING

Jerry Pryor

ARCADIA
PUBLISHING

ISBN 978-0-7385-07664-4

Published by Arcadia Publishing
Charleston, South Carolina

Printed in the United States of America

Library of Congress Catalog Card Number: 00105320

For all general information contact Arcadia Publishing at:
Telephone 843-853-2070
Fax 843-853-0044
E-mail sales@arcadiapublishing.com
For customer service and orders:
Toll-Free 1-888-313-2665

Visit us on the Internet at www.arcadiapublishing.com

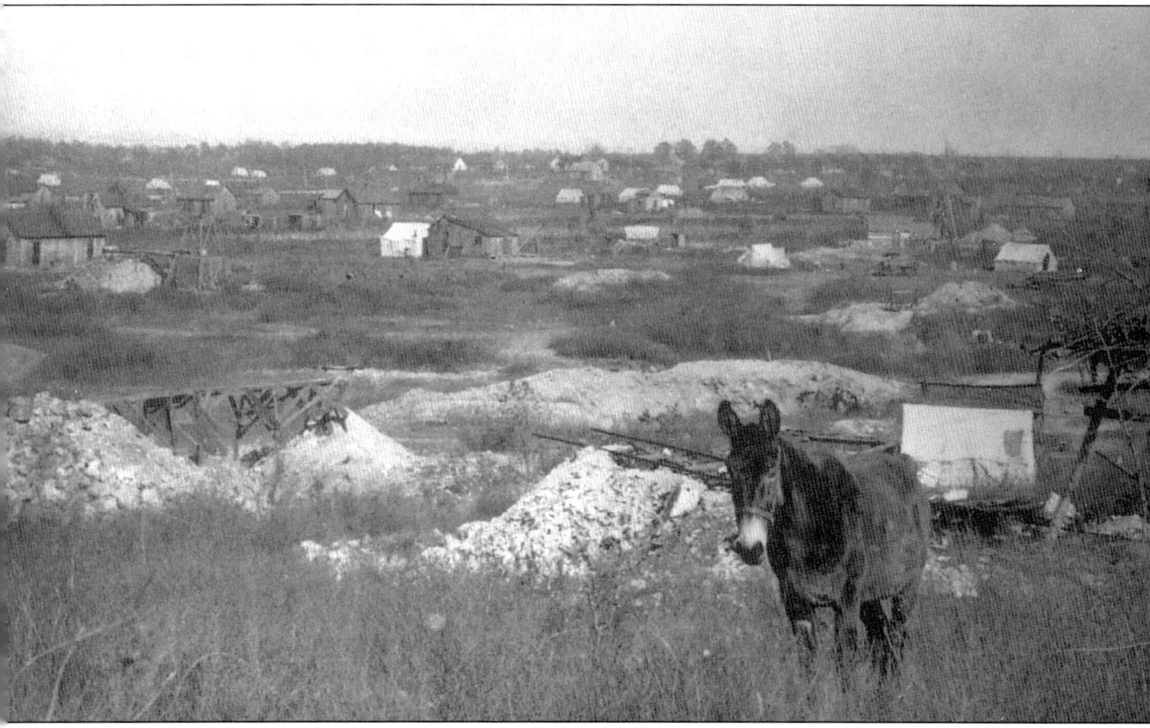

This stolid beast, standing impassively before a landscape scarred by prospecting gouges and the primitive habitations of the early miners, exemplifies two qualities requisite for the first prospectors—patience and fortitude. The use of mules remained an integral part of the underground operations in some mines for decades.

Contents

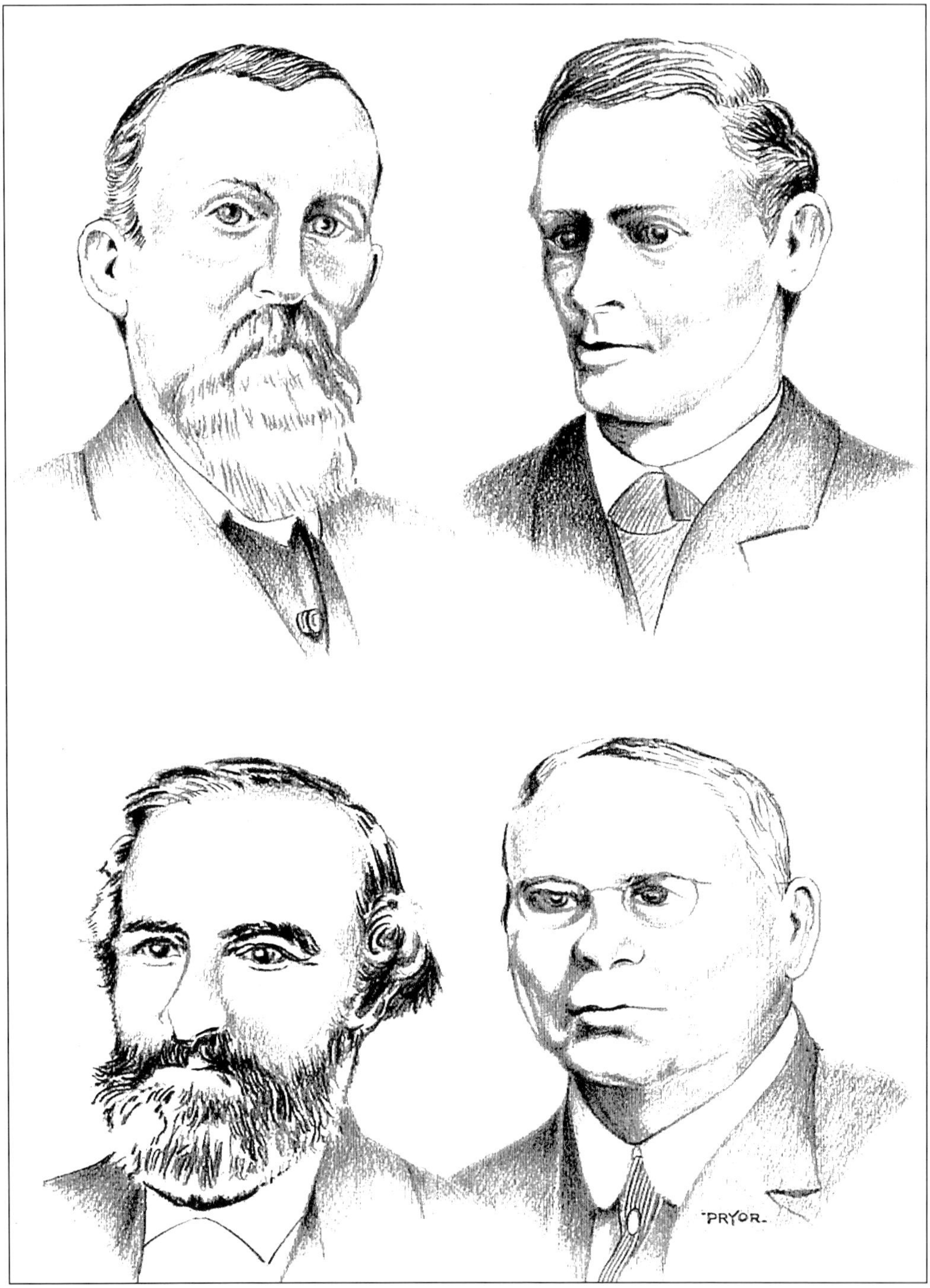

Four of the early pioneers whose names are linked with the history of the region are pictured above. Clockwise from the upper left, they are: John C. Webb, Rev. Harris G. Joplin, Alfred H. Rogers, and John C. Cox. (Sketches by author.)

INTRODUCTION

In the country west of the Mississippi where the mines, towns, and cities featured in this book would grow and prosper, the first explorers found an imposing tribe of native Americans. They were a tall, handsome people who were later called the Osage. They were destined first for poverty, and then to become the wealthiest tribe in the nation—and the chief target of a horde of con men and manipulators—after oil was discovered on their land in Oklahoma. When early settlers in the region found signs of excavation, they were unwilling to credit the digging to the natives, and legends developed that it had been done by de Soto's men seeking ore on his expedition out of Florida from 1539–1542.

Certainly the settlers were aware that the earth contained ore. Primitive mines existed before the Civil War, and the city of Granby in Newton County calls itself "the oldest mining town in the Southwest—1850." But most people were more interested in farming the rich soil than grubbing in the ground for ore-bearing rocks. Still there were those who endured, and every city or town with roots in early mining seems to have its own intriguing story about how the ore, or shines, which triggered its growth, was found.

In 1850, Madison Vickery was digging a well on his property at Granby and discovered a heavy piece of mineral unlike any he had previously seen. William Foster, who was traveling to St. Louis, had stopped to visit in the area and immediately recognized it as ore. He discontinued his journey, sank a shaft, and that was the beginning of organized mining at Granby.

Before that, early wanderers like Daniel Boone, "hearing of the game and Indians in Misoura [sic]," had "pilled up stakes" and gone there. Other men, also infected with wanderlust, heard the tales and were enticed to see for themselves. One such man was John C. Cox, who came with his wife in 1838, built a cabin, and joined the few other settlers already there. The popular story, perhaps true, persists that it was a slave belonging to Cox who later unearthed the first ore on his land while digging for fishing worms.

Apparently no one realized just how important the discovery was—whoever made it—and the region was soon embroiled in the Civil War anyway. Bushwhackers burned Cox's house, and he fled to Newton County for refuge. During the war there was much activity across the region by guerrillas from both sides, as well as by the Confederate and Union soldiers who passed through. The irregulars engaged in wanton killing of the inhabitants to the point that most people moved away, not to return until the hostilities ceased. The Jasper County Courthouse at Carthage was burned, and the town was virtually abandoned for a time. It is estimated that by the end of the war as few as 30 civilians remained in all of Jasper County.

After the war, the Granby Lead and Zinc Company began operations once more—both at Granby and at new locations in Jasper County. One of these was known as Minersville, which later would become Oronogo. In 1870, E.R. Moffett and J.B. Sergeant, using prize money they had won by removing the most ore from a single shaft in a four-month period there, bought 10 acres of land in the Joplin Creek valley from John Cox. Though the land was promising, they had little success at first. Broke and discouraged, near failure, they set off one last blast that uncovered a significant ore deposit and was the spark that ignited the boom in the Joplin area.

John Cox saw the possibilities and platted a town to be called Joplin east of the mines. The new town, as well as the creek, bore the name of Rev. Harris G. Joplin, a Methodist Episcopal minister who had been gone from the area for almost three decades and dead for almost two. Patrick Murphy, an Irishman from Carthage, came over and platted a town named Murphysburg on the west side of the creek. The two towns grew and competed in a sometimes lawless and violent atmosphere until they were finally combined in 1873 under a charter issued by the Missouri General Assembly.

In that same year, John C. Webb was plowing one of his fields a few miles northeast of Joplin. He broke his plow on a large rock that was revealed to be almost pure ore. With the help of a neighbor, W.A. Daugherty, Webb did a little mining but soon gave it up. He leased mining sites, laid out Webbville—which became Webb City—and dabbled in land sales. Royalties and real estate made him wealthy.

G.P. Ashcraft, who had formerly prospected in Colorado, took over Webb's old mine, and with the help of Ben F. Hatcher, overcame flooding problems and made it and other sites immensely profitable, thereby starting a boom which rivaled the one in Joplin.

W.A. Daugherty, perhaps inspired by his neighbor John C. Webb's success with Webb City, bought land from J.G.L. (Leroy) Carter of the pioneer Carter family that had settled earlier on Center Creek near what is now Lakeside. Daugherty platted Carterville, and it soon became another boom town lying to the east of Webb City, separated by Ben's Branch and one of the richest belts of mining land in the area. Unlike their neighbors to the west, however, Webb City and Carterville never combined into one city.

In 1893, Alfred H. Rogers, who had recently moved to the area from Springfield, abandoned his mule-powered trolley line between Webb City and Carterville and organized the Southwest Missouri Electric Railway Company with the offices, car barn, maintenance facilities, and powerhouse all located in Webb City. The service grew to encompass the entire mining district, becoming the largest intercity system west of the Mississippi.

Lakeside Park, also developed by A.H. Rogers, was an offspring of the Electric Railway, which transported its visitors. They both survived until the automobile and the lure of the open road led to their demise.

In addition to a recitation of historical events, there is a human aspect that sometimes gets lost in the telling—the individuals who struggled under difficult and frightening conditions to support their families. At the turn of the century, one such man, Samuel David Robb, moved his family from Pennsylvania, where he had been a foreman on the railroad, to the Tri-State Mining District. He sought a better life for his family. From time to time how they fared will be mentioned—not because that family's experiences are more noteworthy, but because Sam Robb was the author's grandfather and more information is readily available.

As the mines declined and closed, the area experienced difficult times, just like the boom-and-bust reminiscent of other regions that lose their economic underpinning. More recently that has been reversed. Most localities are now prosperous and growing rapidly. Suburban housing developments spread across land once devoted to mining. The scars are disappearing, as have the men who, sitting on shaded benches on summer afternoons, told many of the stories recounted here.

One

Mines and Mining

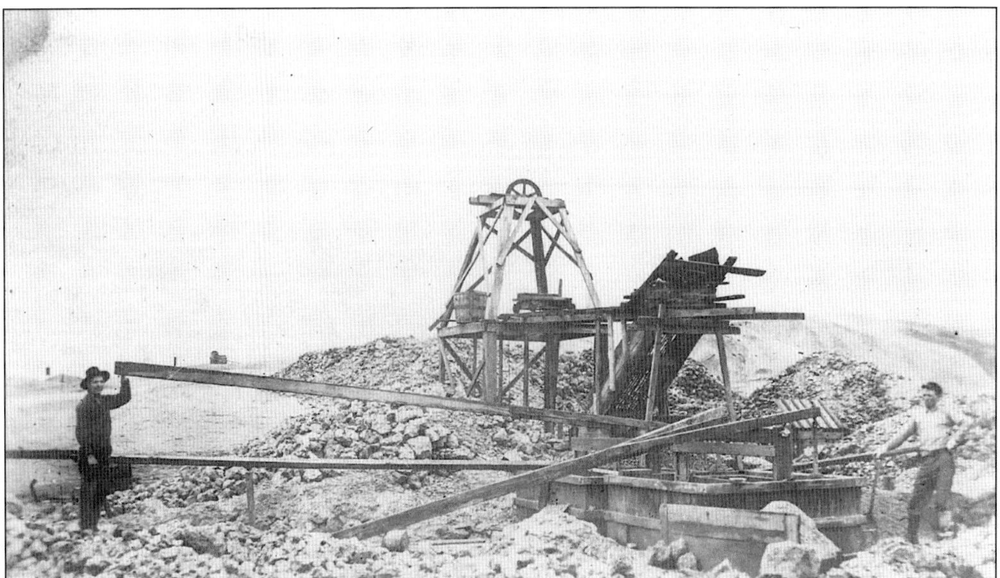

Typical of the first efforts of those who staked everything on the chance of striking a fortune underground, two men struggle to make a primitive mine a paying proposition. One wonders if the initial labors of John Webb and W.A. Daugherty, or E.R. Moffett and J.B. Sergeant did not produce scenes reminiscent of this one.

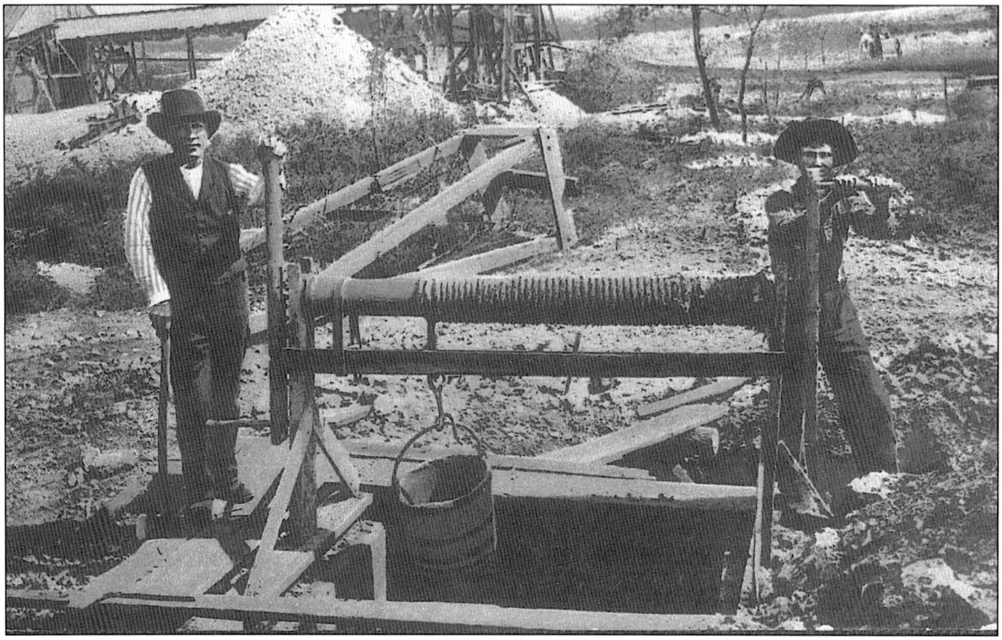

The human hoister mine (pronounced HIGH-ster by the miners) was about as basic as mining got. Every shovel of dirt and rock (ore-bearing if they were lucky) was pulled from the ground by hand. It was a poor man's field, because even a poor man could negotiate a lease agreement and, with hard work, become wealthy.

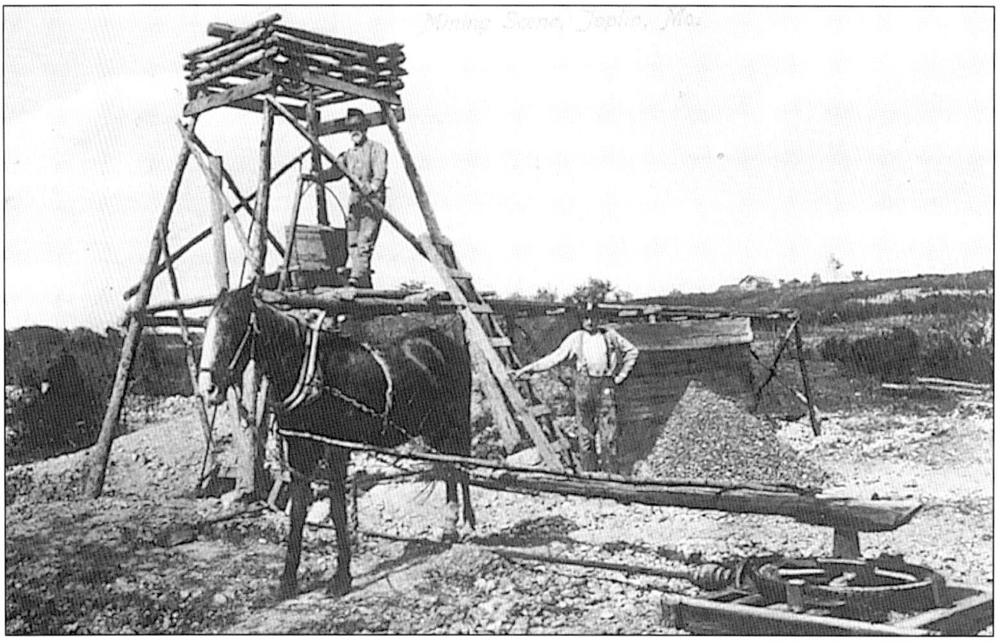

Things have begun to improve dramatically on the mining scene. Someone has replaced the human hoisterman (HIGH-ster-man) with a horse, an innovation calculated to increase efficiency and to moderate physical labor. At first, a whip hoister was installed with the horse pulling directly away from the shaft using a simple pulley. Later it was replaced by this circular-geared device.

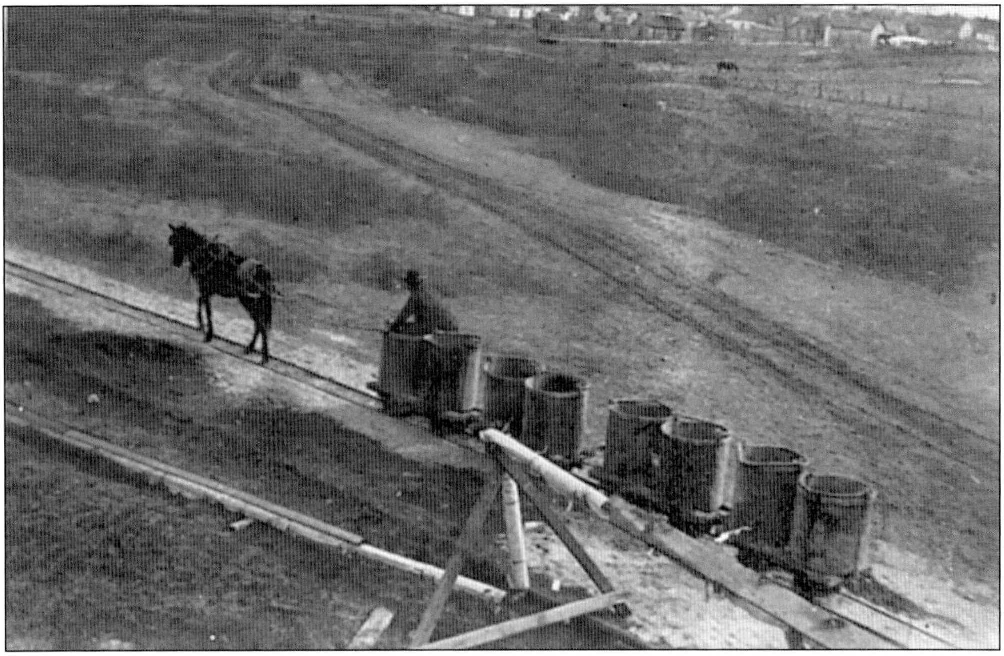

This "train" of eight tubs (also called cans or buckets) being pulled by a horse is indicative of the valued role horses and mules played in the mines. In 1973, during Joplin's centennial celebration, some of the small rails on which the ground or rock cars rolled were cut into short sections and sold as souvenir paperweights.

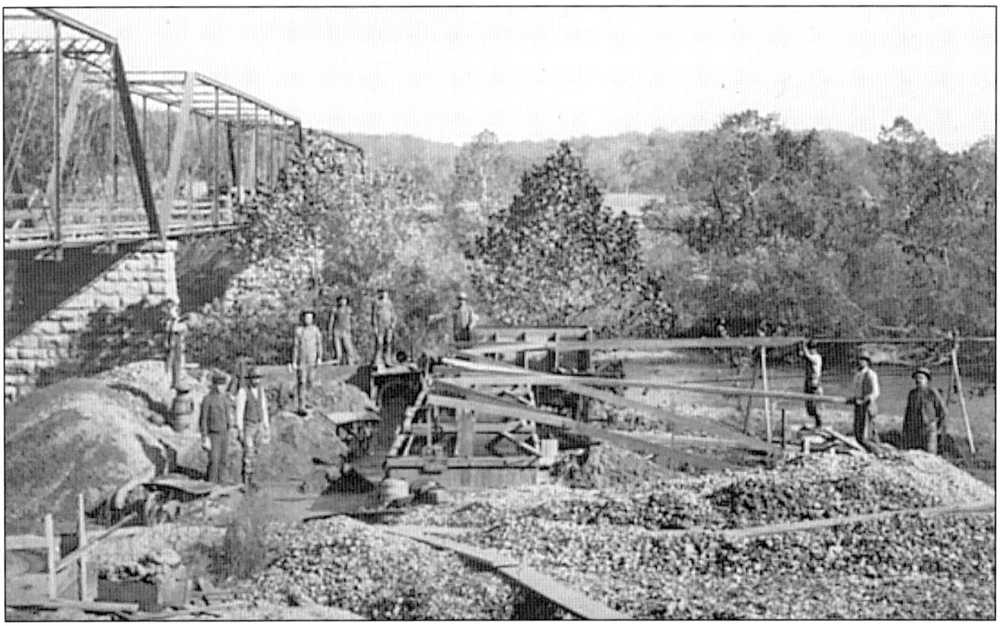

A complex of five hand jigs has been set up on the bank of a creek. The hand jig was an early method of separating the ore from the waste. Men shook the submerged material in a box with the long lever-like handles, causing the heavier ore to settle out. A proficient jig man might average as much as 1 ton of ore per hour. The process was later replaced by steam-powered jigs in the mills.

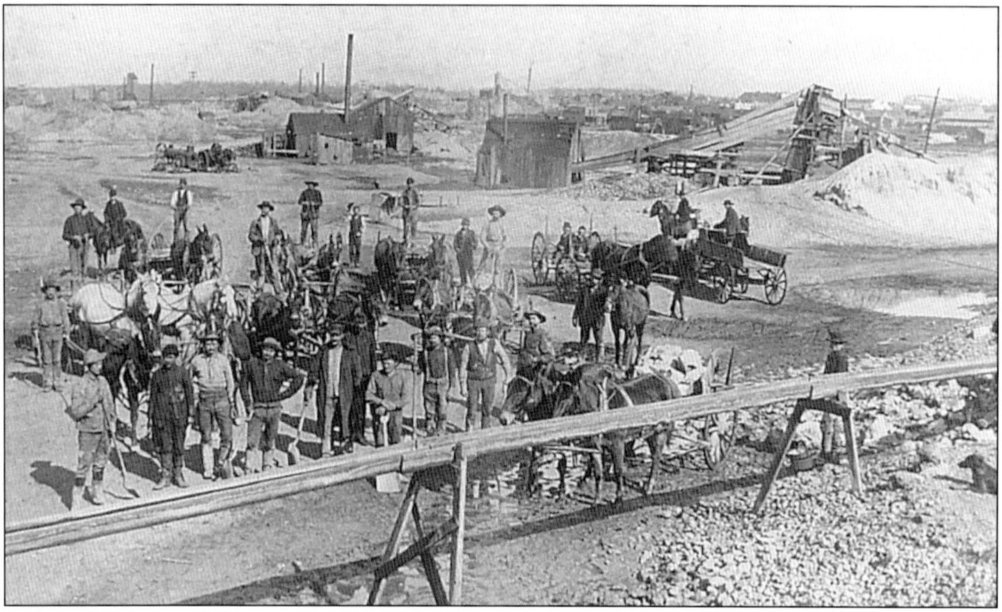

Before the dawn of the twentieth century, the Sucker Flat area of Webb City was a warren of prospect "diggings" and ramshackle mine structures. It was clearly a promising location. Notice the steam engine in the background to the left of the photo. Steam would become the chief source of power for the mines and mills. (Courtesy of Webb City Historical Society.)

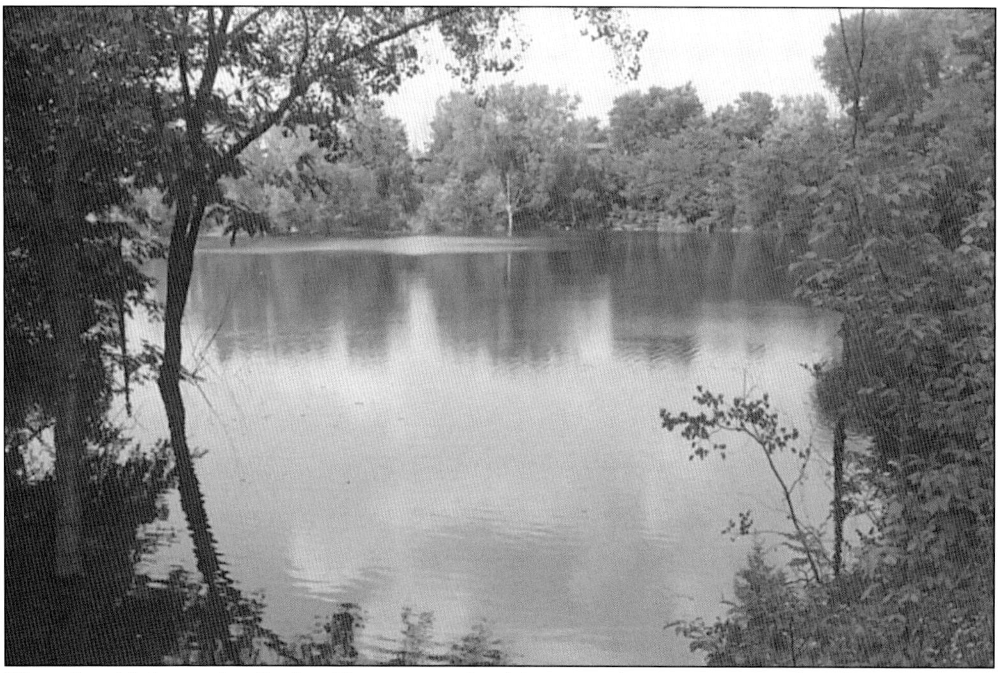

This placid lake in the Sucker Flat mining land has now become a part of Webb City's King Jack Park. Hidden in the trees atop the cave wall in the background is the reconstructed Prosperity Junction Station, which was originally located at the east edge of Carterville as part of the Southwest Missouri Electric Railway system. The building is now occupied by the Webb City Chamber of Commerce. (Photo by author.)

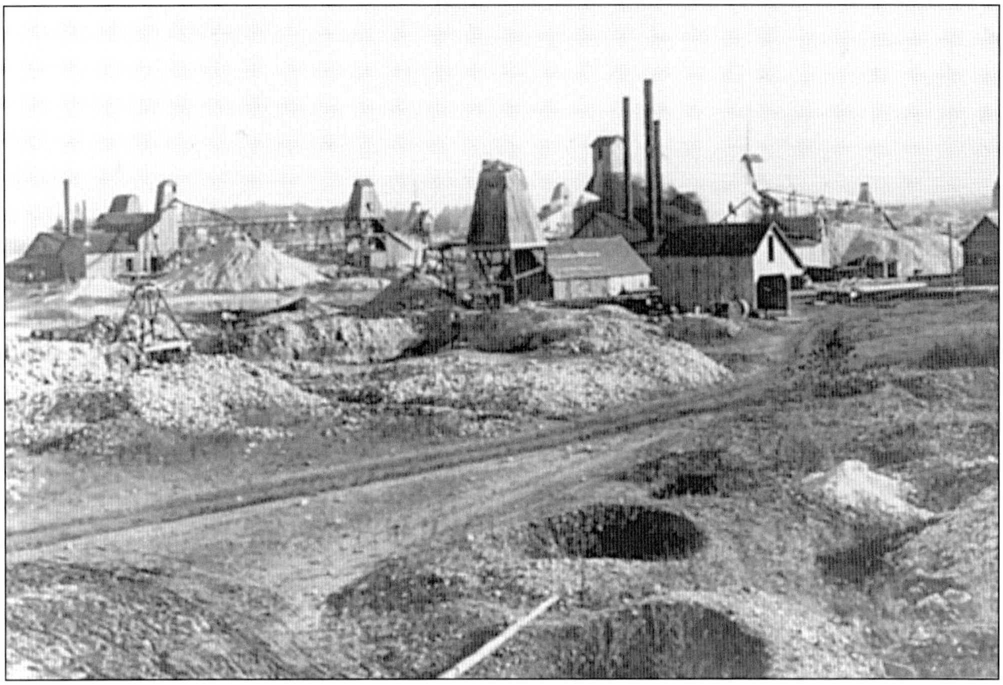

Most people in the region would now be unlikely to think of Aurora as an early mining town, but this photo shows otherwise. A 1902–1903 nine-month tabulation of mining production ranks the top producers as follows: Webb City-Carterville, Joplin, Galena-Empire, and Aurora. Aurora's output of zinc was nearly 39 million pounds and nearly 1 million pounds of lead.

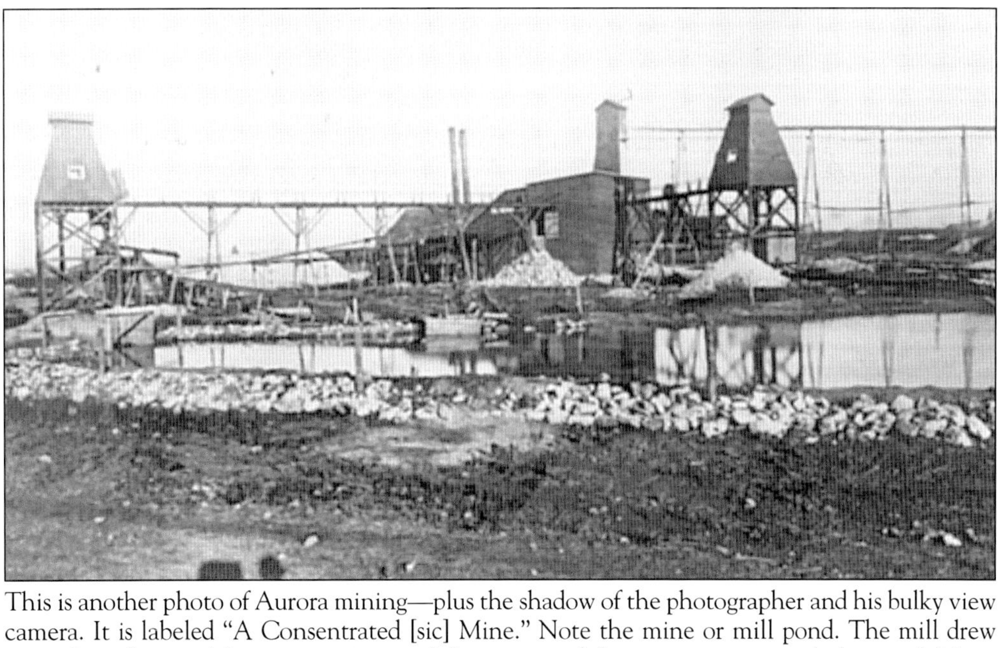

This is another photo of Aurora mining—plus the shadow of the photographer and his bulky view camera. It is labeled "A Consentrated [sic] Mine." Note the mine or mill pond. The mill drew water from the pond for its operation, and then returned the waste water to a sludge pond. Many mines had a more than adequate supply of water, as it was pumped from underground to avoid flooding. These ponds and their runoff were sources of ground and stream contamination.

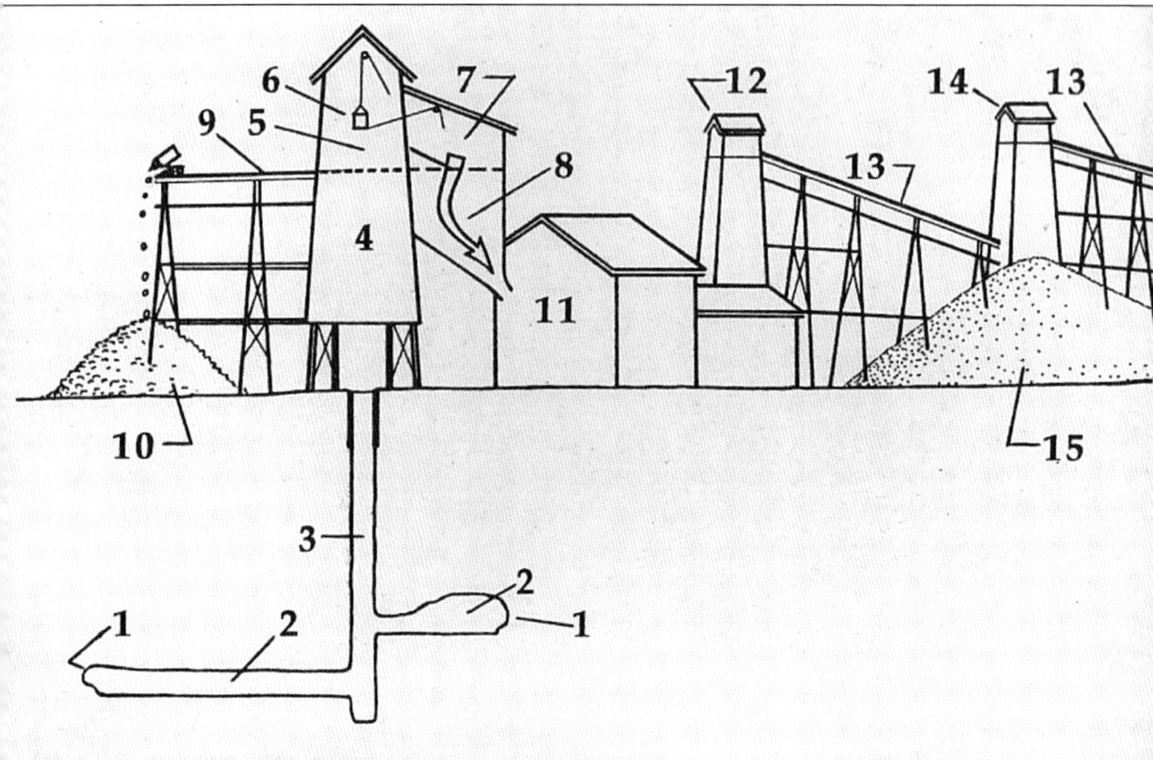

This diagram identifies various elements of the mine with a brief description of the mining and milling operation: 1. The ore or mine face where the actual drilling, blasting, and removal of material occurred. 2. The drift or tunnel along which the excavated material was moved to the shaft to be raised to the surface. 3. The shaft that was sunk into the ground to intercept the vein or pocket of ore. It was cribbed with timbers in the soft earth at the top. 4. The derrick, an enclosed timber structure built above the shaft and supported by spilings. 5. The hoister room where the hoisterman and his equipment were located. 6. The tub, can, or bucket used to pull the excavated material from the mine and to transport the miners and their tools. 7. The screen room into which the filled tub was drawn by a tail chain, dumping its contents onto a screen, often called the grizzlies (grizzly bars). Men called "screen apes" used sledgehammers to reduce the rocks to chunks small enough to pass the screen. 8. The hopper through which the ore-bearing rock moved out of the screen room. 9. A tramway along which the waste (non-ore-bearing rock) from the screen room was pushed by hand in rock cars along an open track, a particularly unpleasant job in inclement or windy weather. 10. The waste rock pile. 11. The mill where the ore-bearing rock was crushed and where the ore was separated from the waste using powered jigs, the "fines" being further separated with flotation using water and chemicals. (Froth flotation to capture zinc fines was developed in the United States at Mascot, Tennessee, and at Carterville.) The resulting concentrate from the mill went to smelters where large furnaces using heat and chemicals removed the impurities. The molten lead was cooled and shaped in molds for shipment. The smelting of zinc was a more difficult and costly process than was the smelting of lead. 12. An elevator raised the waste from the milling process to be removed for disposal. 13. Using water and gravity, flumes carried the waste away from the mill. 14. Dummy elevators transported the tailings to greater distances as the initial pile grew too large. The Davey tailing pile in the Johnstown area of Carterville was considered at one time the largest in the world. It is said to have had five dummy elevators. 15. The tailing (sometimes called chat) pile.

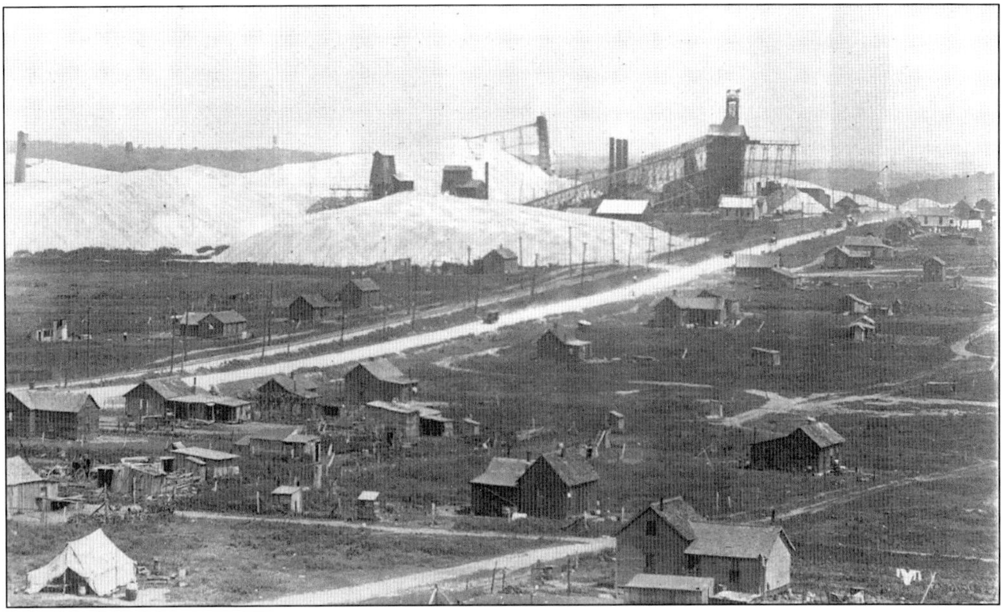

The Underwriter's Mine, better known as the Yellow Dog, was located between Webb City and Oronogo, and was at one time considered the largest zinc mine in the district. Once again, how close the houses were located to the mines is apparent.

According to the Webb City Genealogical Society's publication "Mine Accidents and Deaths, Volume 1," in May of 1892, a mother whose home had been built near the Troup Mine at Prosperity heard the chilling sound of the mine whistle. She quickly wrapped her three-month old baby and went to check on the safety of her husband. When she returned home, she found she had smothered the baby. It was said to be the fourth death at the Troup cave-in that day.

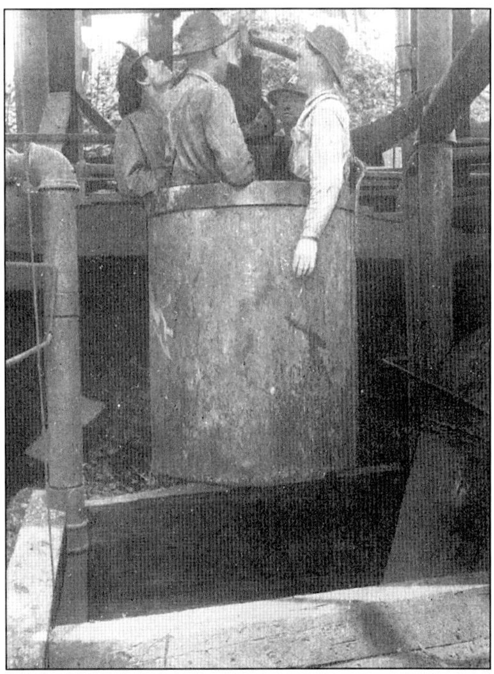

These miners are being lowered into the shaft in a tub. In shallower tubs, they sometimes rode with only one leg inside, the other dangling over the rim. This led to injuries if the tub happened to swing, scraping their legs against the sides of the shaft.

In May of 1890, two men riding in that manner were severely injured when the gearing failed, as recorded by the Webb City Genealogical Society. Their legs were crushed and mutilated in the fall. In another incident, a bystander in the hoister room became an instant hero when an ice-covered brake failed to hold. He grabbed a bar and jammed it against the spool, thereby saving the miners from serious injury or death.

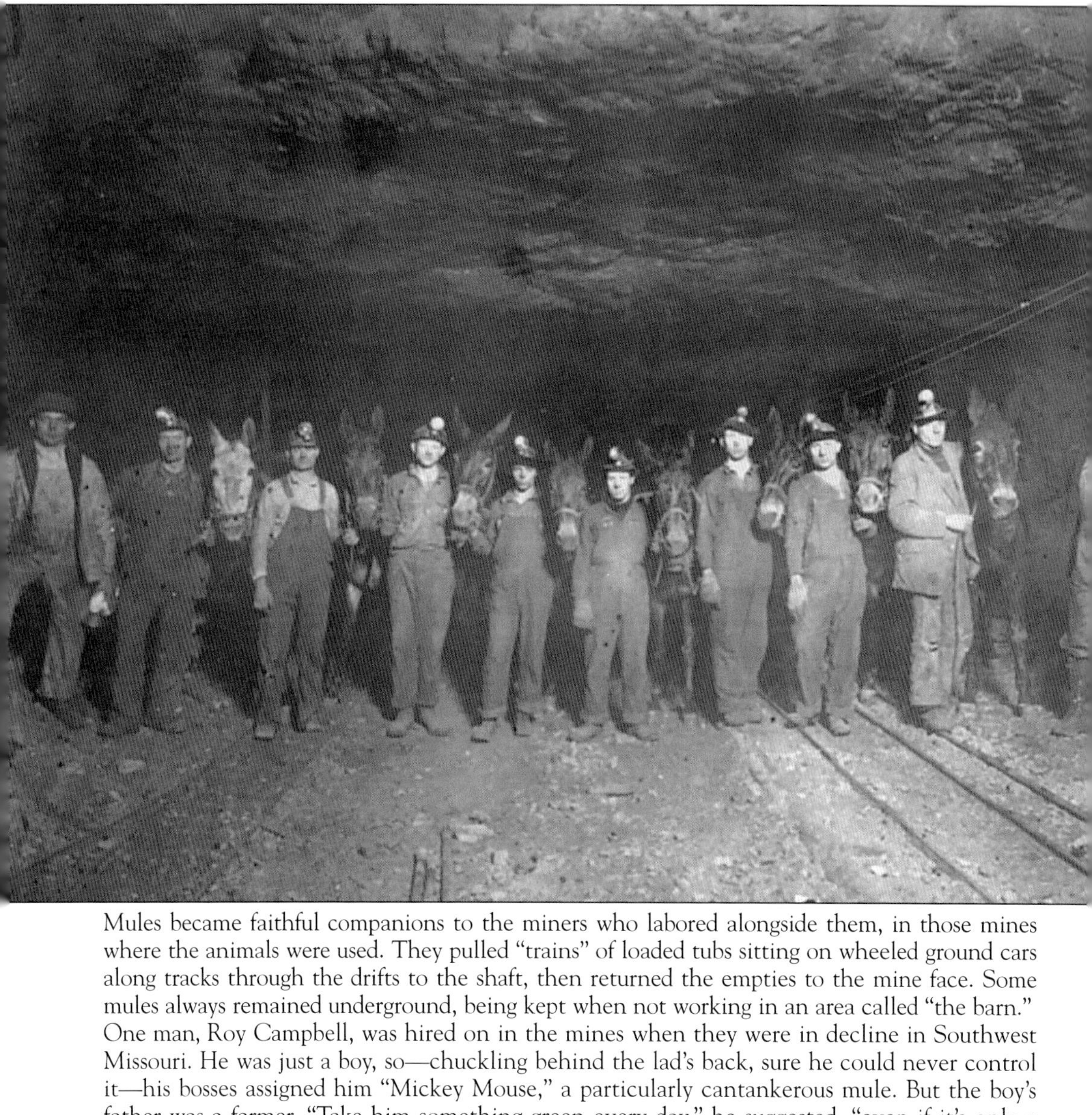

Mules became faithful companions to the miners who labored alongside them, in those mines where the animals were used. They pulled "trains" of loaded tubs sitting on wheeled ground cars along tracks through the drifts to the shaft, then returned the empties to the mine face. Some mules always remained underground, being kept when not working in an area called "the barn." One man, Roy Campbell, was hired on in the mines when they were in decline in Southwest Missouri. He was just a boy, so—chuckling behind the lad's back, sure he could never control it—his bosses assigned him "Mickey Mouse," a particularly cantankerous mule. But the boy's father was a farmer. "Take him something green every day," he suggested, "even if it's only a hedge ball." To the chagrin of the other miners, Roy soon had Mickey Mouse working as well as any other mule in the mine.

Years later as Roy was driving in the country, he saw the animal grazing in a field alongside the road and stopped the car. When he walked to the fence, Mickey recognized his old friend, trotted over, and raised his foot for the hoof to be picked, something that had qualified him for a treat in the mine.

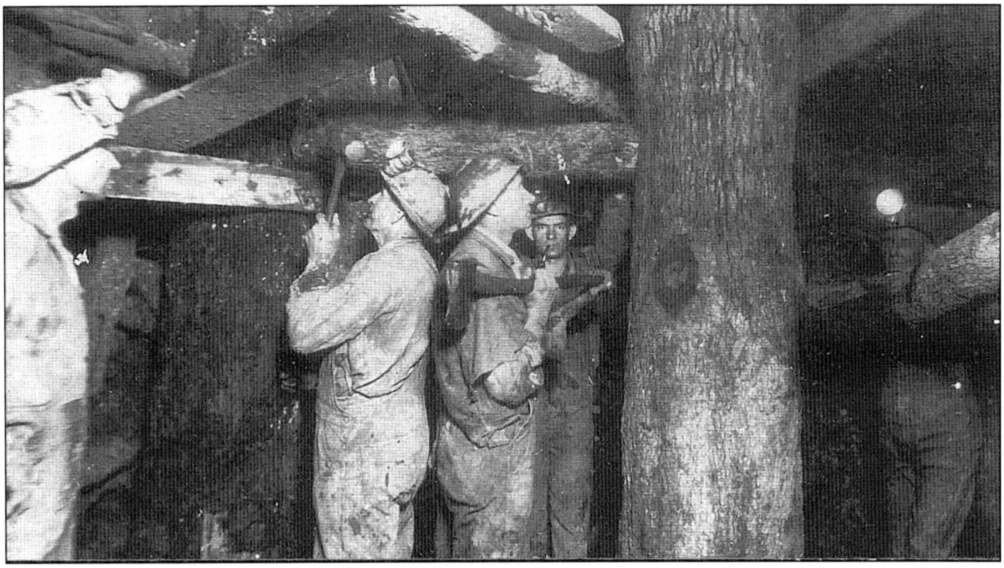

A crew installs timbers in the Gresselli Mining Corporation's No. 3 Shaft. Drifts in earth or in unstable rock required timber supports to avoid cave-ins. Even with the bracing, however, there were no guarantees of safety. In January of 1902, though it had timbers in place, the Aida C. Mine southeast of Carl Junction caved in, trapping six men. One body was freed and taken to Knell's Undertaking Rooms in Carterville. Another young man, Don Onstatt, was pulled out alive but died soon after. At the end he took his watch from his pocket and handed it to another man. "Give this to my mother. It is all I got," he said. The Webb City Genealogical Society states that the hardened miners and Dr. Chenoweth, who would become a veteran of such accidents, wept at the scene. (Photo Courtesy Carl Junction Archives.)

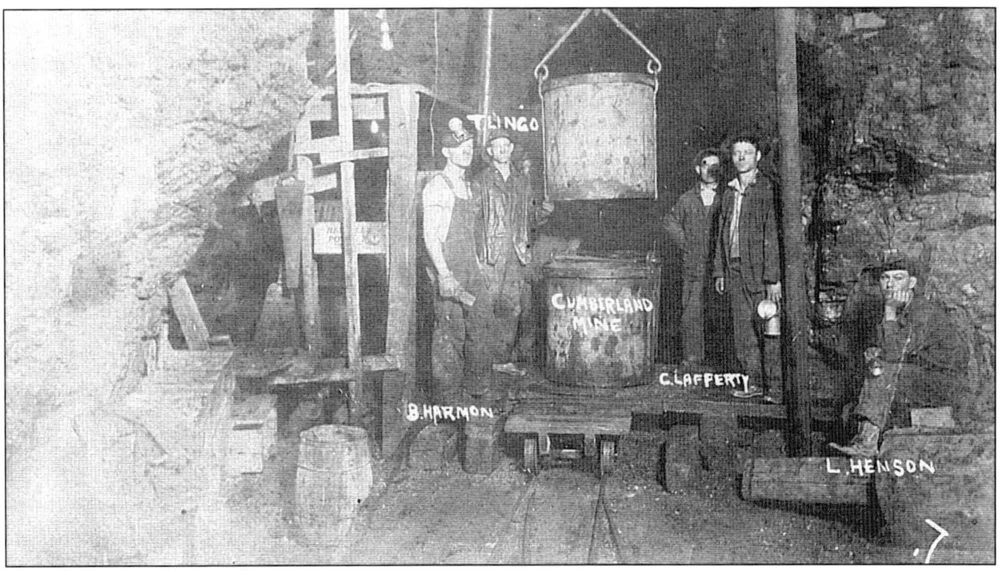

Tub hookers pose with a partially hoisted tub while a ground car waits to return the empty to the ore face. (Photo Courtesy Webb City Historical Society.)

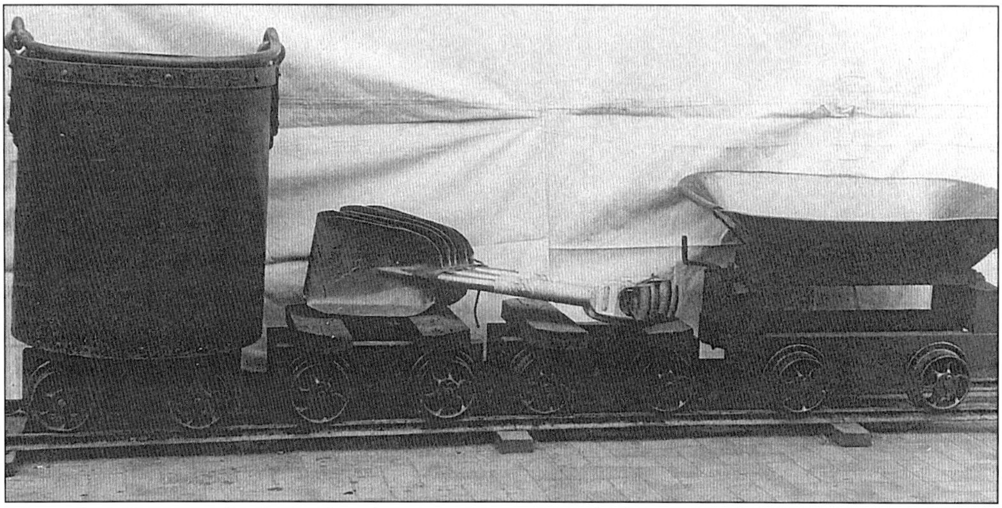

The tub, left, ground cars, and jack cars were items sold by Weeks Hardware in Carterville. Established in 1900, Weeks manufactured and sold mining equipment throughout the district. Zinc ore was called jack. Output from the Webb City area outpaced any other location, and it called itself "The Zinc City." To this day, the name of the high school yearbook is *King Jack*. (Photo Courtesy Joan Johnson.)

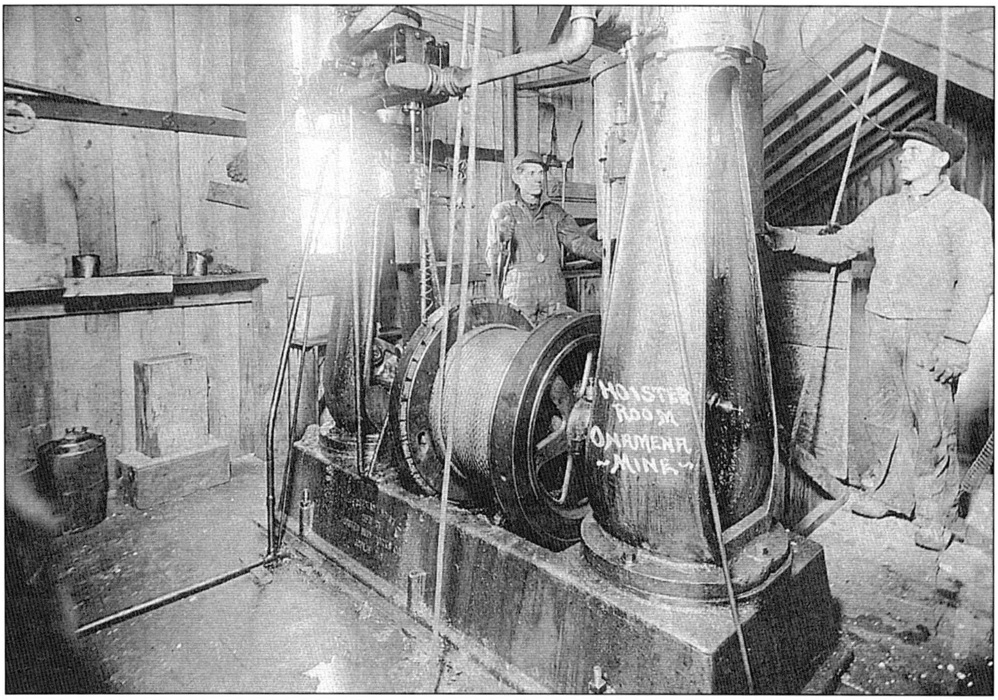

On the left is hoisterman Welford Hamer at work in the hoister room of the Onamena Mine. Visible behind is a raised tub and the screen room. Hoistermen became very adept at their job. Some, in a bit of devilish humor, would occasionally drop a can full of miners down the shaft at great speed, braking in the last few feet to avoid splattering them at the bottom. While this may have given the hoisterman his jollies, it was ill-calculated to endear him to the miners involved. (Photo Courtesy Webb City Historical Society.)

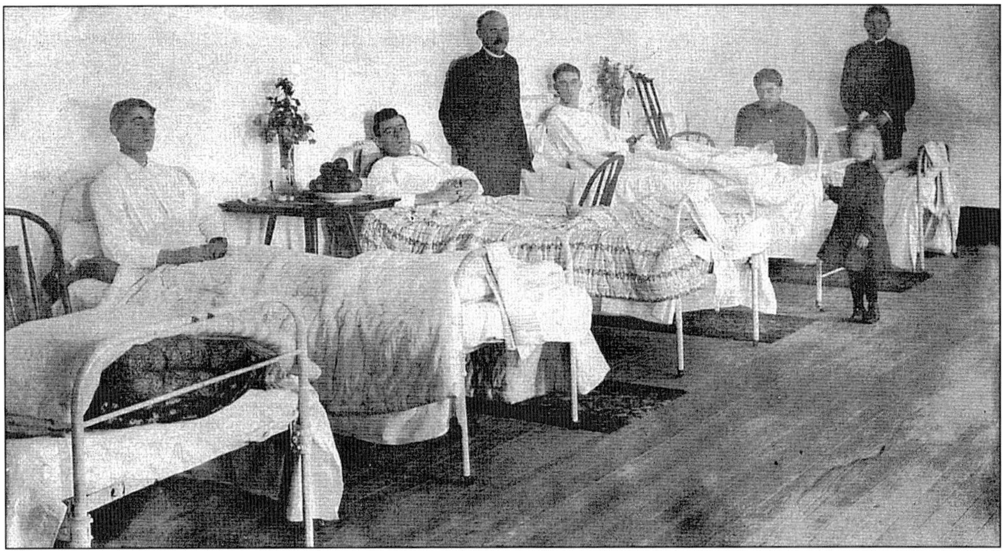

As previously mentioned, some hoistermen on occasion found it amusing to give miners "fast rides" down the shaft. According to the Webb City Genealogical Society, on December 12, 1905, the man operating the hoister at the Yellow Dog Mine decided to frighten four men in that manner but failed to stop the descent in time. Three of the miners sustained broken legs and the fourth a severely injured ankle. The men were Ward Byerly, John Cellner, Milton Pridemore, and Frank Turner. Pridemore, of Carterville, had a broken left leg. His brother-in-law was the hoisterman, yet Pridemore blew the whistle on him, stating that though he felt bad about it, it had been a deliberate act. The men were taken to the Salvation Army Hospital in Webb City. Captain (or Ensign) Vredenburg ran the hospital, which was said to be the only one in existence. (Photo Courtesy Henry Hulett.)

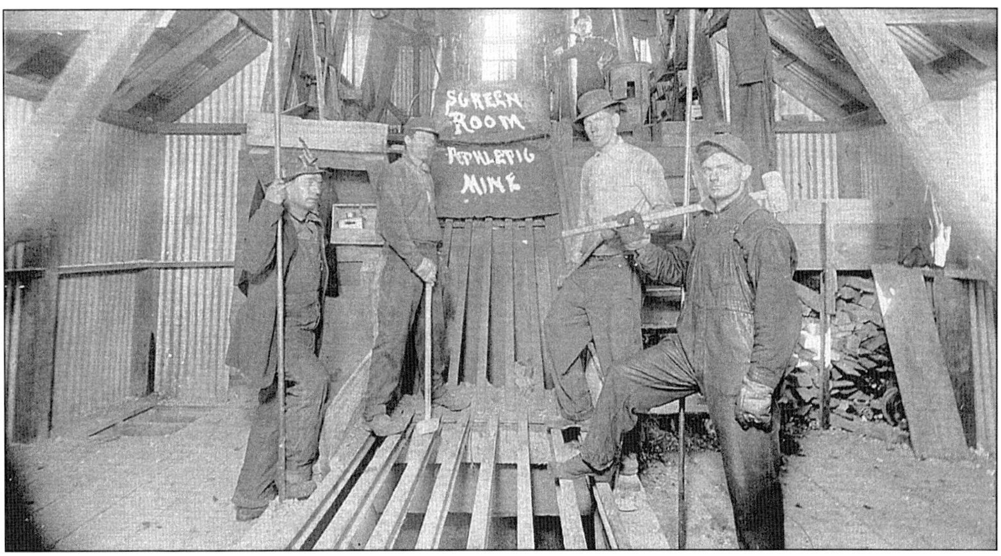

Once the full tub was raised out of the ground, a tail chain was attached to it, and it was pulled into the screen room where its contents were dumped onto the grizzlies. The screen apes, here at the Athletic Mine near Duenweg, used sledgehammers to reduce the rocks to pass the screen. (Photo Courtesy Brad Belk, Joplin Museum Complex.)

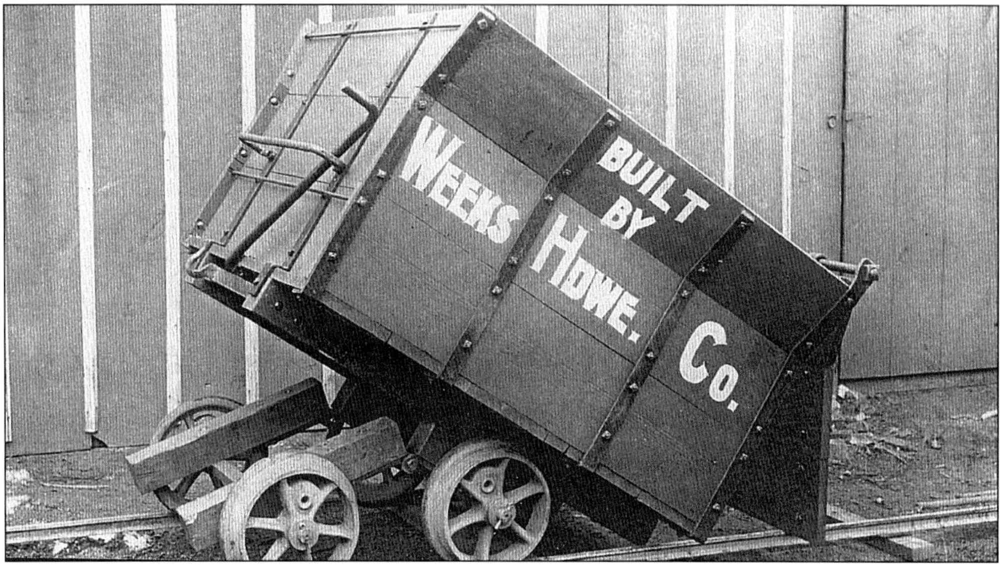

This is an example of the Rock Cars built and sold by Weeks Hardware Company. The waste rock from the screen room was pushed along a tramway supported by a spidery wooden framework. At the end of the track, the contents were dumped on the waste pile below. (Photo Courtesy Joan Johnson.)

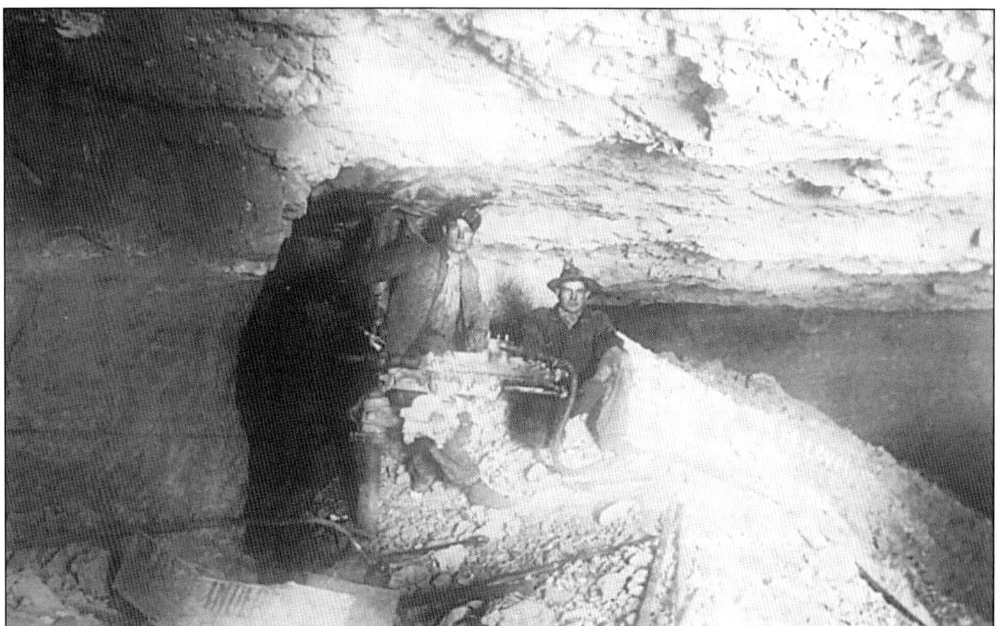

A stope is a step-like excavation formed by the removal of ore from around a shaft. According to the inscription on the photo, on December 20, 1912, these two miners were stoping at the 174-foot level in the Samson L&Z Mine at Joplin.

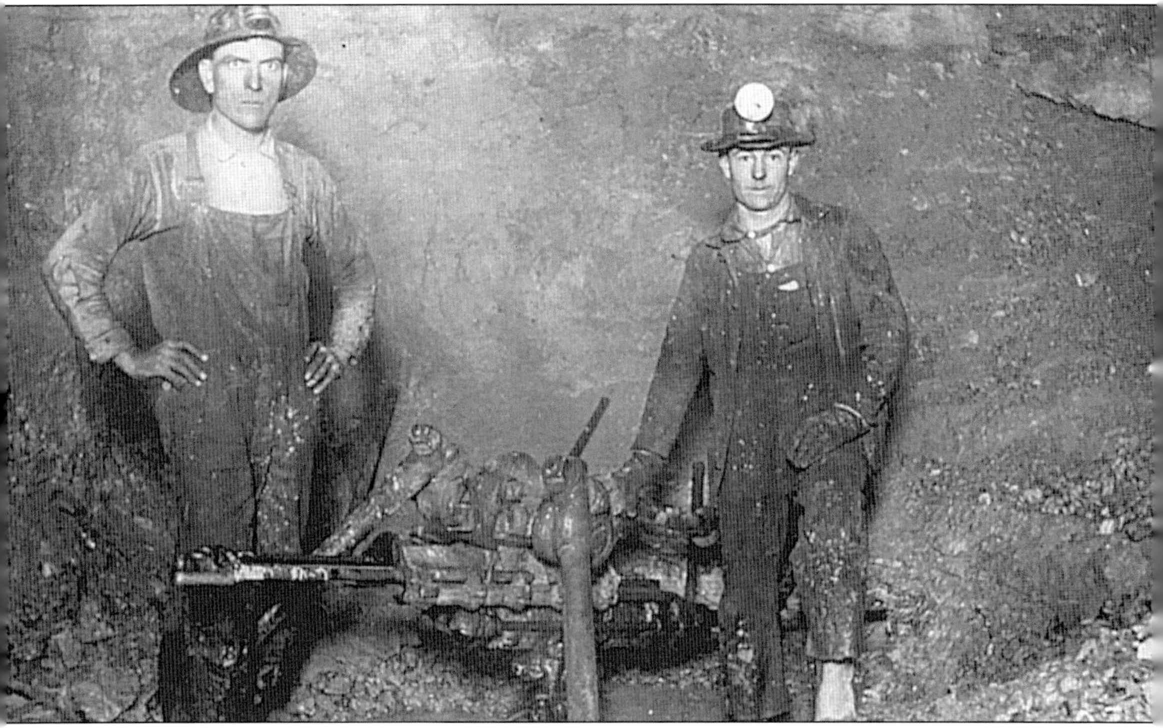

These men were operating a drill for the Imperial Mining Company in 1915. Drilling the holes for blasting powder was originally done manually with one man holding a bit which was driven by another man wielding a sledgehammer, with the bit being turned a little between strikes. The use of steam-driven machine drills was much more efficient, but it was a filthy job. The bits had an opening down the center through which water was forced for cooling and to flush the drill hole. It is apparent that these miners are wet from the chest down. One man recalled with a chuckle that, after his shift when he would come above ground on bitterly cold winter evenings, his clothes would freeze like a shell so that when he changed in the "dog house," he could just lean them against the wall.

It was common in the mines to call the bits "drills." The Webb City Genealogical Society reports that on January 9, 1904, John W. Canfield was working for the J.W. Durby and Co. Mine south of Carterville. He was killed when a newly sharpened drill fell down the shaft and passed through his body. He was 21 years old, single, and lived with his parents on South Fountain Street, Carterville. He was insured for a few hundred dollars. (Photo Courtesy Webb City Historical Society.)

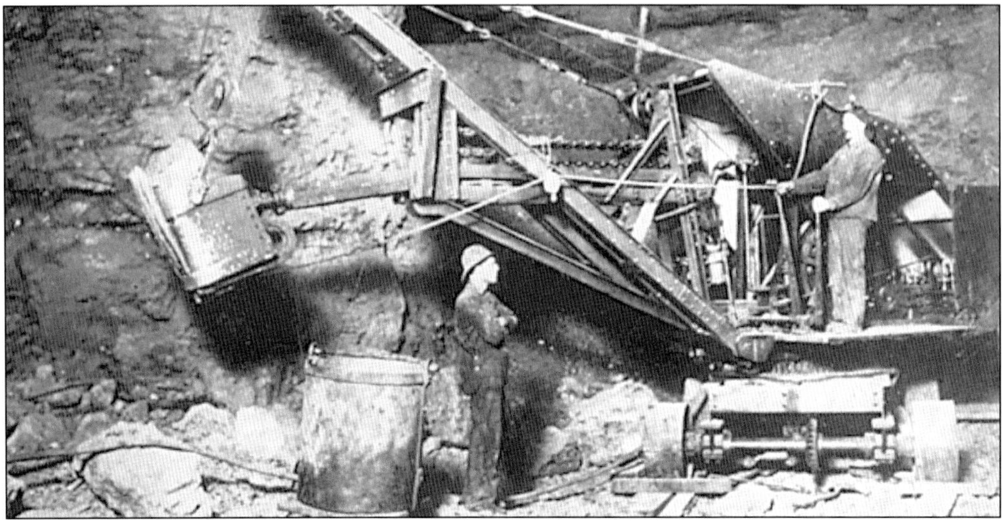

In a large cavern, a steam shovel is being used underground in one of the Davey mines. The American Davey Company's mining operations were extensive in the area. So many of their shafts were located between Prosperity and Johnstown that they were designated by numbers.

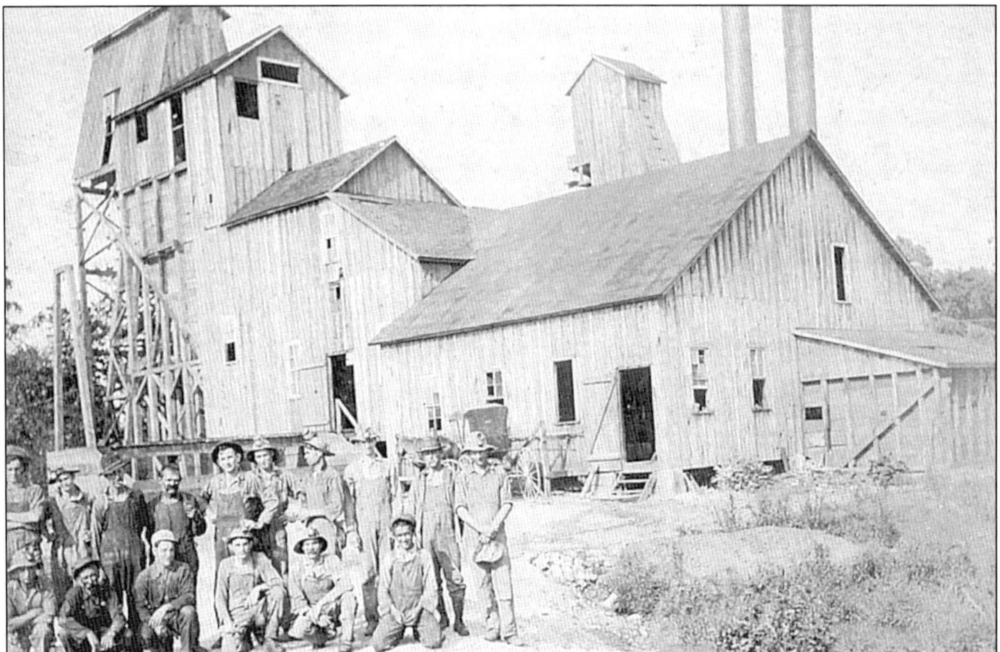

This photo from the Webb City Historical Society archives shows a group of miners posing in front of the Fighting Wolf Mine. Giving a mine a woman's name was considered bad luck, but the names used were colorful, as were the men who worked in them. The Webb City Genealogical Society declares that on Friday, May 28, 1895, the entire night shift of the Annie D. Mine walked off the job at midnight after seeing a ghost in the drift. The ghost, which was reported to be over 9 feet tall, with eyes that shone like "two balls of fire," walked into the wall and disappeared. The men "may refuse to work in it longer."

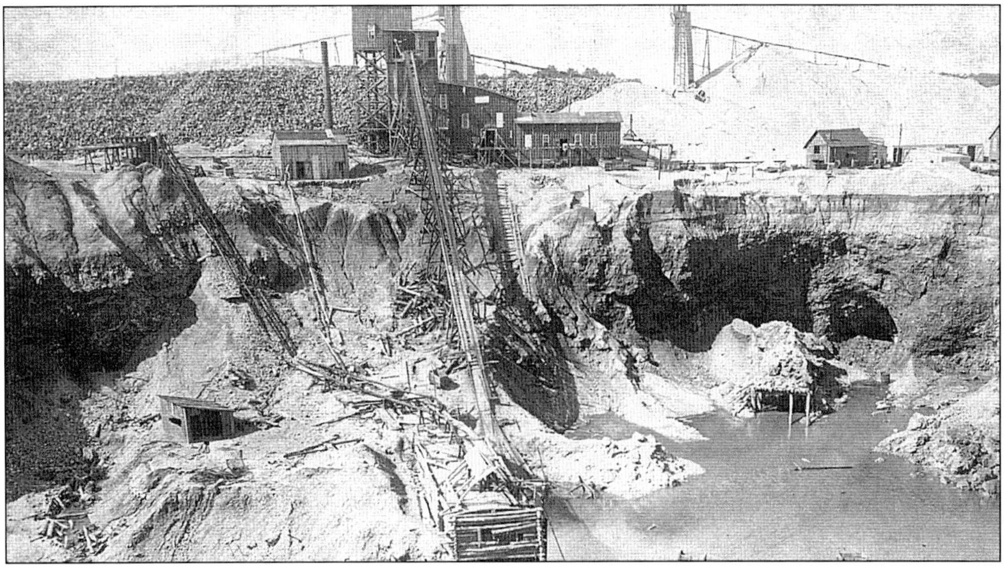

Carrying ore up the face of an open pit mine was called a skip operation. In an obvious state of disrepair, the abandoned Chicago-Lehigh Mine near Carl Junction begins to fill with water. In later years, the water-filled caves became "attractive hazards." Reciting tales of previous drownings, mothers tried with scant success to persuade their children not to play in or around them. In addition, there was the intriguing mystery of vehicle tracks that were sometimes visible leading up to and over the edge. One man once admitted to "having a couple of car bodies in the Orange Cave." (Photo Courtesy Carl Junction Archives.)

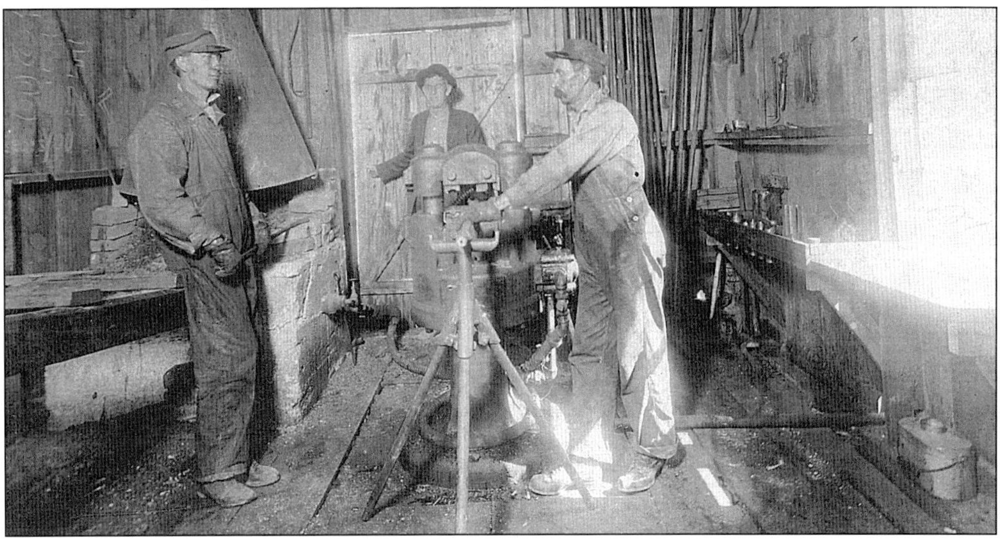

The blacksmith shop, such as this one at the Panther Mine, was an important facet of the mining operation. There the drill bit was reconditioned, the center opening was blown out, and it was resharpened by heating it in the forge and then hammering a new point—not by grinding as might be imagined. Other tools such as picks were sharpened, and tempering after heating was done as required. Machinery in the mine and mill was maintained and repaired. Some old timers swore that a good blacksmith could forge weld better than any of the new-fangled acetylene or electric welders, which later came into use. (Photo Courtesy Webb City Historical Society.)

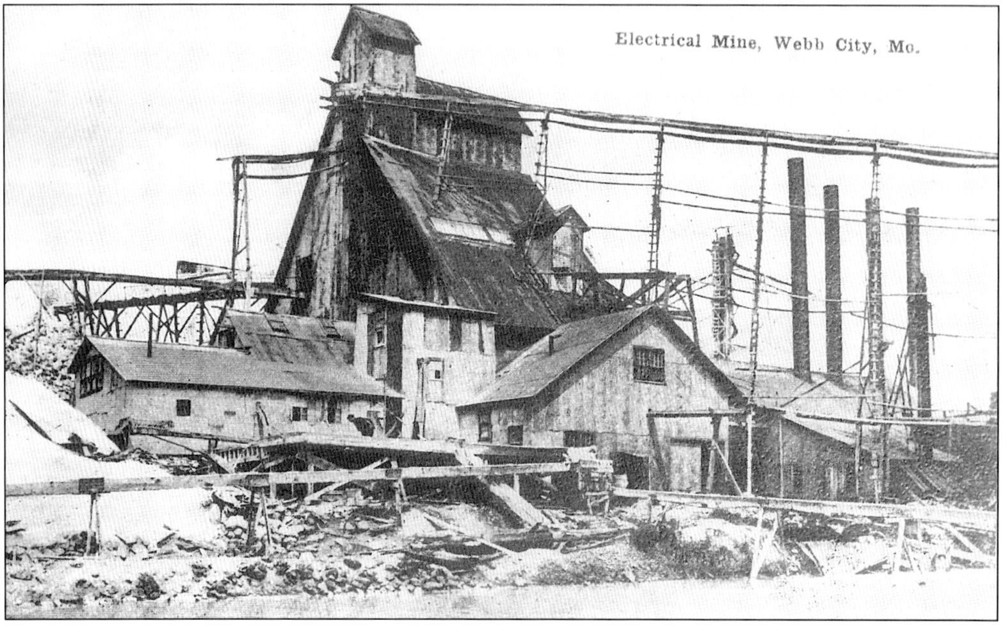

The melange of buildings at the Electrical Mine at Webb City depicts the apparently haphazard growth caused by the expansion of a successful mine. New buildings were grafted piecemeal to existing buildings as needs dictated.

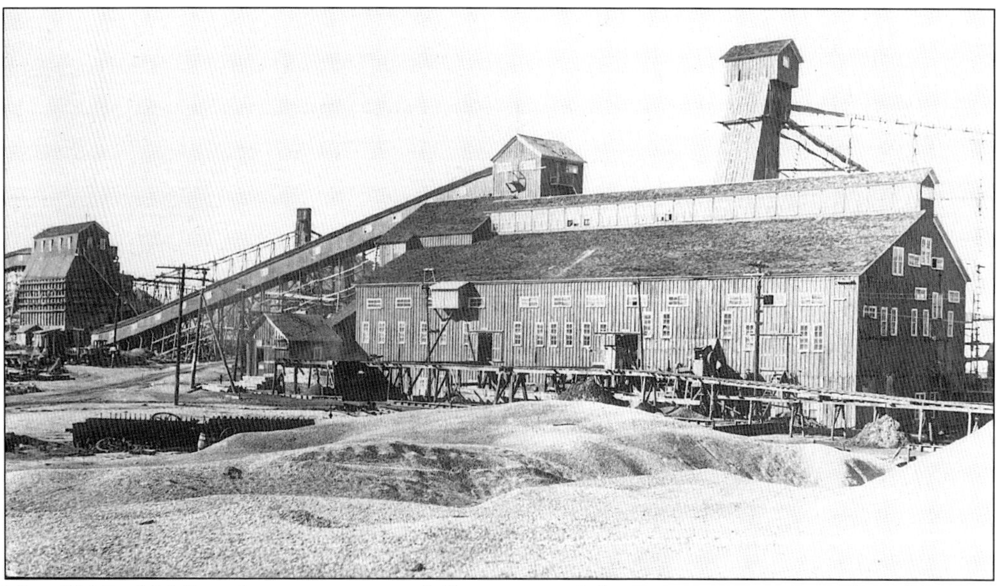

The Old Red Mill at Oronogo is a good illustration of the impressive size a mill could attain when it was combined with a producing mine. In most mills, machines operating at each stage of the process derived their power through a belt turned by a central power shaft running the length of the building. The belt was engaged or disengaged at each location to start or stop the operation.

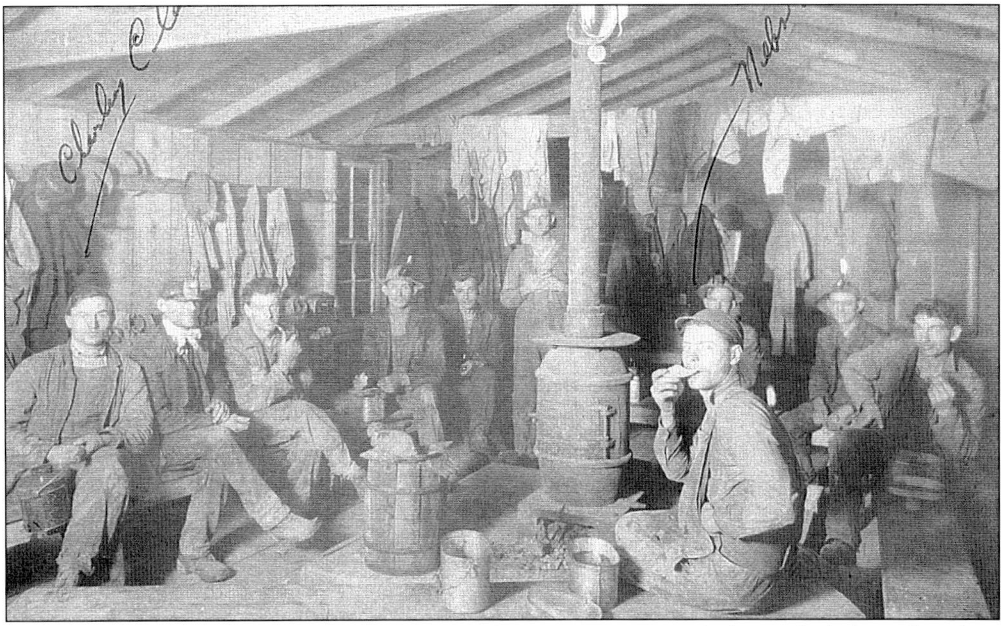

This is the interior of a typical "dog house" where miners ate and changed clothes after their shift. Note the pail-type lunch buckets in which they carried their food. Years ago, an elderly man recalled that when he was a young boy, he would hitch his goat to a cart and haul his mother's baked goods to miners who would tip him a few pennies for his efforts. (Courtesy Webb City Historical Society.)

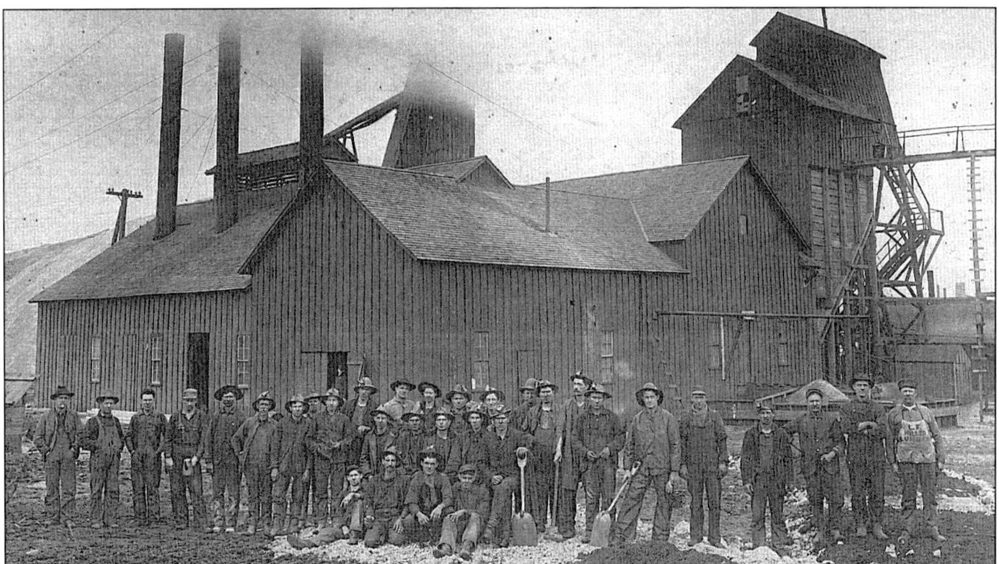

A group of miners poses in front of the Ground Floor Mine near Joplin. Standing sixth and seventh from the left are Samuel Robb and his son, Will. Samuel moved his family from Pennsylvania before the turn of the century seeking a better life. By 1915, Harry Robb was dead at age 28, the first son to be claimed by miner's consumption ("the con," a form of tuberculosis). Samuel followed at age 61 in 1924, dying in the Jasper County Tuberculosis Hospital at Webb City. Will succumbed in 1926 at age 38, leaving a wife and two small daughters.

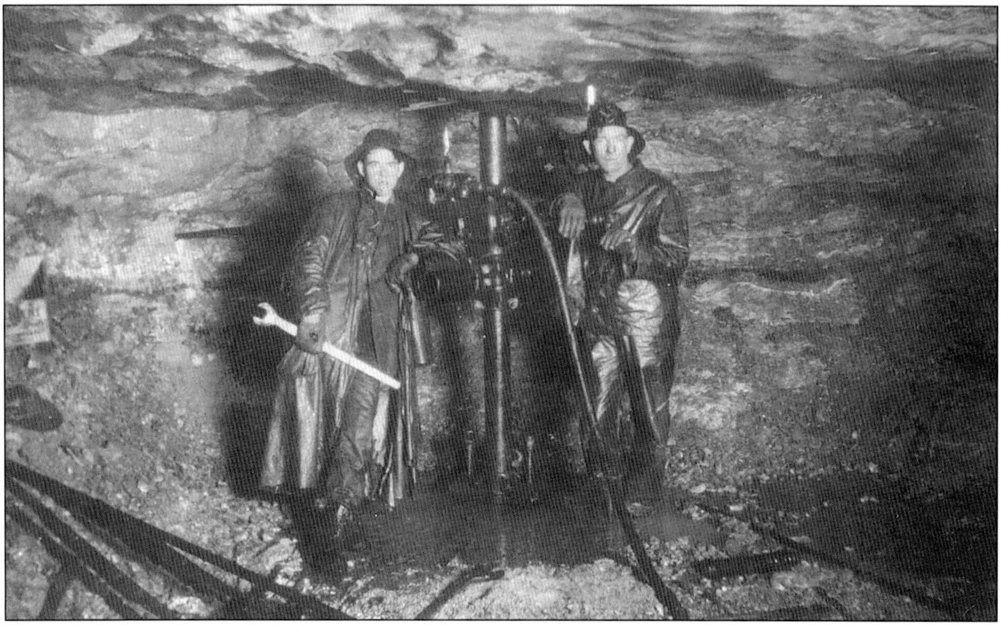

Slicker-clad Will Robb, holding the wrench, poses for the photographer by his drill. As mentioned before, using a drill was a wet job. Notice the pool of water in which the men are standing. They were apparently using sunshine lamps (wickers) burning lard oil for light, one of which has been transferred to the top of the machine. Candles were also used, as were paraffin and carbide lamps. The man with Will may have been his friend "Rassie" Patton, who was born the same month and died the same month as Will, also leaving a wife and two small daughters. (Personal Collection.)

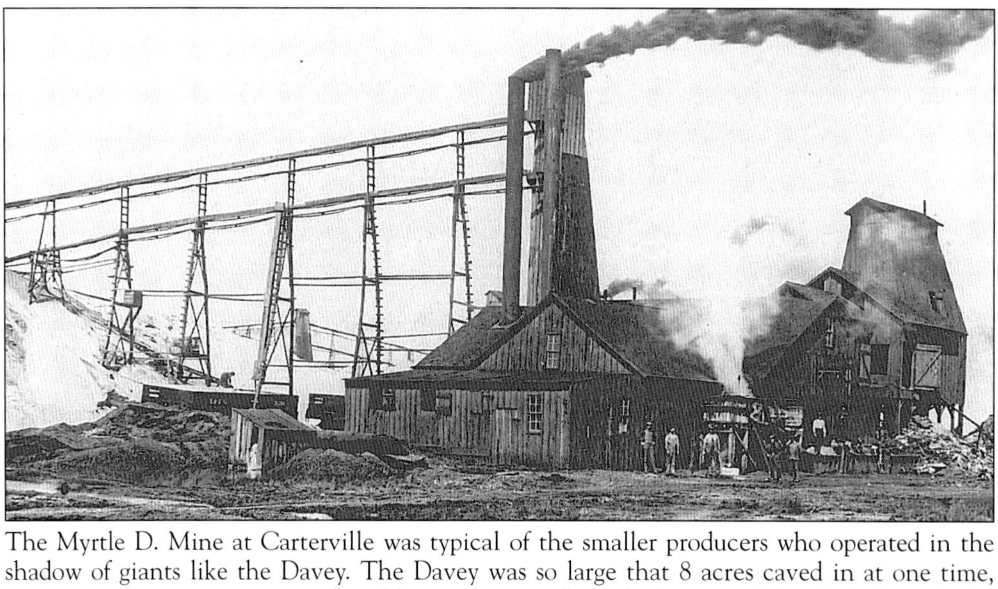

The Myrtle D. Mine at Carterville was typical of the smaller producers who operated in the shadow of giants like the Davey. The Davey was so large that 8 acres caved in at one time, generating pressure great enough to blow equipment from the bottom of the shaft through the top of the derrick. One old miner said you could always tell when there'd been a cave-in: "Wind came a-whooshing outta every shaft and drill hole on the lease." (Courtesy Joan Johnson.)

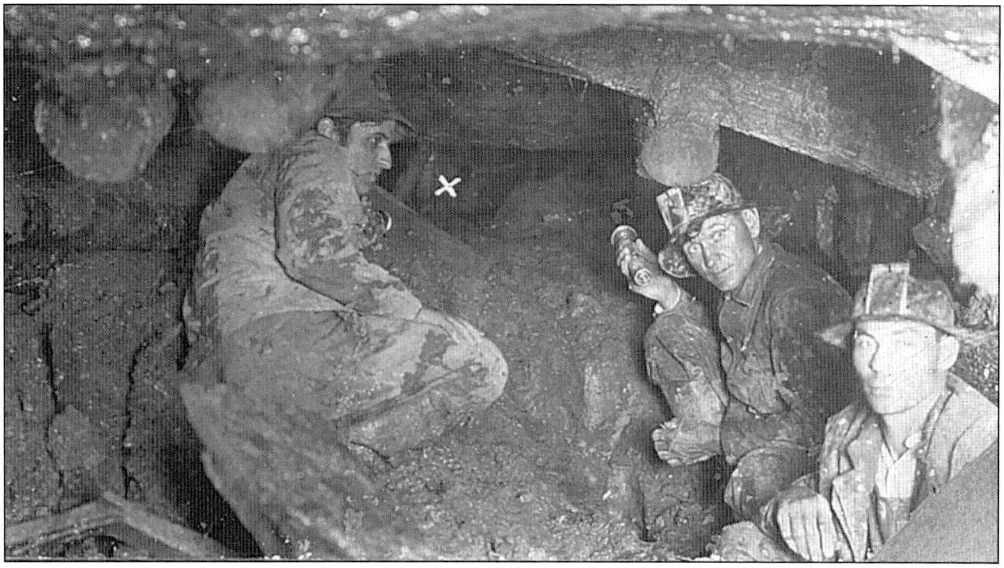

This old postcard of an unidentified cave-in purports to show three unidentified miners attempting to free a trapped companion, marked with the "X." Often when there was a cave-in, nearby mines would cease operations so their workers could help with the rescue efforts. At one, it is said, a hundred men worked to save those caught underground. The possibility of cave-ins was ever present. An old man once swore he never went on the job a single day without fear as a companion. (Photo Courtesy John Burrington.)

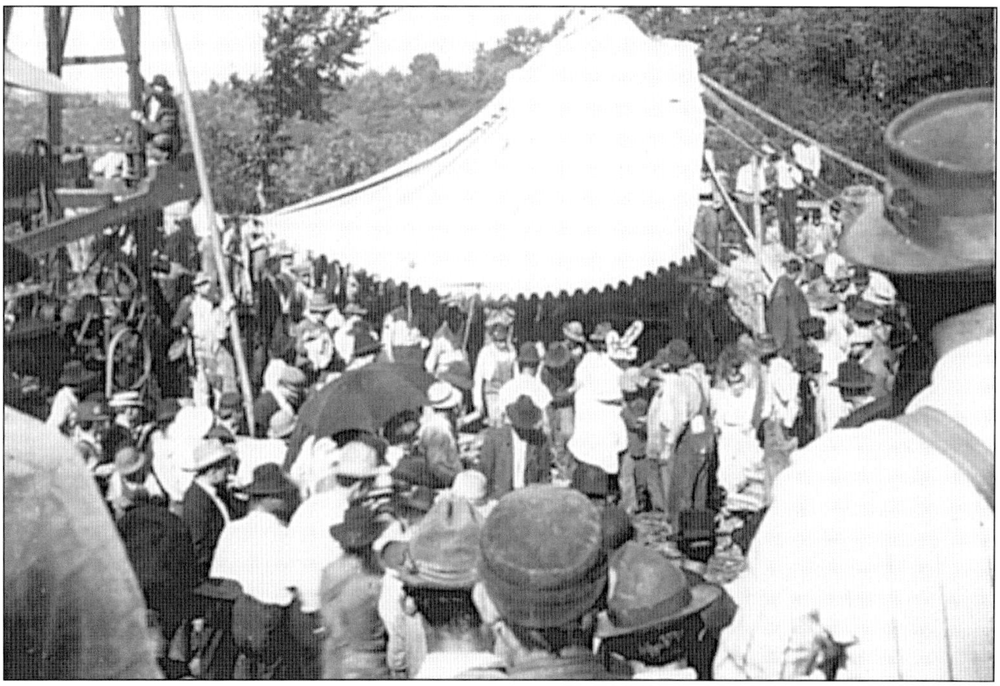

On August 2, 1911, a large crowd gathered to watch rescue efforts when John Clary was trapped underground in the White Oak Mine at Joplin. After three days and three nights, Mr. Clary was pulled from the ground alive. He became a local hero for a time and was often invited to speak about his harrowing experience.

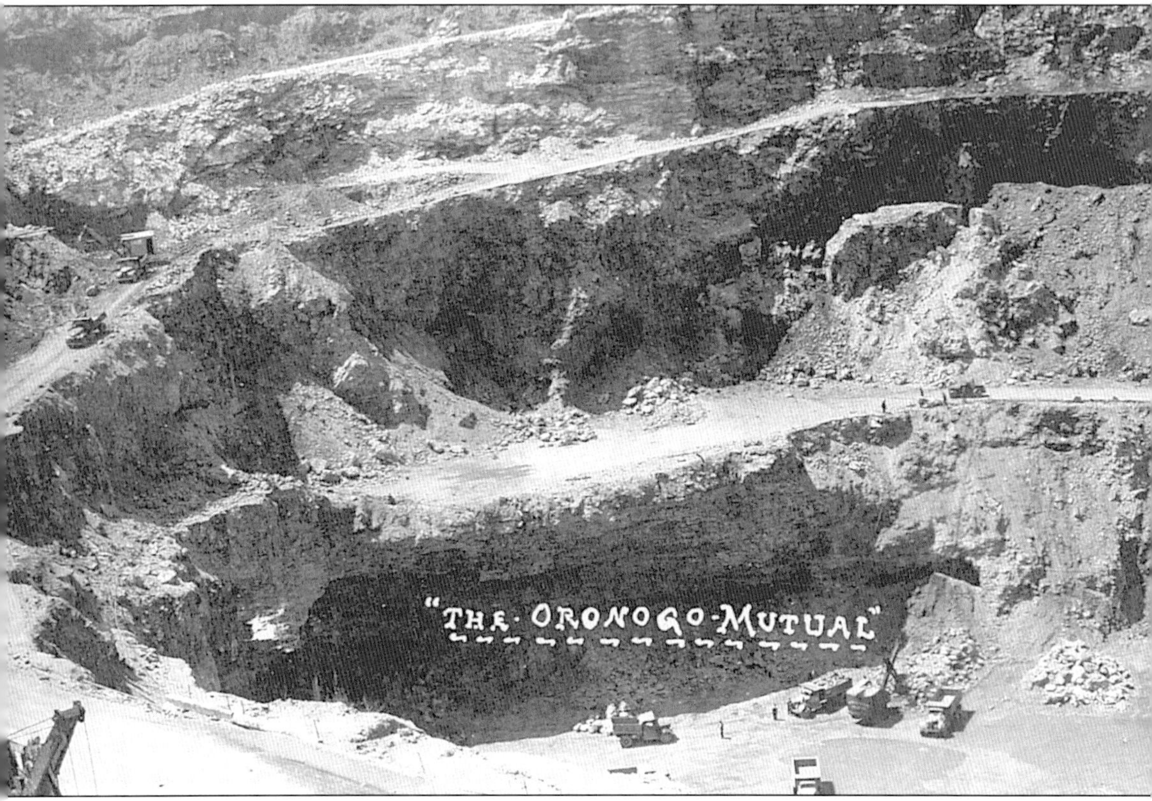

A panoramic view of the Oronogo Mutual Mine, commonly called the Oronogo Circle, shows what was considered at one time to be the largest open-pit lead and zinc mine in the world. It was reported that from 1850 to 1915 the Great Circle furnished 70 percent of the production of lead and zinc in the United States and well over 35 percent of the production in the entire world. Water, always a problem, was continually pumped to maintain the operation. After World War II the pumping ceased, and the water was allowed to fill the pit. It became a popular destination for divers from throughout the country who came to explore the submerged passages and rooms. On one tragic day in 1970, three of the divers failed to surface. Many other divers rushed to the location to search for their bodies in the great cave. One volunteer recalls that the first man was found at the 130-foot level. The mask and tank were still on the body as if he had just gone to sleep. The other two were located deeper, and their tanks and masks had been ripped off as if they had suffered some sort of distress. (Rescue information Courtesy Les Sage, Harlen Hampsten Photo Courtesy Roy Campbell.)

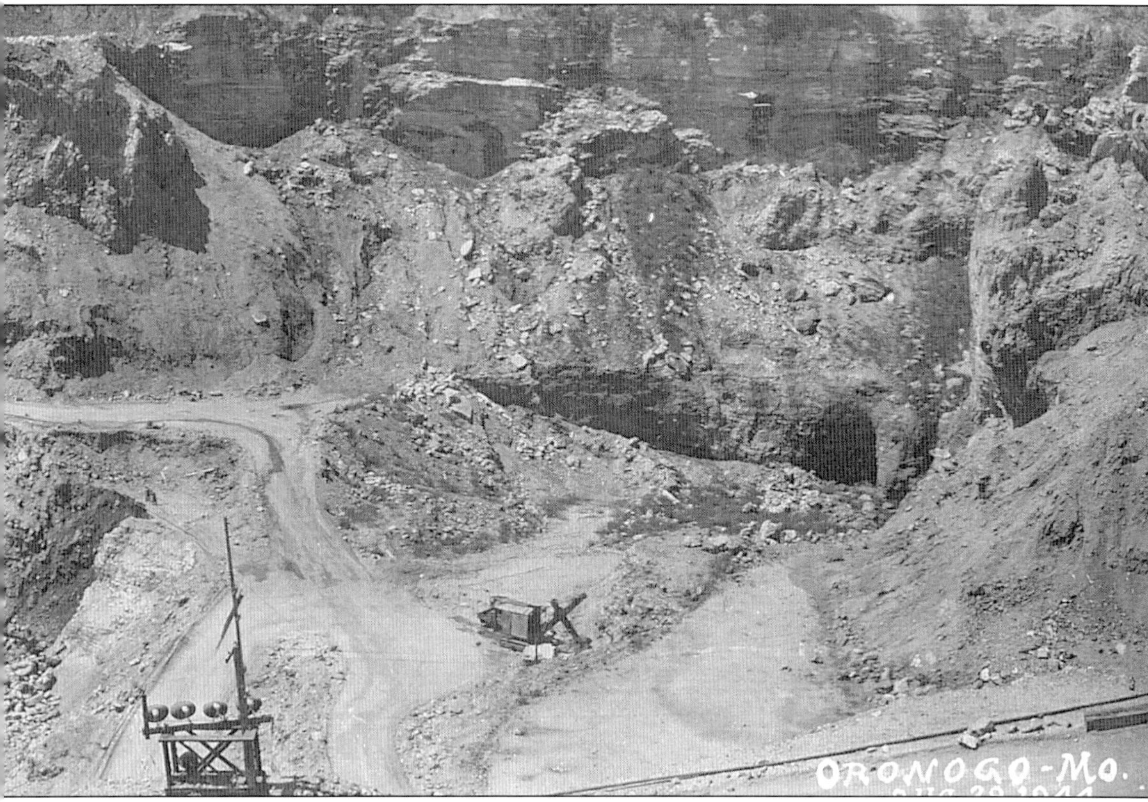

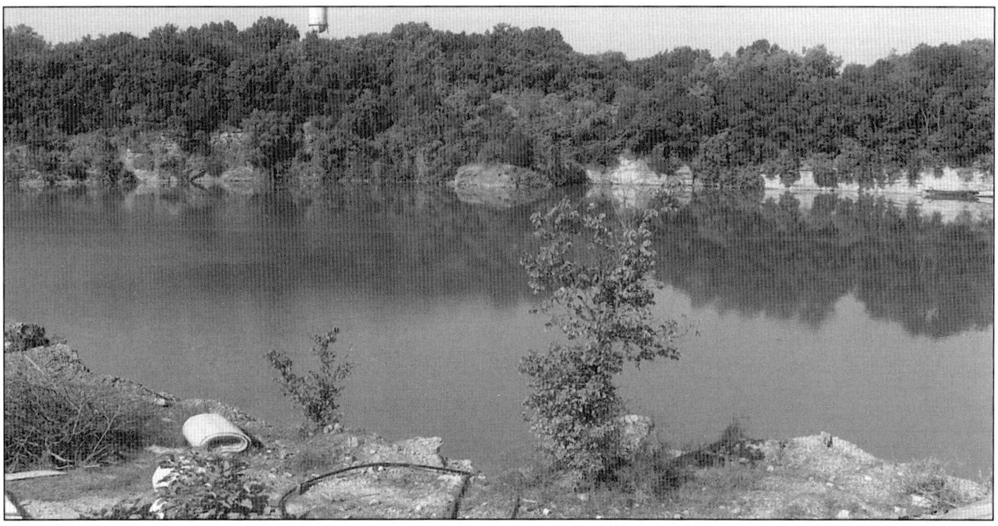

The hidden mysteries of the Oronogo Circle Cave, shown as it appears today, continue to draw divers who come to explore the hidden depths. Some of the water-filled drifts and caverns hold machinery and equipment left just as it was when the mine flooded, including two "new" steam locomotives assembled underground early in the century in the northwest drift. They were never used. (Access to the cave Courtesy Captain John's Diving, photo by author.)

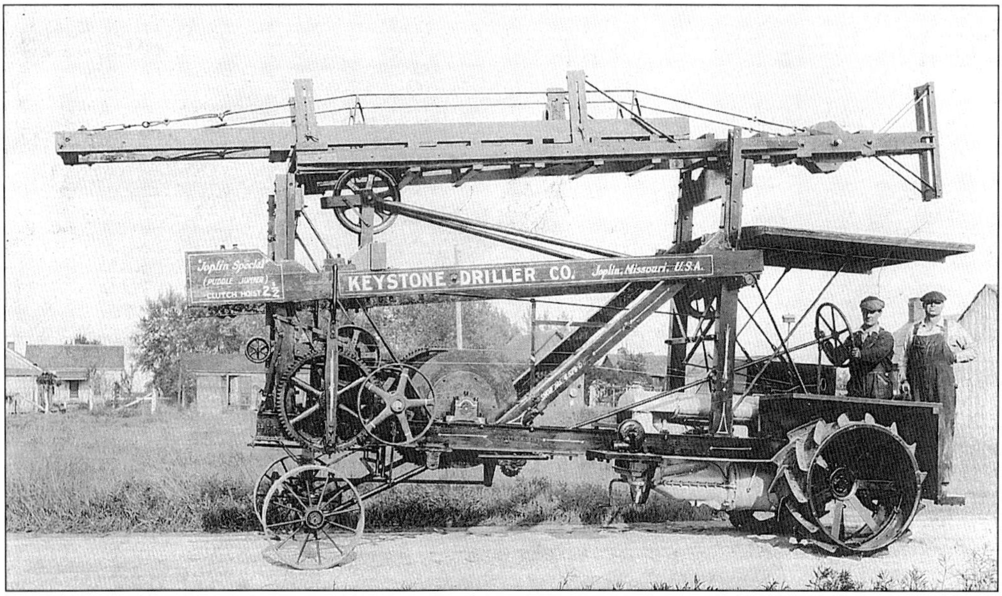

Keystone drillers, built by a company in Joplin, were a welcome addition to the mining scene. Drilling enabled the miners to know what was beneath the surface without the wasted time and effort of "dry-hole shafts." Frank Pryor is at the wheel. He married one of the Robb daughters, Marguerite, and worked only briefly in the mines, avoiding the health problems associated with the occupation. (Personal Collection.)

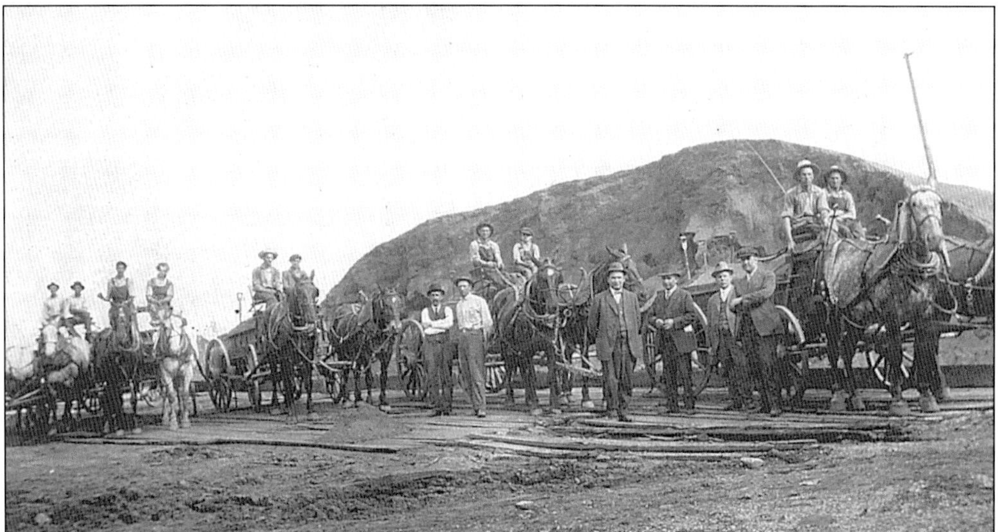

Bill Rhea, owner of the Mont B Mine, was highly regarded by the miners. One man who worked for him said Mr. Rhea became a friend who would visit his modest home and roll around on the floor playing with his children. He once remarked that he was making $20,000 per week. When the relater of that information was asked, a bit sarcastically, what he was paying his employees, the reply was three or four dollars a day when other operators were paying only one or two. When the price of zinc ore declined, he didn't lay off workers, but continued mining—piling the ore adjacent to his mill. The pile shown contained 6,850 tons and was sold on June 3, 1915, to George W. Moore for $111,700.46. (Photo Courtesy Henry Hulett.)

Two

JOPLIN

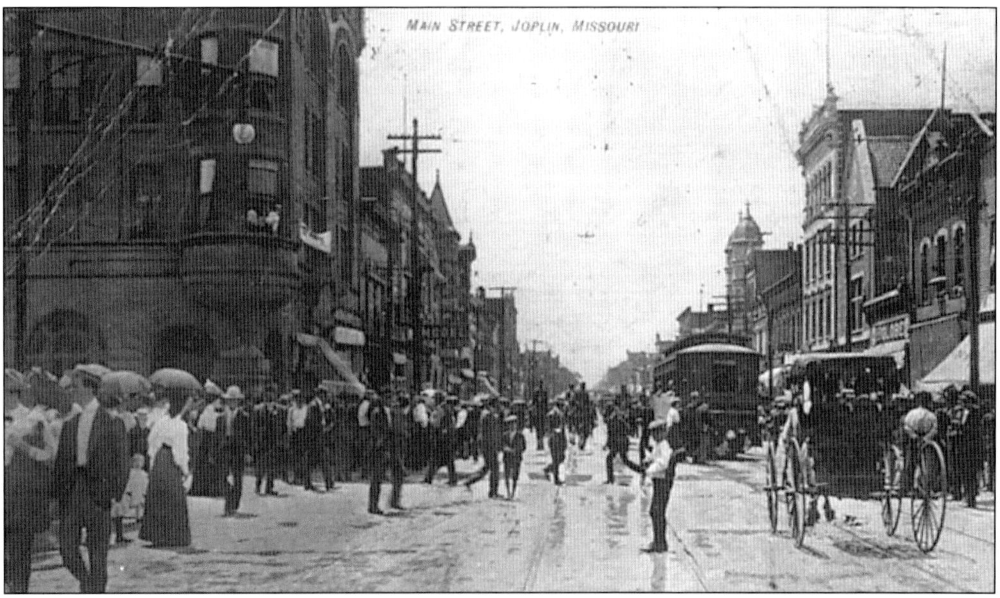

Civil War maps illustrating the war in Missouri in the September 7 and October 19, 1861, issues of *Harper's Weekly*, make no mention of Joplin, but do show a landmark, Grand Falls, on "Neh-hah" River. However, by 1874 a Jasper County map shows Joplin to be one of the dominant cities in the area. This early view of its bustling, though muddy, Main Street is demonstrative of the rapid growth following its birth—as two feuding mining camps on either side of Joplin Creek. The Keystone Hotel, on the left, was built in 1892, less than two decades after Joplin officially became a city in 1873. (All photos Courtesy Ron Mosbaugh unless otherwise indicated.)

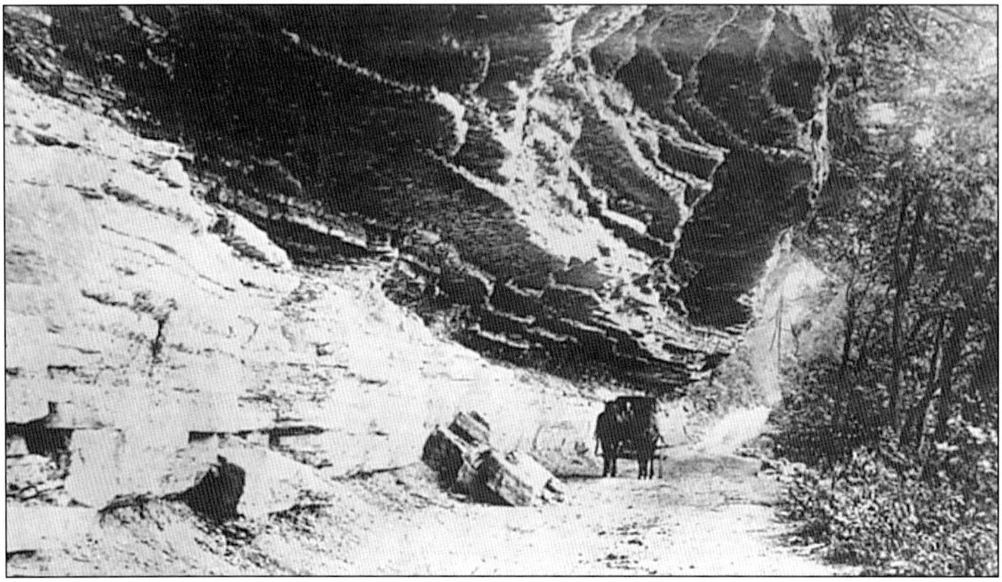

The Joplin mining boom occurred only 30 miles north of the spectacular Ozark scenery that began around Noel and Pineville and extended deep into Arkansas. Busy modern highways now carry heavy traffic beneath these overhanging rock bluffs.

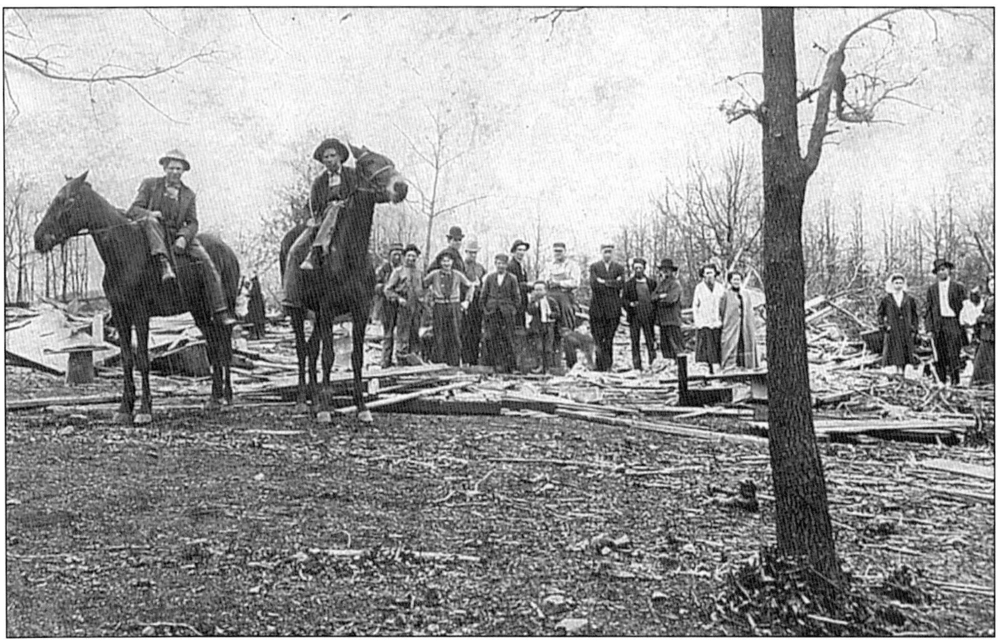

The Joplin boom also occurred in the heart of "Tornado Alley," and the region has been visited frequently by the destructive storms. In 1883, a "cyclone" virtually destroyed Oronogo. Here a group of men and women view the devastation wrought by one such early twister. (Photo Courtesy John Burrington.)

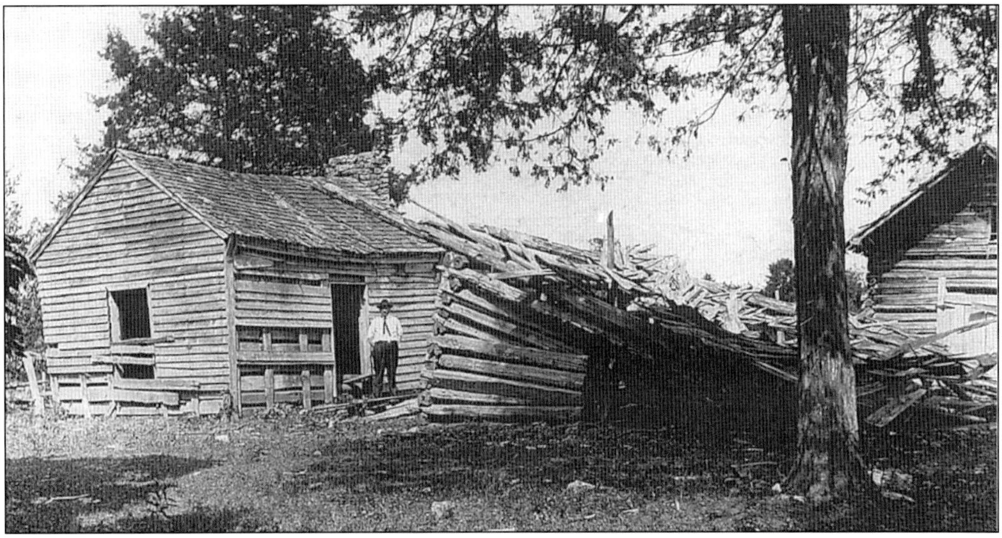

This turn-of-the-century photograph was identified as being taken just to the northeast of the current Range Line-Zora Street intersection and is a good example of early building materials—split logs or whole logs (sometimes with the bark left on) and clapboards. Range Line was little more than a country lane then. In 1999, it carried close to 25,000 vehicles per day. (Photo Courtesy Webb City Genealogical Society.)

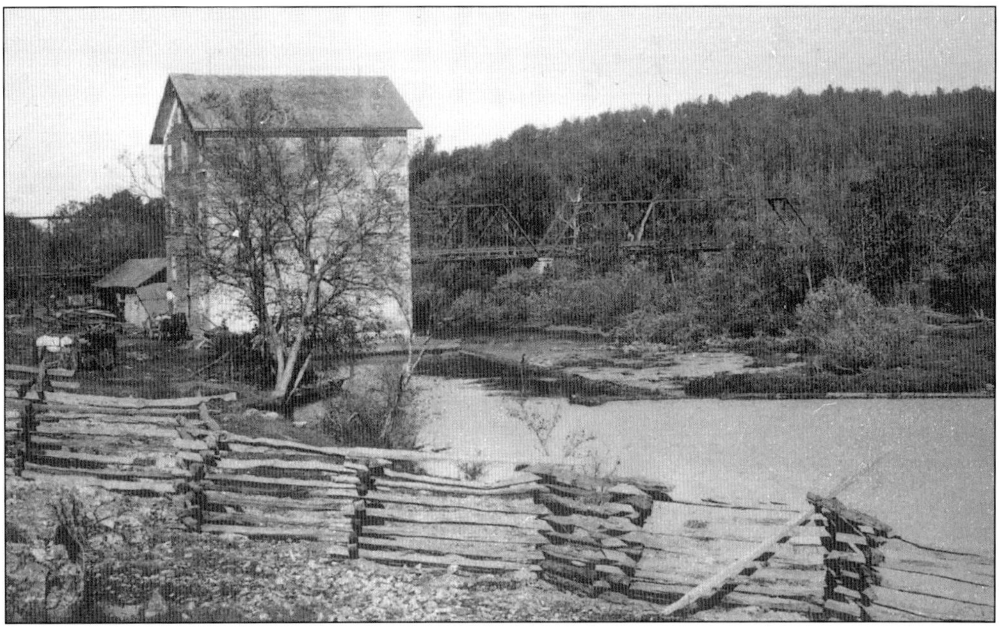

Redings Mill, established in 1844 by John S. Reding on the bank of Shoal Creek, grew to be one of the largest in the region and became a Joplin landmark before burning in 1932. Mills were omnipresent. The 1874 Jasper County map shows them scattered along the banks of almost every free-flowing stream. This view from upstream shows the old truss bridge barely visible in the distance. The name Reding seems to give everyone troubles—even the Missouri Highway Department. At one time the sign at one approach to the village bearing the same name showed Redings Mill, while another listed it as Reddings Mill.

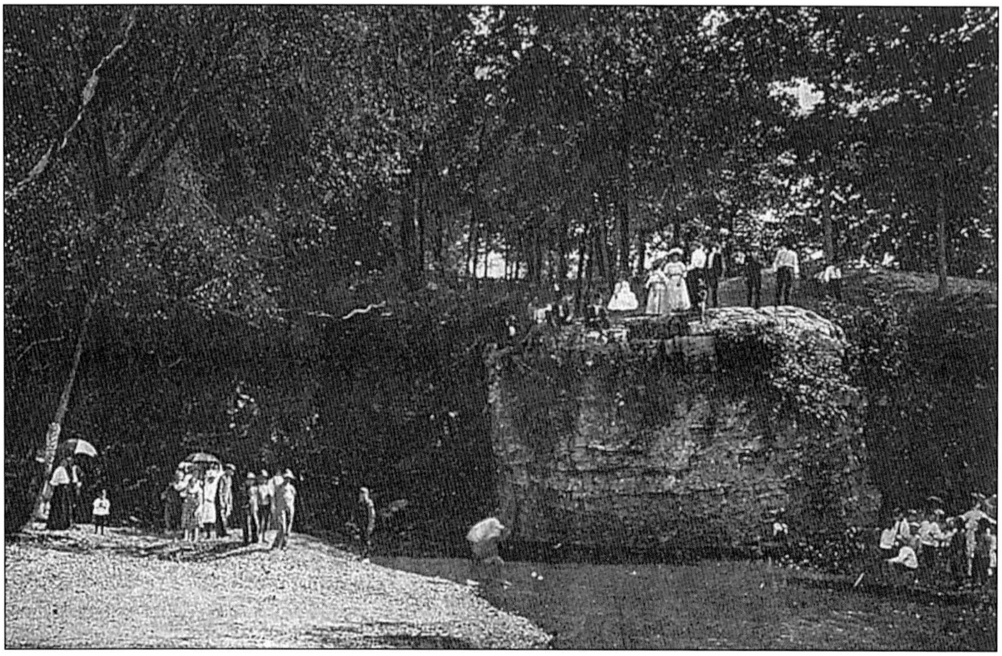

In 1908, these people gathered at Castle Rock, an unusually symmetrical rock formation at Turkey Creek on the north edge of Joplin. Charles Schifferdecker, an early German immigrant, built a beer garden there, chilling the beer in a cave located a few hundred feet east of the rock. He also built a house nearby, which still stands, cooling it with drafts from the cave, an early form of air conditioning. (Courtesy Helen Chickering, John Burrington photo.)

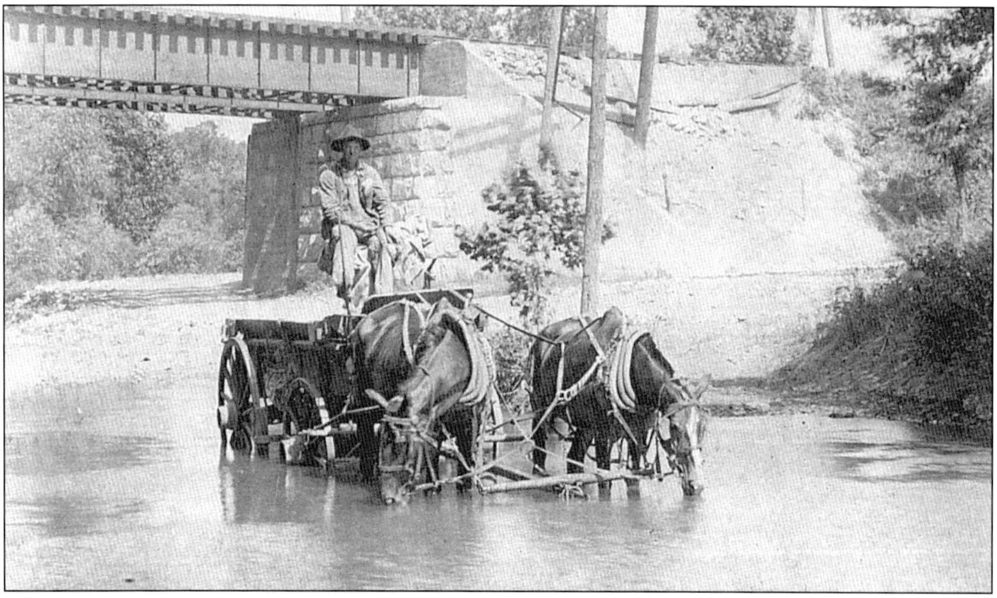

A teamster pauses to water his horses near the Castle Rock streetcar bridge spanning Turkey Creek. On the back of the card is written: "Frank B. Williams used this rig to haul groceries for the Interstate Grocery Co...sold to Roy this fall." The Interstate Grocery Company opened an immense five-story warehouse on Tenth and Main Streets in Joplin near the Missouri Pacific Depot in February of 1915. (Photo Courtesy John Burrington.)

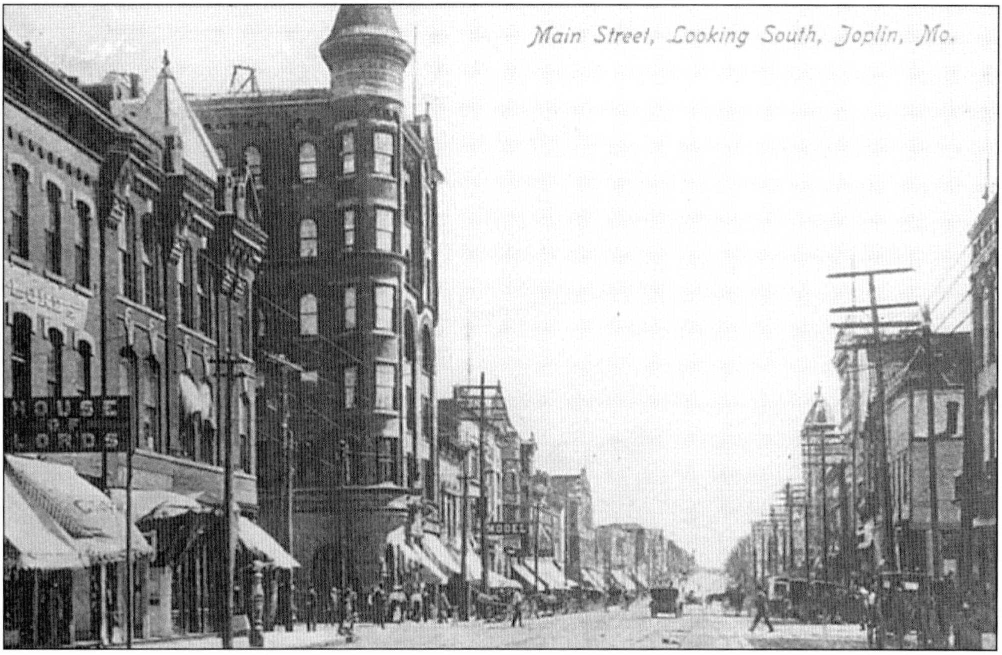

Another view of Joplin's Main Street looking south shows, on the extreme right, the old Joplin Hotel which would later be razed and replaced by the Connor. To the left, across Fourth Street north of the Keystone Hotel, stands the House of Lords—the notorious saloon, gaming house, and brothel. The House of Lords was to contribute to Joplin's image as wide-open and hell raising.

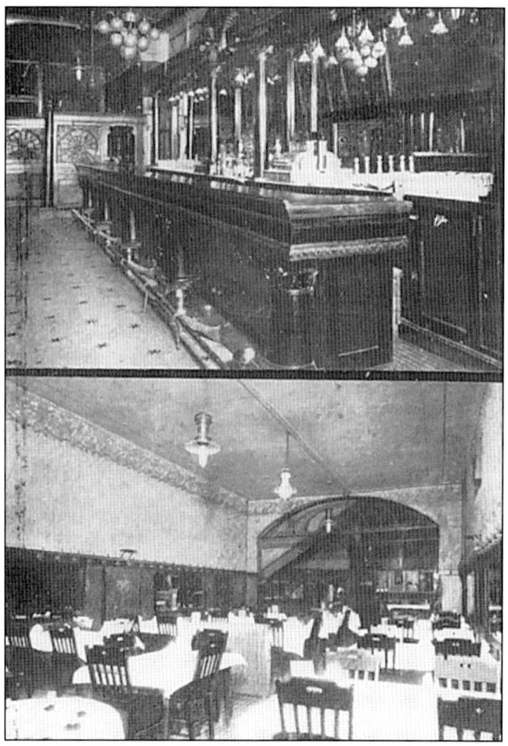

Legend has it that a visitor from England, impressed with the elegant interior, suggested that the new saloon be called the House of Lords. Be that as it may, its fame soon spread far and wide. The bar and cafe, pictured here, shared the second floor with various games of chance. Influential musicians such as Scott Joplin gave it a reputation as one of the founding venues of ragtime music. One local musician who felt that influence was Jimmy Scott of Neosho, who later became a noted songwriter and performer. He penned the music to "Take Me Out to Lakeside" in honor of Lakeside Park, another area attraction where he performed.

Joplin's picturesque Spiva Park is now located where the House of Lords once stood. In it a marble statue of a miner stares, perhaps in wonder, out across the city.

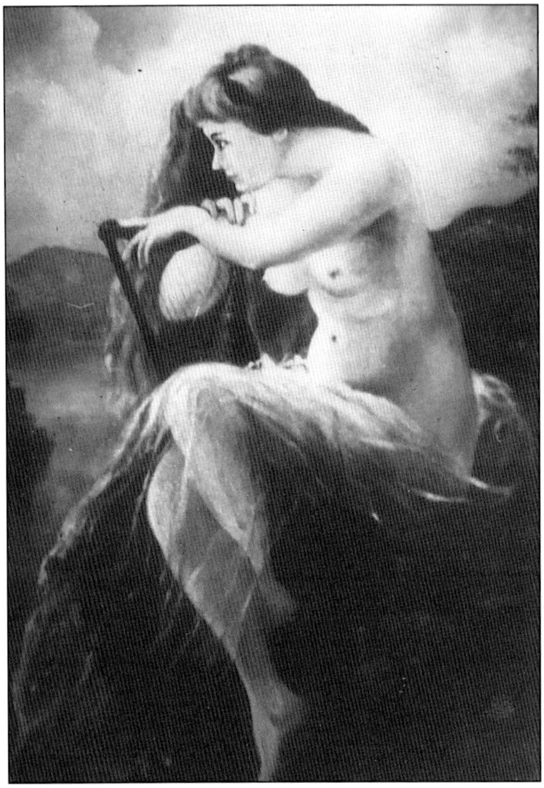

Another young man from nearby Neosho who frequented the bar at the House of Lords was Thomas Hart Benton. He later remarked that when he worked as a young cartoonist for the *Joplin American*, the jibes of the older Lords patrons, plus the paintings displayed on its walls, had a strong influence on his decision to take up painting as a career. Mr. Benton eventually became one of the country's most noted muralists. One of his murals was specifically painted for the Joplin City Hall, which stands only a few blocks from the former location of the House of Lords. Shown here is one of the paintings he would have seen on his visits to the bar.

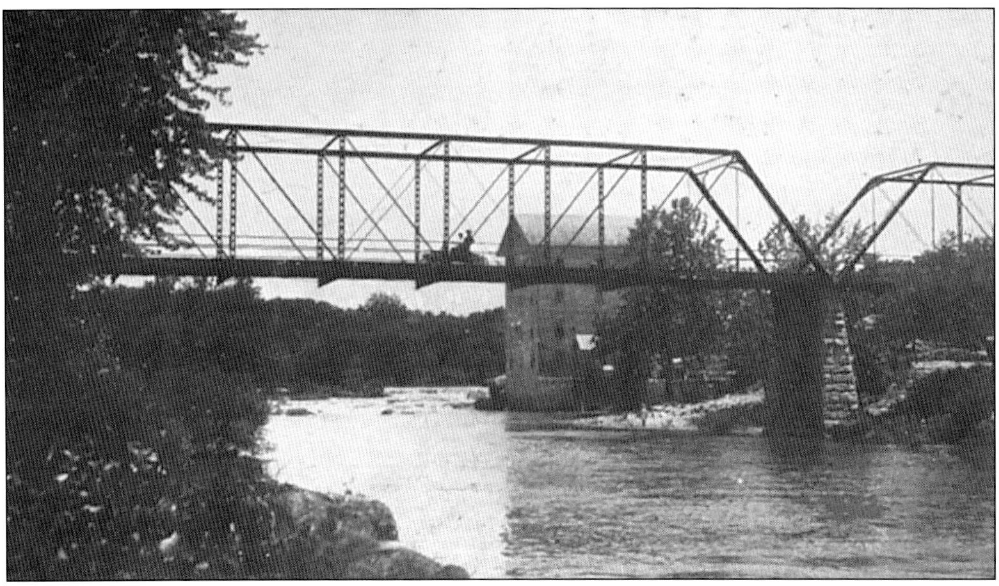

Another of Joplin's scenes of infamy was this Redings Mill landmark. This location on Shoal Creek became famous in 1933 as the high-speed escape route of Bonnie Parker and Clyde Barrow after their deadly shoot-out with local lawmen in Joplin. That escapade, along with one similar in Carterville in 1927 involving the Ma Barker gang, contributed to the area's freewheeling reputation as typified by the House of Lords. In 1930, the truss bridge was replaced with a multi-arch concrete structure across which the fleeing bandits escaped.

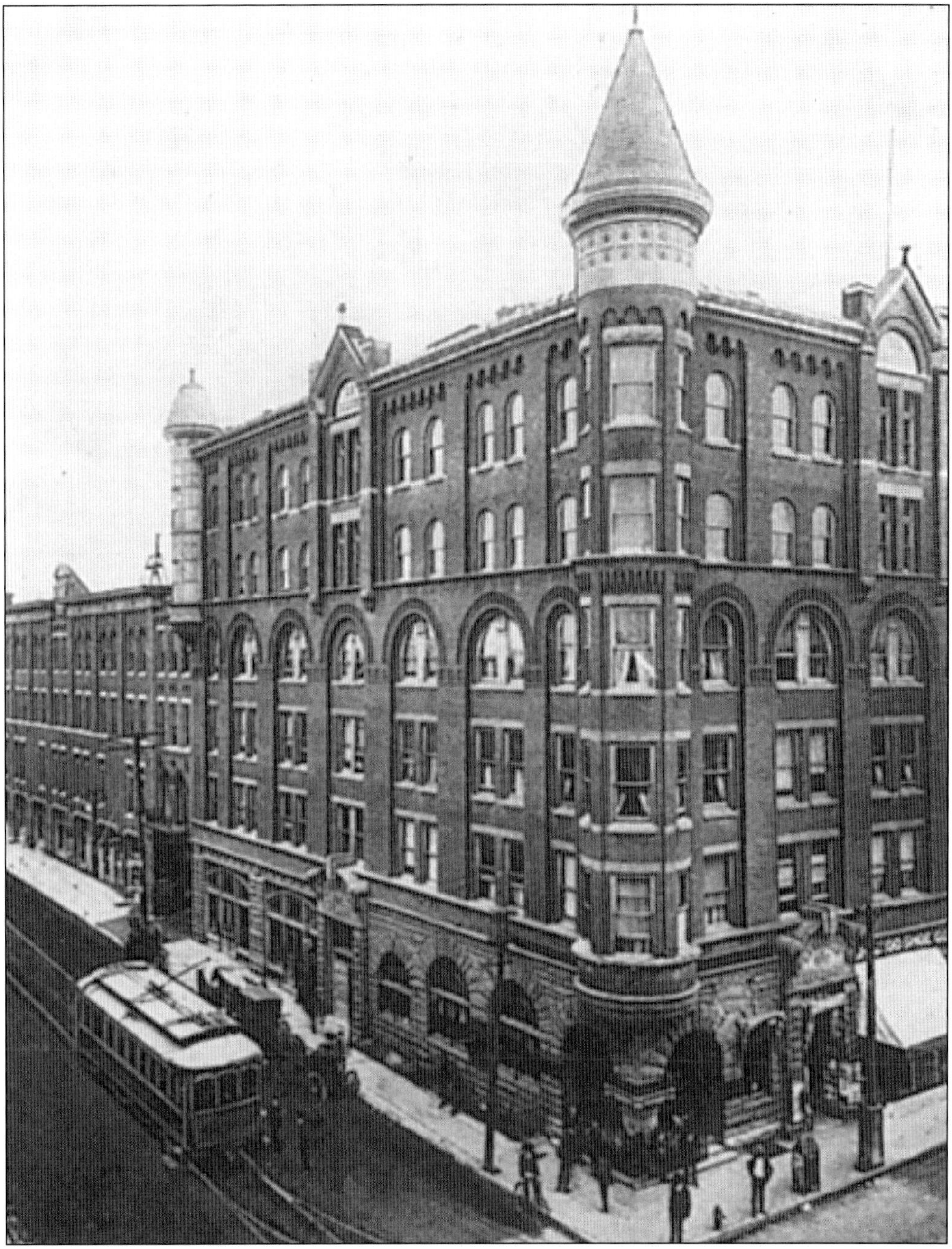

The Main and Fourth Street intersection in Joplin proved to be a magnet for impressive hotels. The first, the 50-room Joplin Hotel, was built on the northwest corner in 1874, only a year after Joplin became a city. The Keystone Hotel, pictured above, followed on the southeast corner 18 years later. Its builder, Elias Z. Wallower, named it "Keystone" in honor of the commonwealth of Pennsylvania, from which he hailed. For most of its existence, it stood in the shadow (both literally and figuratively) of its more imposing neighbor, the Connor, which replaced the Joplin Hotel. However, there were those who believed the Keystone had more architectural character by comparison.

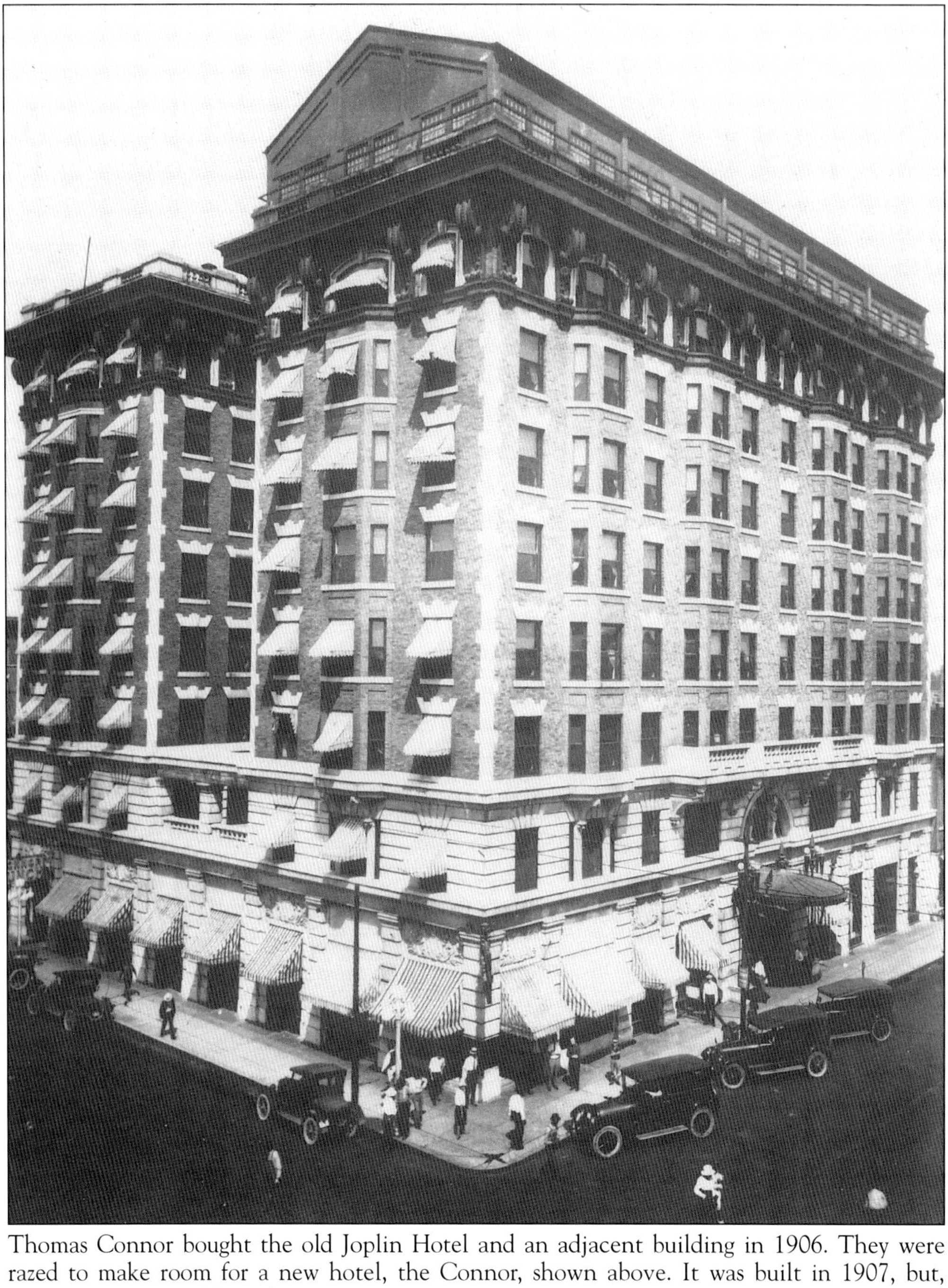

Thomas Connor bought the old Joplin Hotel and an adjacent building in 1906. They were razed to make room for a new hotel, the Connor, shown above. It was built in 1907, but, sadly, Mr. Connor did not live to see its opening, dying that year shortly before its completion. The impressive structure survived until 1978 when, though listed on the National Register of Historic Places, it was being prepared for demolition. It made headlines nationwide when it collapsed prematurely, trapping three workmen beneath tons of debris. One was pulled from the rubble alive.

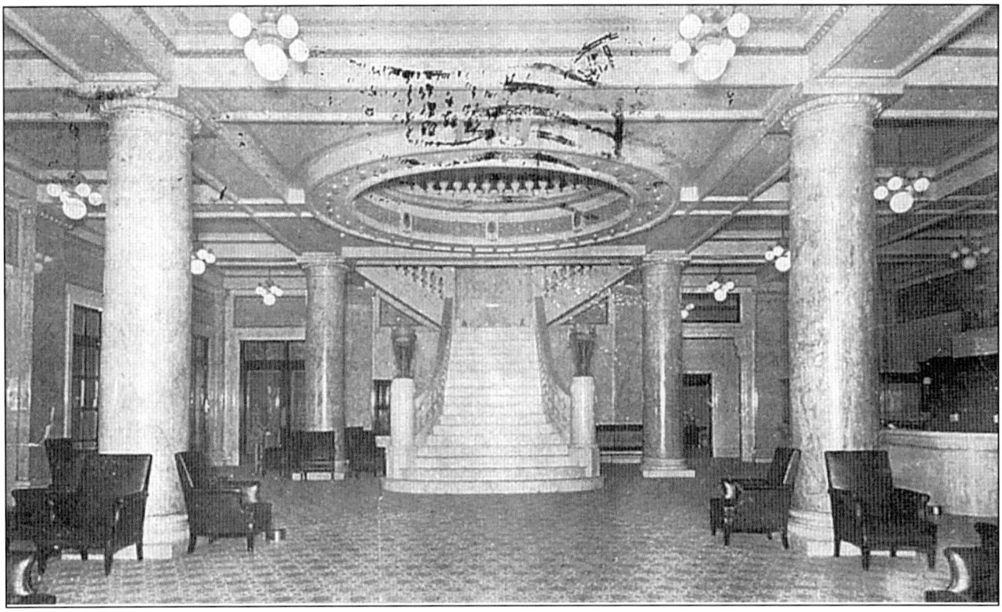

An old postcard, with the cancellation stamp visible on the back, depicts the elegant interior of the Connor Hotel. Visitors from other parts of the country were surprised and impressed with the marble splendor of the lobby, which doubtless would have been acceptable in the finest hotels in the world. (Photo Courtesy John Burrington.)

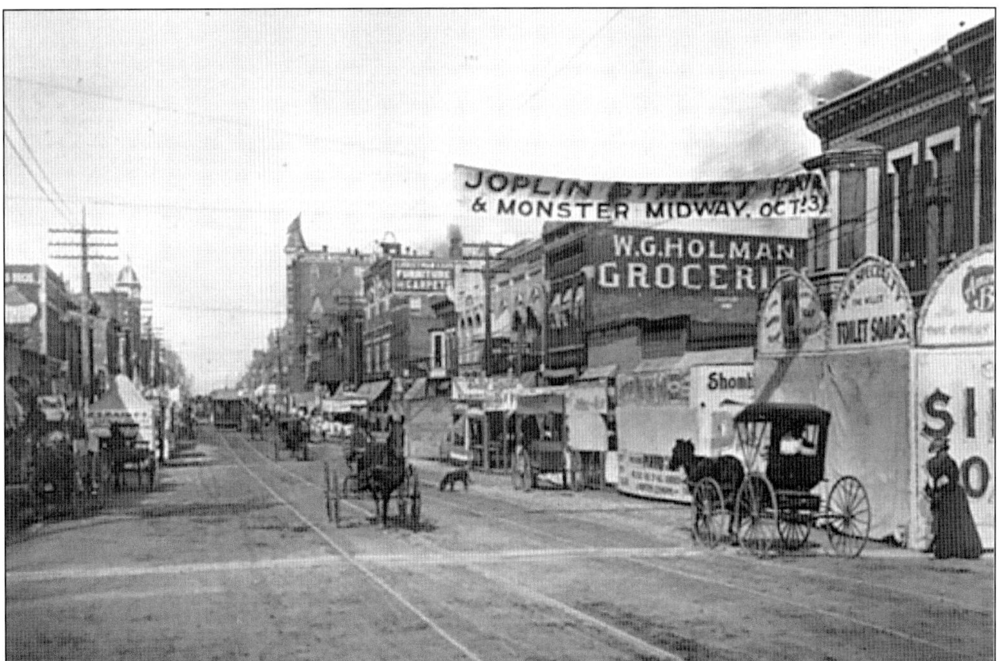

An early view of Joplin's Main Street during a street fair looks to the north. A tower of the Keystone Hotel is visible in the distance on the right side of the street, with the twin domes of the Columbian Block on the left. The Connor was yet to be built.

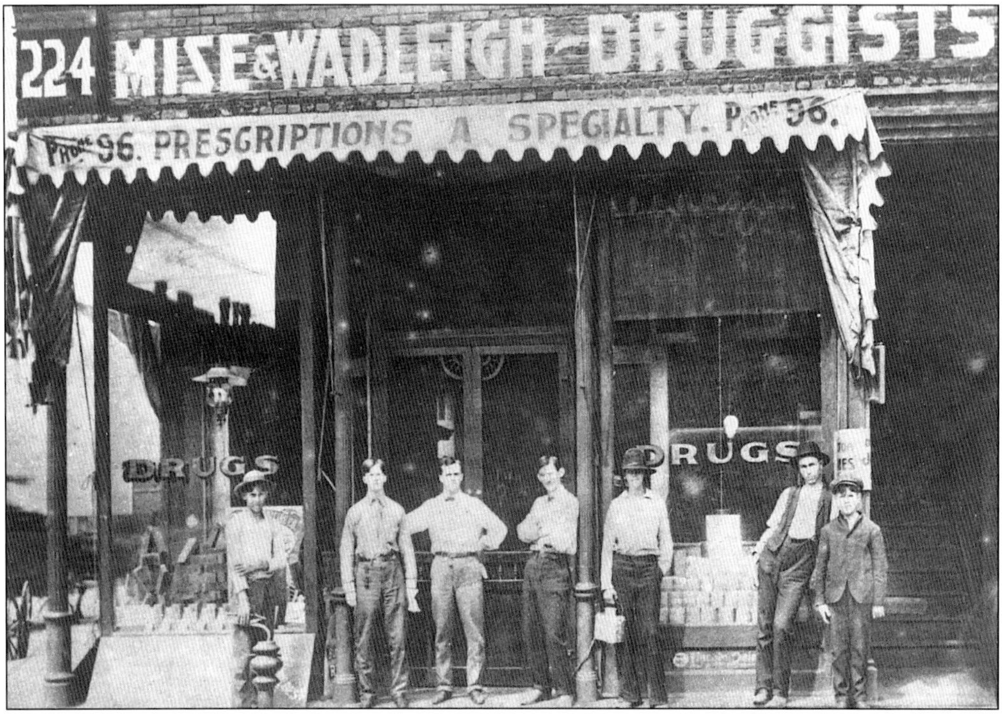

In 1898, bystanders pose in front of the Mise & Wadleigh Drug Store in Joplin. Much like today, early drug stores sold more than the apothecary's drugs and nostrums, but also vending sundry items from wallpaper to jewelry.

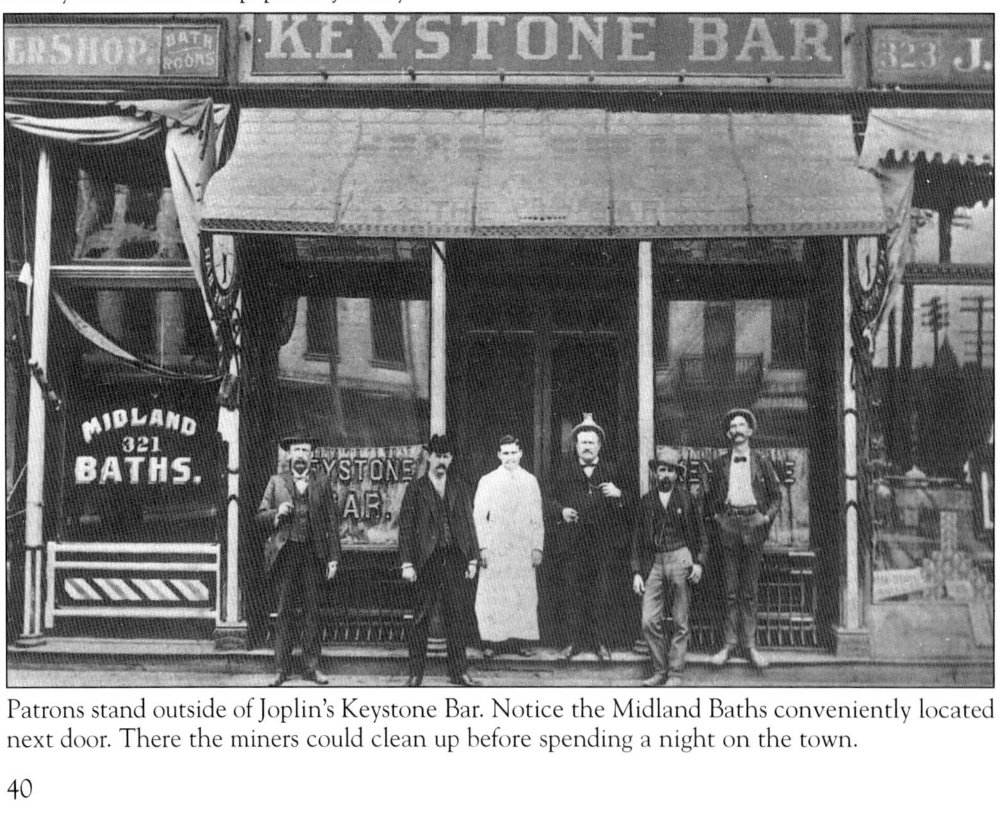

Patrons stand outside of Joplin's Keystone Bar. Notice the Midland Baths conveniently located next door. There the miners could clean up before spending a night on the town.

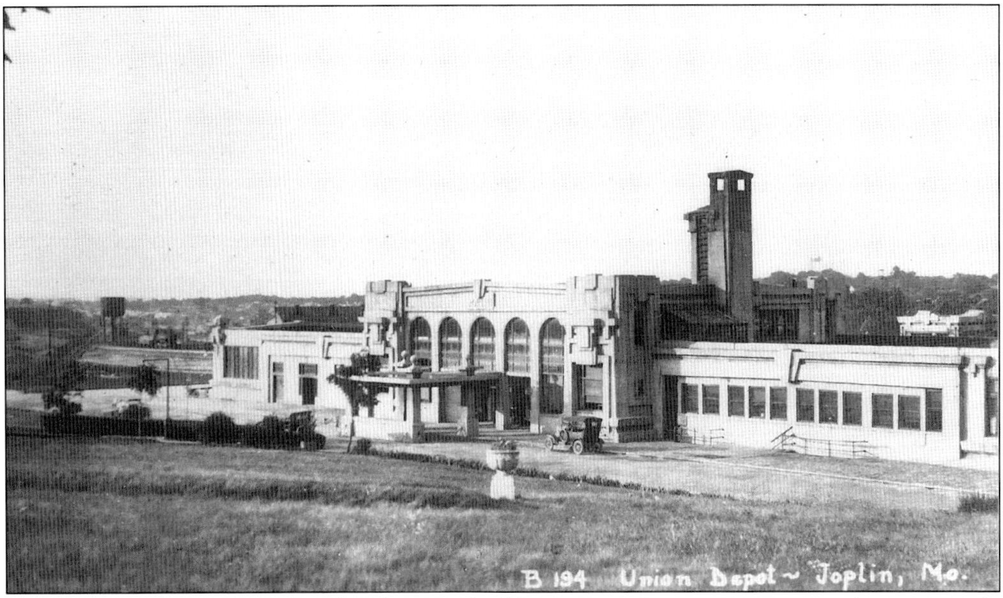

The Union Depot is one of Joplin's historic buildings. Its architecture is unique, being an early example of structures built with reinforced concrete. Train service began in 1911 and was in use until 1969. Now an empty shell, the old depot is almost hidden in the valley behind the post office. Efforts have persisted for years to save it for commercial or public use.

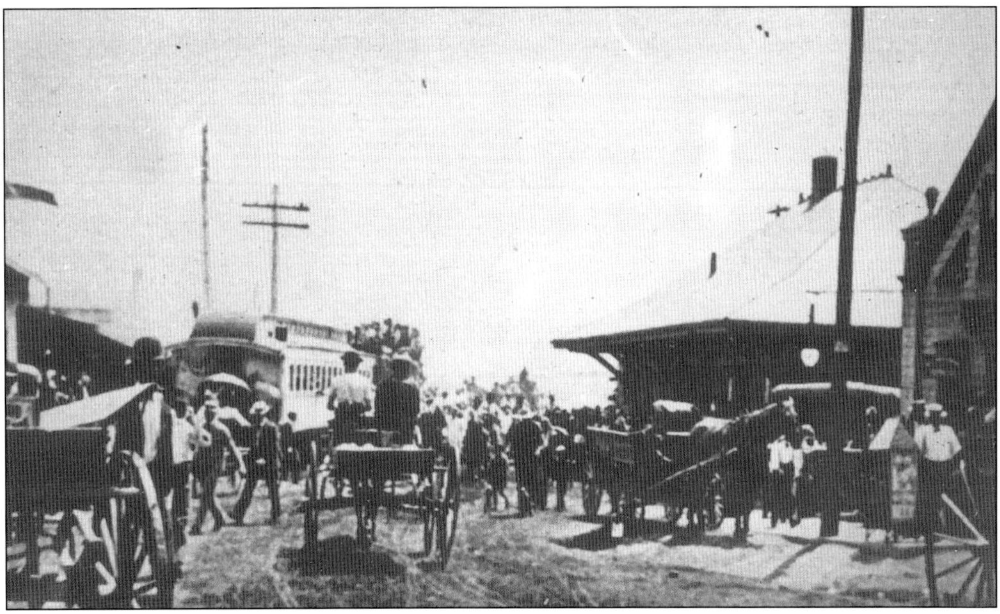
The Frisco Depot (shown) at Sixth and Virginia was a busy station, as was the Missouri Pacific at Tenth and Main. Late in the twentieth century, the MoPac building was dismantled stone by stone by a contractor, David Glenn, and rebuilt on Joplin's busy Range Line where it became the Iron Horse Restaurant and Brew Pub.

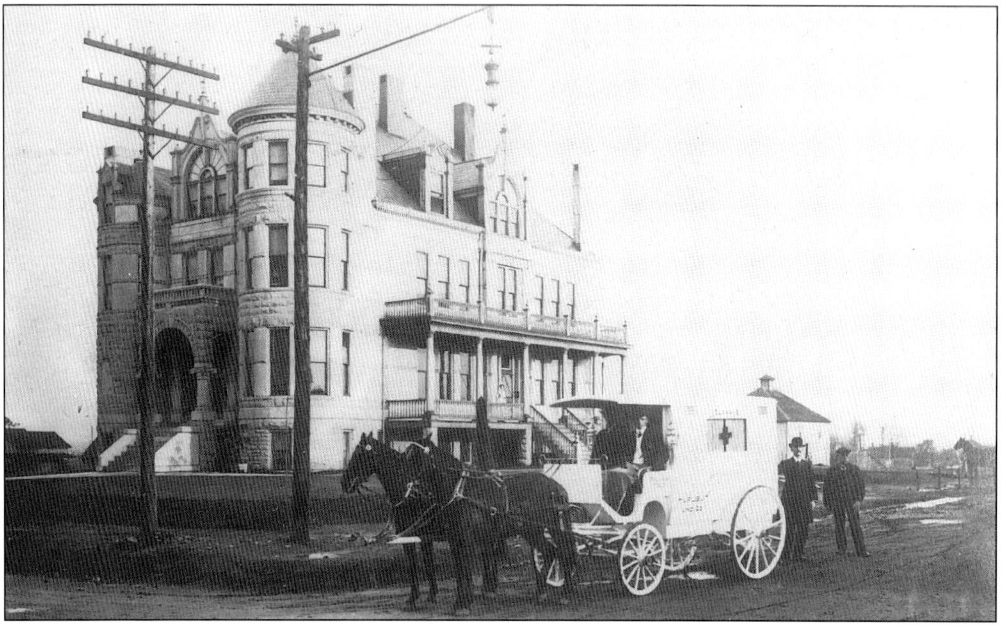

Because the need was so apparent as the city grew, the original St. John's Hospital was opened in 1900 by the Sisters of Mercy at 22nd and Connor Streets after locations in temporary quarters elsewhere. It was designed by Joplin's own self-taught architect, August Michaelis, and was built of Carthage stone. The structure cost $40,000, with the money for the building and furnishings coming from donations and fund-raising events, and was considered a modern and well-equipped institution. Mr. Michaelis, son of a German immigrant baron, was noted for the impressive structures he designed.

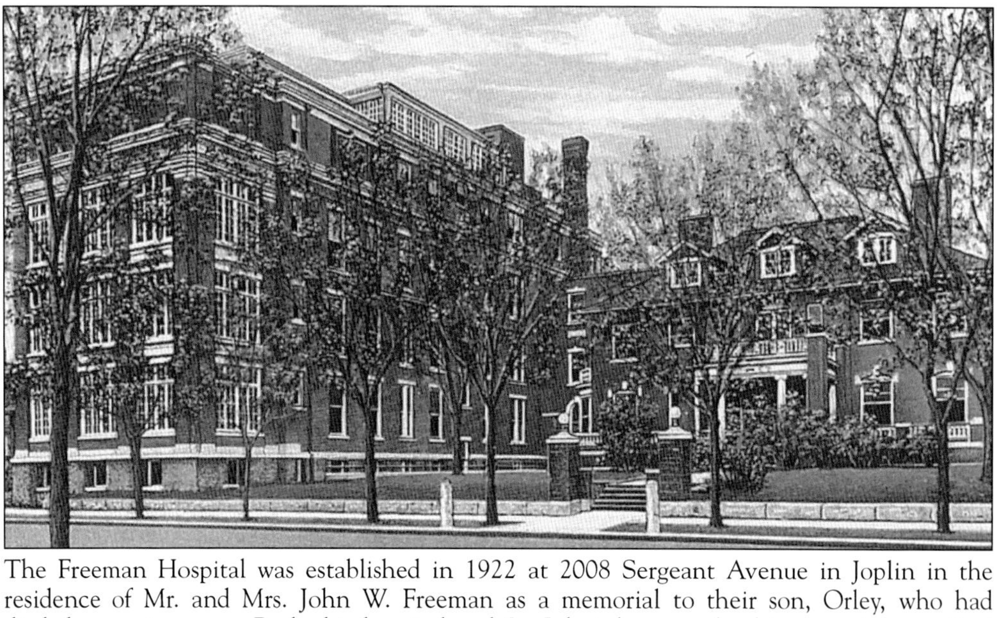

The Freeman Hospital was established in 1922 at 2008 Sergeant Avenue in Joplin in the residence of Mr. and Mrs. John W. Freeman as a memorial to their son, Orley, who had died the previous year. Both this hospital and St. John's have evolved into modern medical complexes, which draw patients from throughout the region and surrounding states. (Photo Courtesy John Burrington.)

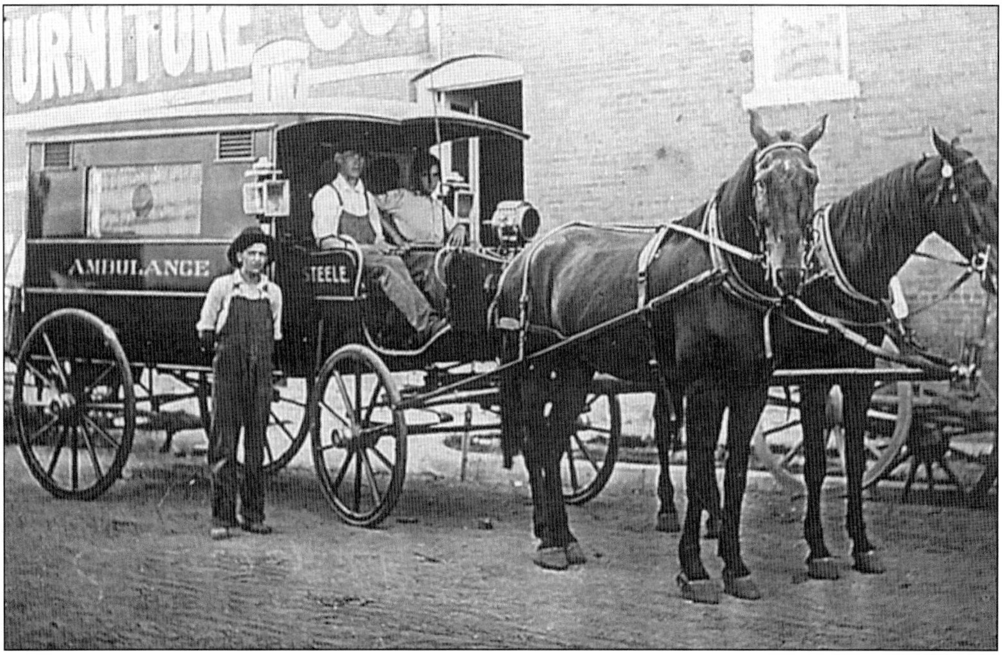

In the early 1900s, there was no organized service to transport sick or injured people, so mortuaries began furnishing ambulances which often were in competition to see who could arrive first at an accident. This practice continued into the second half of the century. (Photo Courtesy John Burrington.)

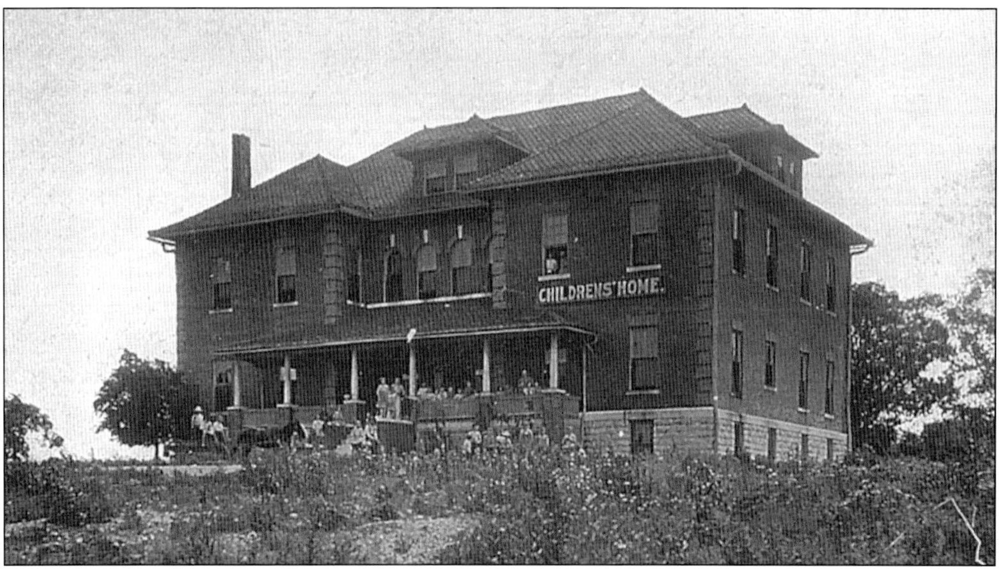

The WCTU founded an orphanage in Joplin in 1900. After early moves, the Children's Home settled in the building pictured above, which had been erected for that purpose at Third and Comingo Streets. A number of the nearly six thousand people who had lived there as children—the earliest from 1916—were drawn back to Joplin from across the nation to a reunion held there in 1991. The facility ceased operation in 1957. (Photo Courtesy John Burrington, information Doyle White.)

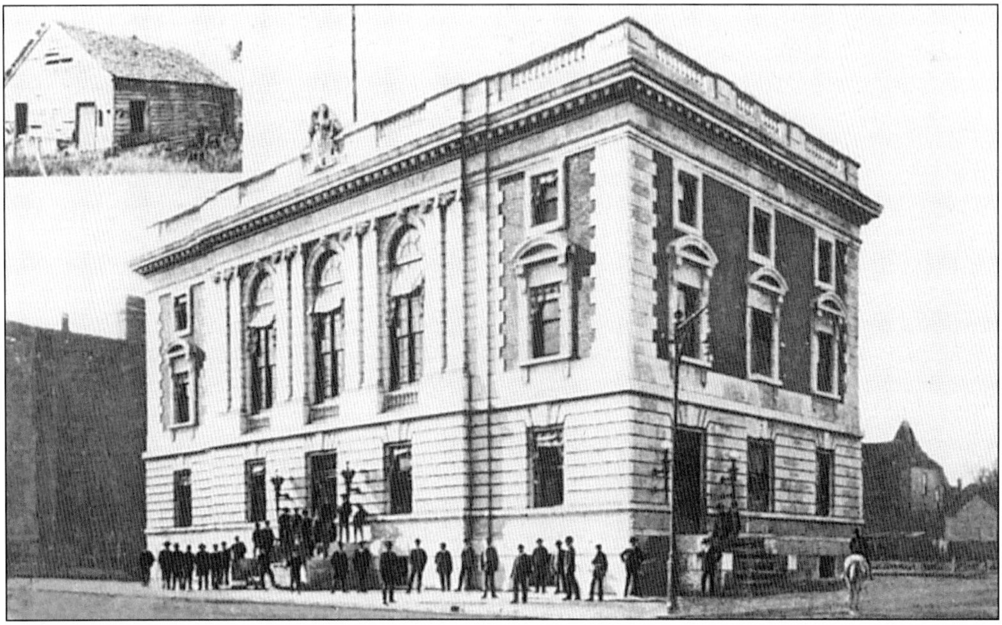

This post office later became the Durward G. Hall Building for federal offices, and later continued use as the federal courthouse. Media attention was focused on the court in 1984, when a change of venue moved the Christine Craft discrimination trial there. The inset shows Joplin's first post office.

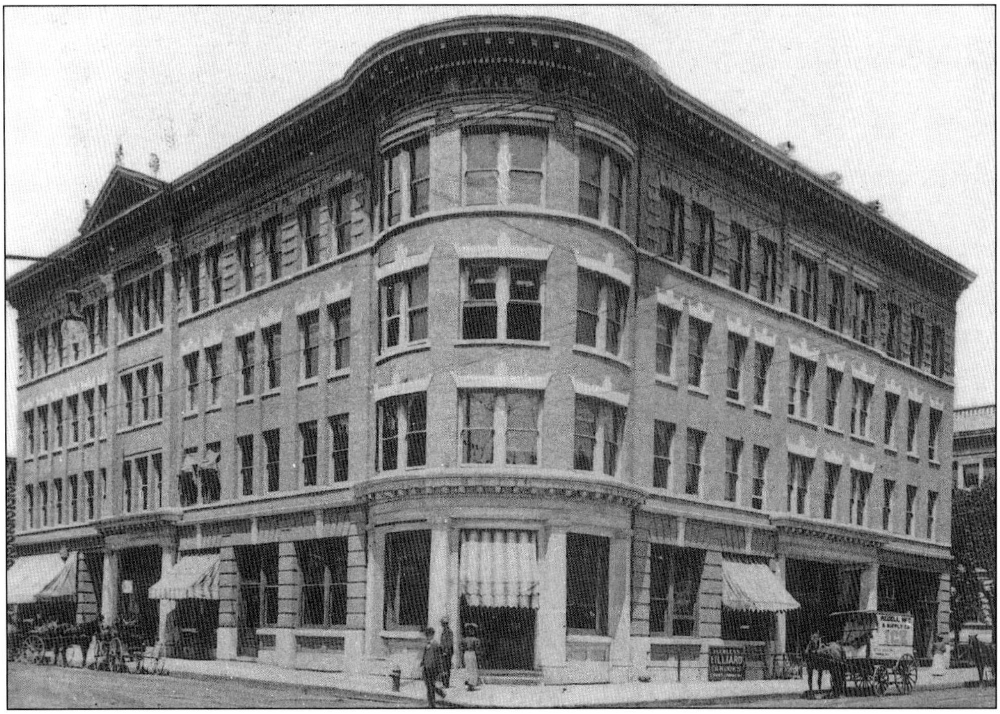

The Miners Bank first opened in 1877, only four years after Joplin was officially chartered. It relocated into this building, another Michaelis design, on the northwest corner of Fourth and Joplin Streets in 1905. The building no longer exists.

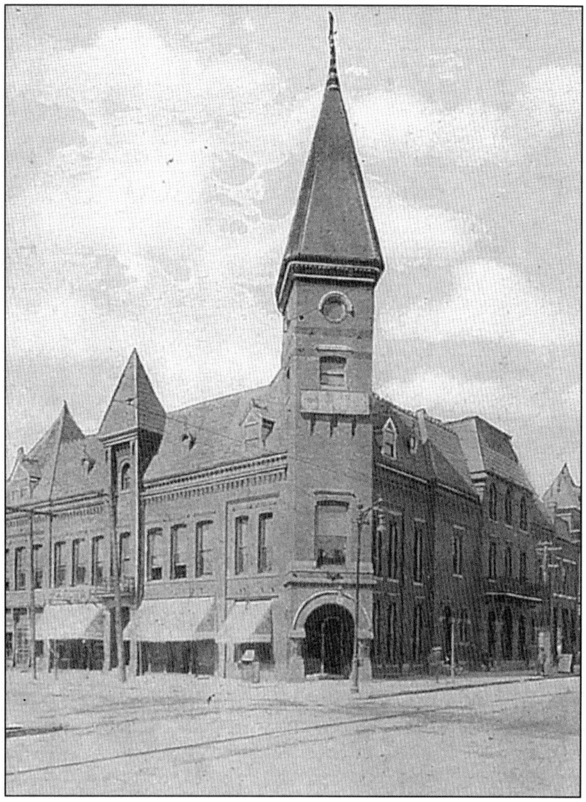

The Club Theatre opened in 1891 south of and across the street from the future location of the Miners Bank Building. It remained a popular destination for audiences as well as the touring performers who appeared there. The theater underwent major remodeling and was reopened in 1917, but in December of 1918 was tragically destroyed by a spectacular fire, which for a time threatened the bank across the street. Other portions of the building survived, one of which housed the Joplin *News Herald* newspaper until 1922. (Photo Courtesy John Burrington.)

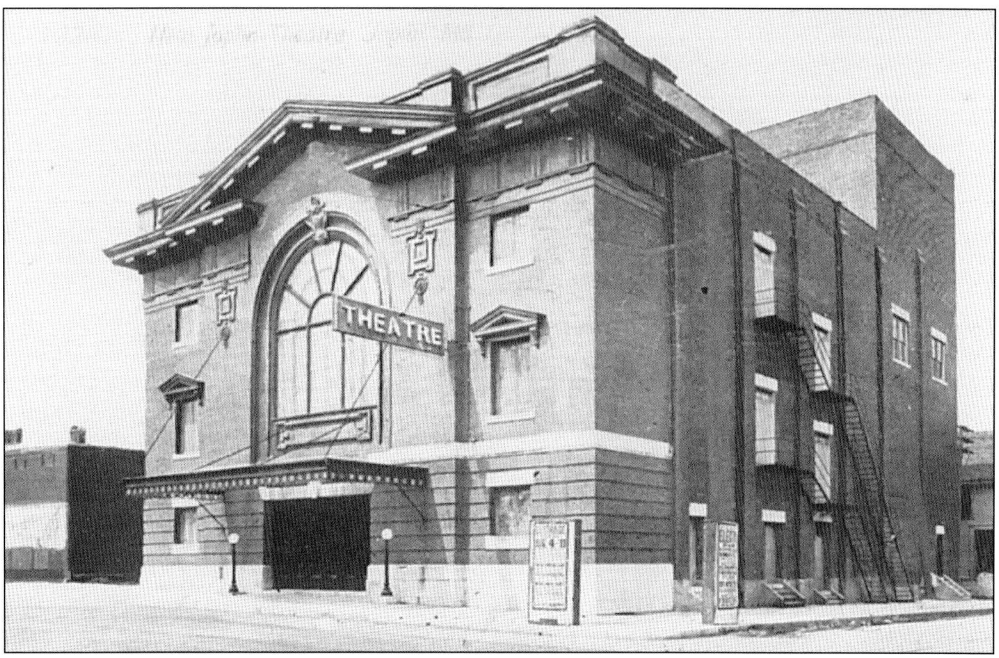

This is another popular Joplin theater of which no trace remains. It was variously called the Shubert Theatre or the New Joplin Theatre. It opened in 1908, seated over 1,500 people, and was located at the southwest corner of Seventh Street and Joplin Avenue.

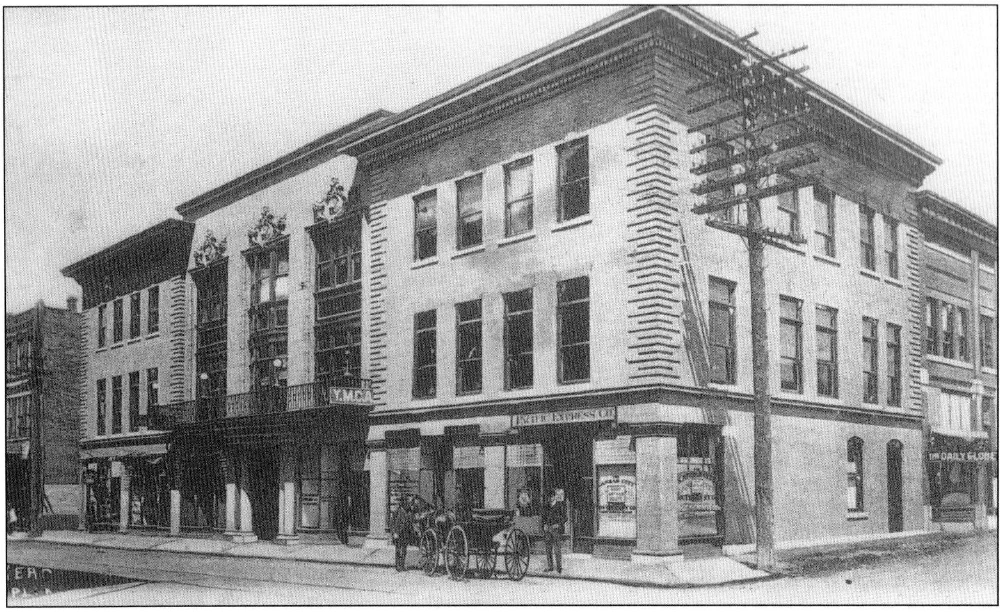

Shortly after the turn of the twentieth century, the YMCA was located in this building at the northwest corner of Fourth Street and Virginia Avenue on the site of the old Haven's Opera House. At the far right the Daily Globe sign is visible. The Pacific Express Company was located on the corner.

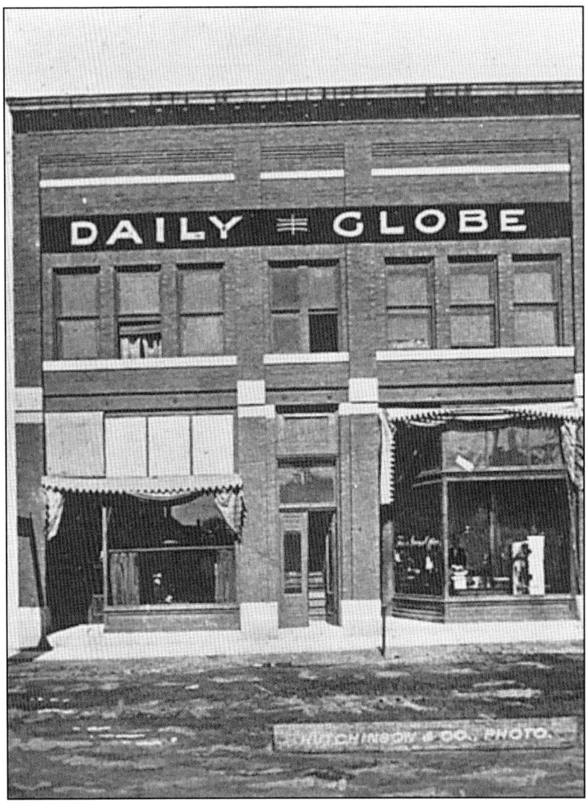

The *Joplin Globe* newspaper has been a part of Joplin's history since 1896. In the early years it was located just north of the YMCA building, which stood at the northwest corner of Fourth Street and Virginia Avenue.

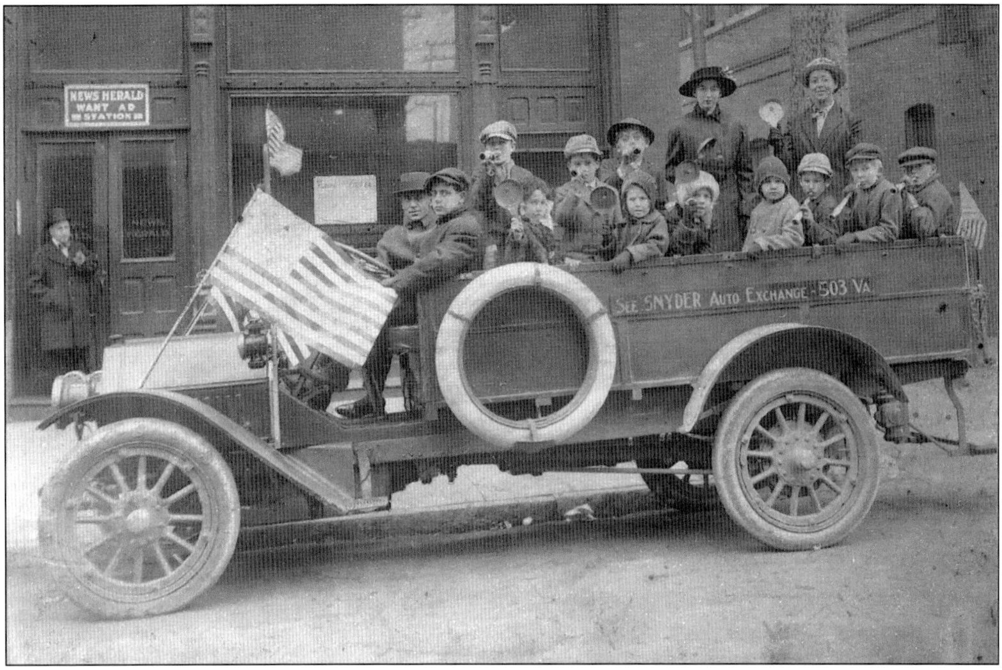

The occupants of this vehicle—bundled against the cold and holding toy horns and other noisemakers—were apparently prepared for a parade, perhaps the first Armistice Day celebration. Behind them is a want ad station for the *Joplin News Herald*, Joplin's afternoon paper. Older people remember the shout of newsboys on the corners of Joplin's busy Main Street—"JOPLIN NEWS HEREEAULLD."

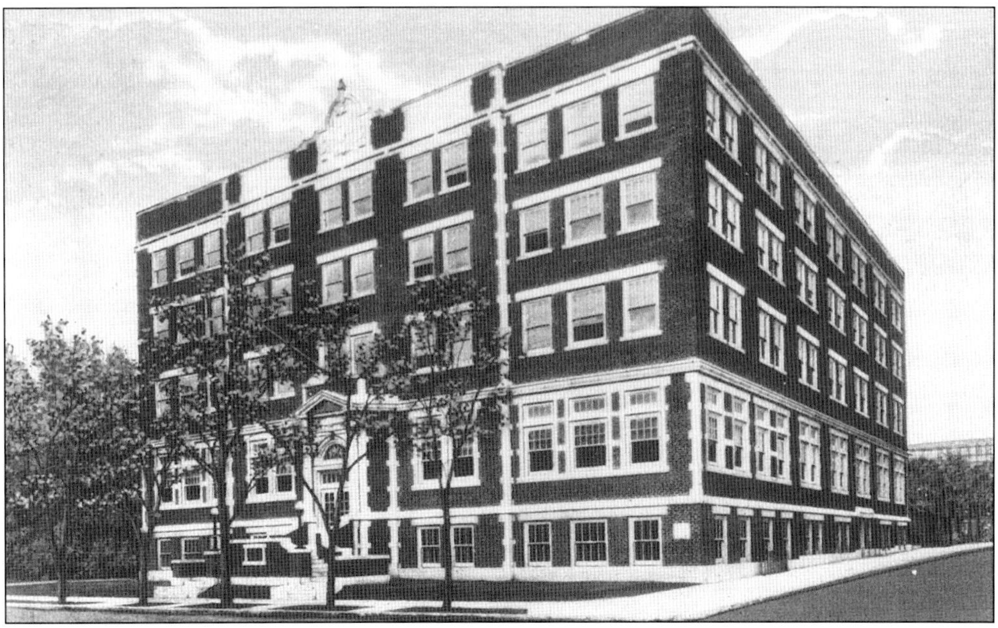

This building, erected specifically for that purpose at the corner of Fifth Street and Wall Avenue, became the home of the YMCA in 1920. It has been remodeled and continues in use to this day as the Joplin Family Y.

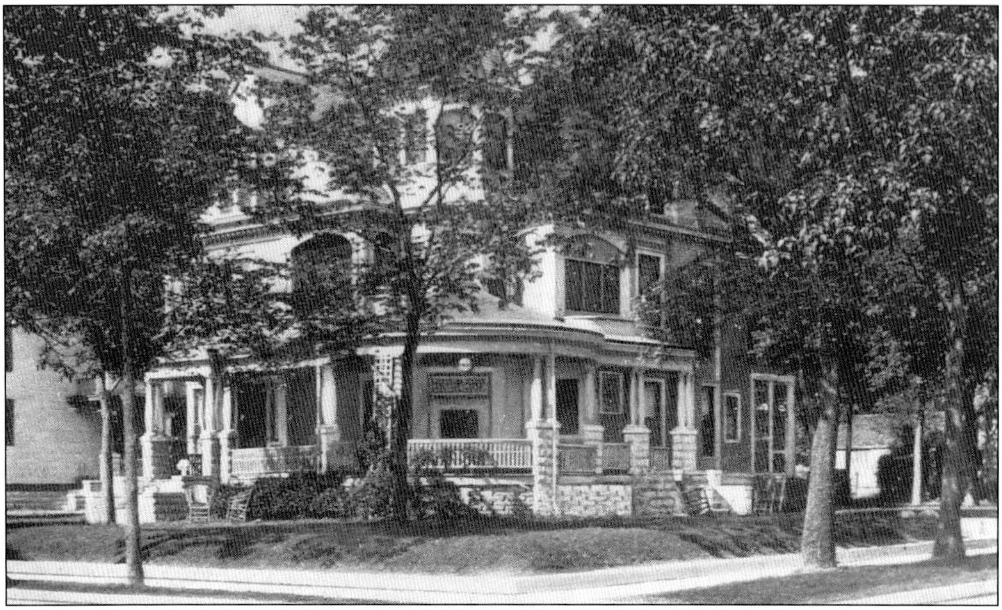

The Joplin YWCA was located in this beautiful Victorian residence at 504 Byers Avenue from 1920 to 1974. Later an artist, Ken Eberts, used it as the snowy setting for a Leanin' Tree Christmas card depicting a World War II soldier returning home in a 408 Cab. Older people will recall the throngs of soldiers who packed the sidewalks of Joplin's Main Street during the war when Camp Crowder was in operation at Neosho.

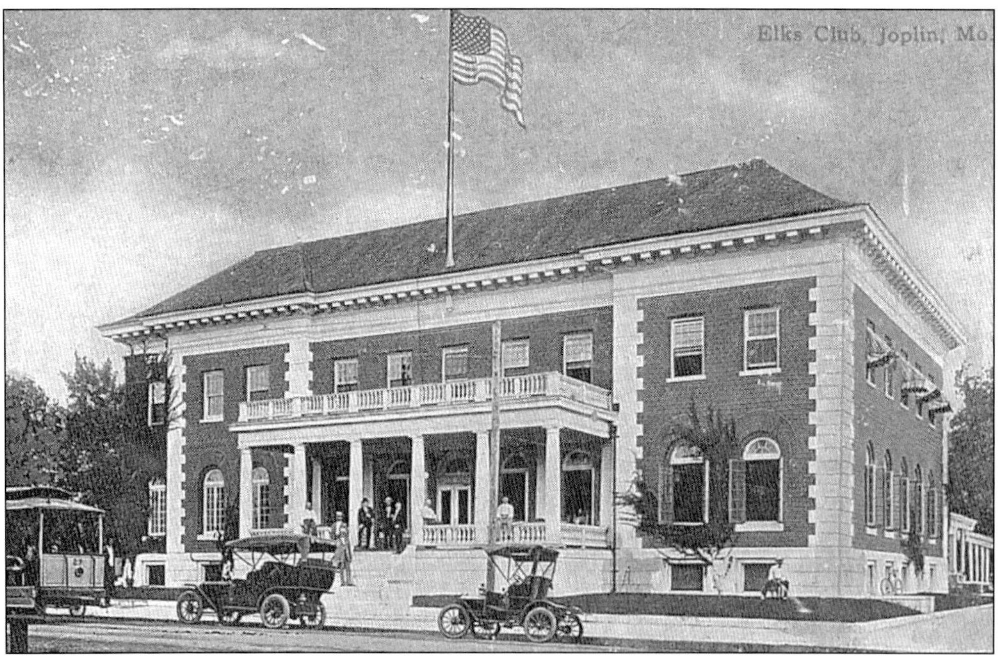

The Benevolent and Protective Order of Elks opened their handsome new clubhouse building in 1905 on the southeast corner of Fourth Street and Pearl Avenue. The building remains, but the Elks have moved to another location. (Photo Courtesy John Burrington.)

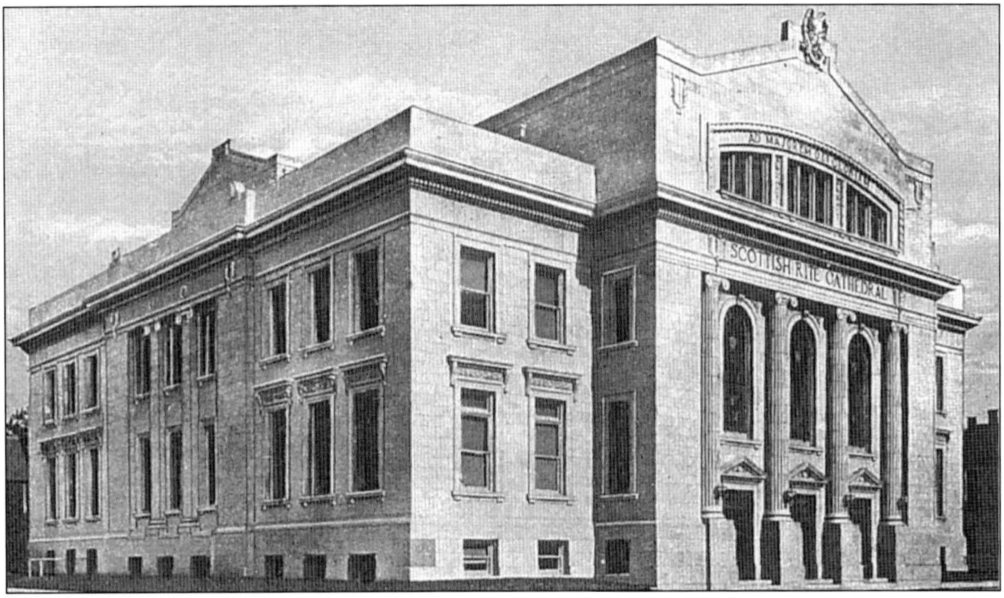

Opened shortly after the turn of the century, the first Scottish Rite Cathedral in Joplin was also said to be the first in the state outside of St. Louis. The Cathedral pictured here later replaced it. Located at the corner of Fifth Street and Byers Avenue on land donated by Charles Schifferdecker, it was dedicated in 1923 and continues to be used today. (Photo Courtesy John Burrington.)

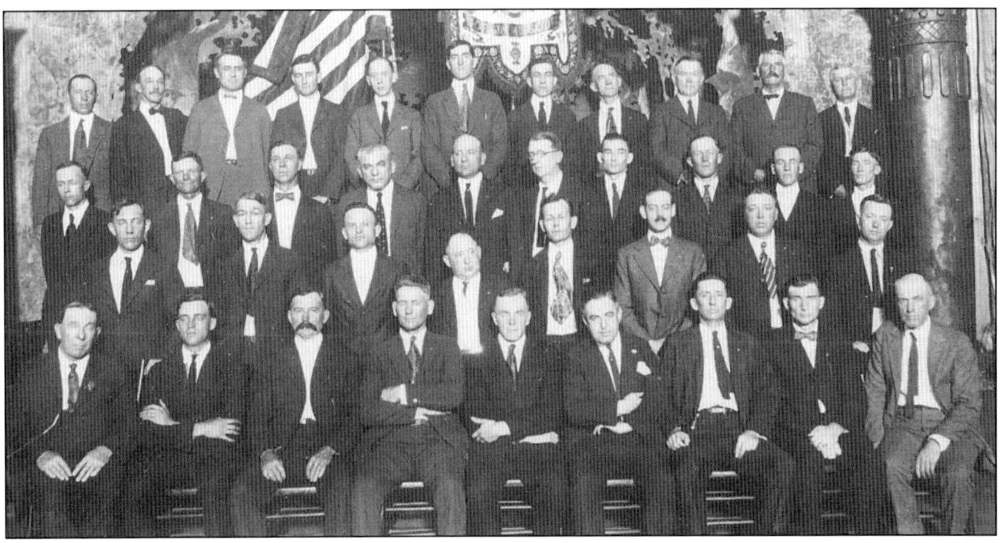

The 1921 Scottish Rite New Era Class poses for a photograph. The men are identified from left to right as follows: (front row) A.J. Elliott, R.G. Bryant, J.S. Merdon, V.W. Anderson, H.H. Wallower, M.J. Ettinger, H.A. Gardner, B. Harnish, V.A. Wilbur; (second row) C.S. McKee, J.N. Maddy, G. Ross, S. Brown, G.C. Butler, K.C. Johnson, B.W. Rogers, B.A. Cartwright; (third row) W.A. Roby, T.D. Stephens, C.O. Williams, M.S. Jankowsky, A.F. Smith, E. Hawkins, H. Turner, T.B. Lowe, F.M. Jones, C.W. Bellden; (fourth row) D.C. Kellog, T.C. White, J.E. King, E.J. Gaines, H.B. Sanders, S.N. Douglas, T.N. Murray, O.A. McDill, F.W. Tuckwood, G.H. Kelley, and J.R. Woodfill Jr. (Photo Courtesy Webb City Genealogical Society.)

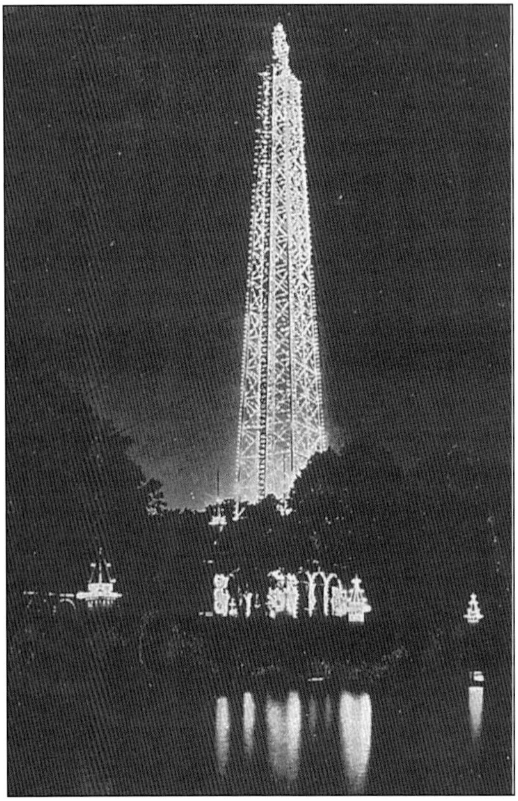

With the advent of electricity, an entire new world of magical possibilities emerged for people accustomed to the dim flickering illumination of candles, kerosene lamps, and gas jets. Nothing drove this fact home better than the impressive displays in Schifferdecker Electric Park, which opened in 1909. It was called by some the "Coney Island of the mining district," and drew crowds from throughout the area. The tower became a landmark at night with its thousands of incandescent bulbs. The park also featured amusement rides, concessions, swimming, and a theater. It eventually became part of the city park system. (Photo Courtesy John Burrington.)

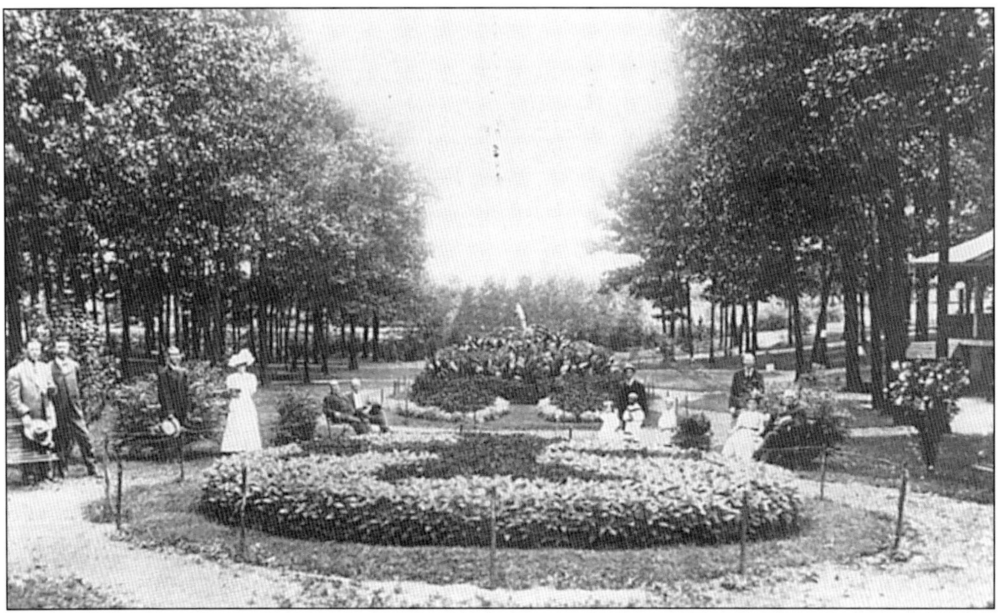

Joplin is blessed with many parks. Some have been left in a state of natural beauty, while others have been carefully manicured and maintained. This is an early view of one of the latter, Cunningham Park at 26th Street and Maiden Lane.

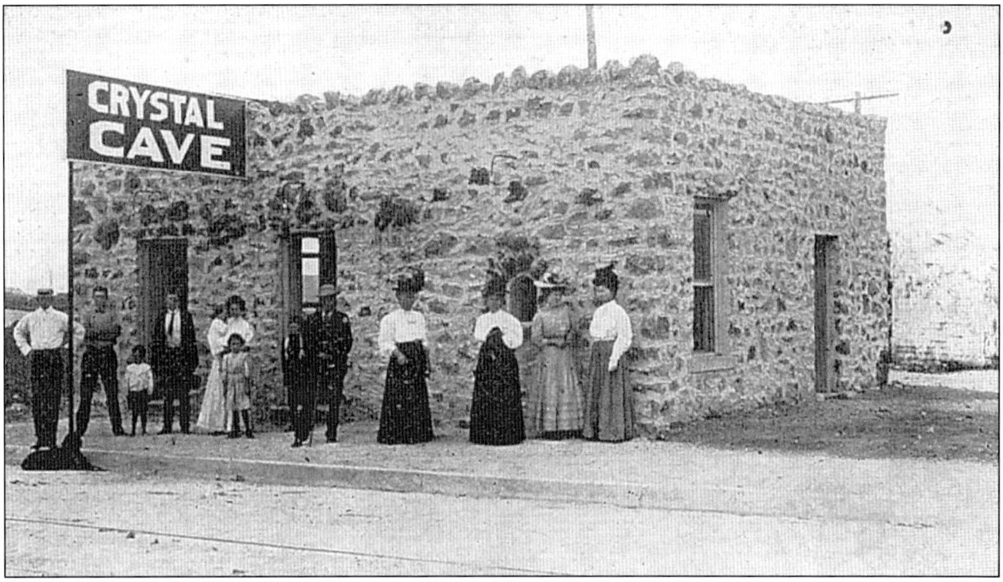

Crystal Cave, so named because of its sparkling crystal formations, became an early Joplin attraction. Its cool depths were doubly appealing in an age before air conditioning. Flooding later closed the cave. (Photo Courtesy John Burrington.)

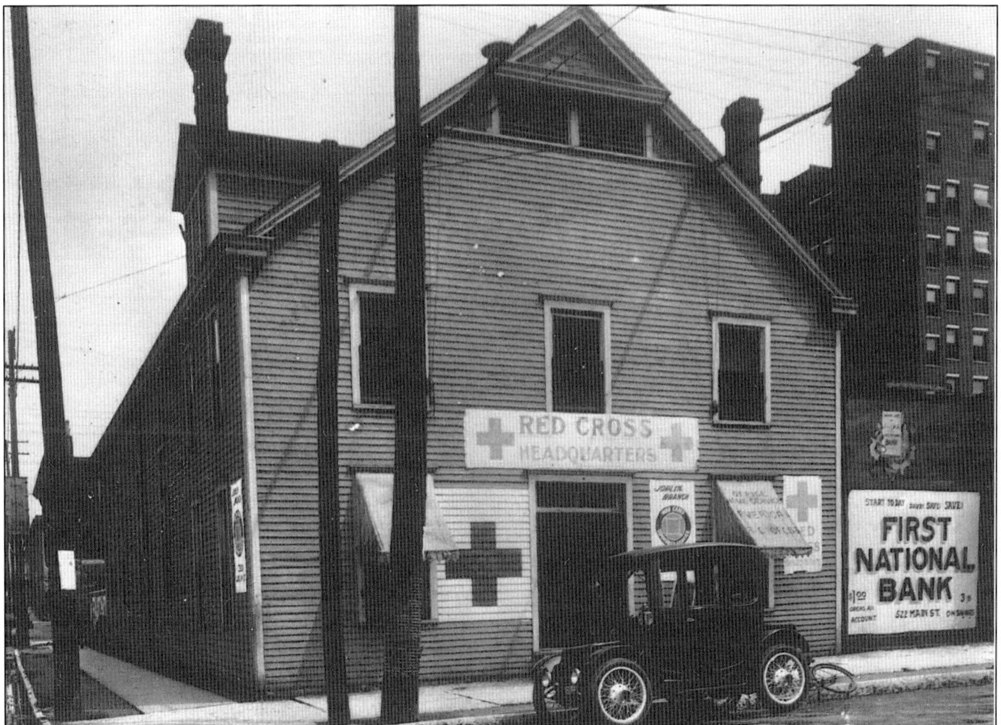

One of the early locations of the Joplin Red Cross was in this frame building located at Third Street and Joplin Avenue, a structure donated by a German club when anti-German sentiment rose at the beginning of World War I. At the curb stands what appears to be an early Detroit or Milburn Electric Coupe.

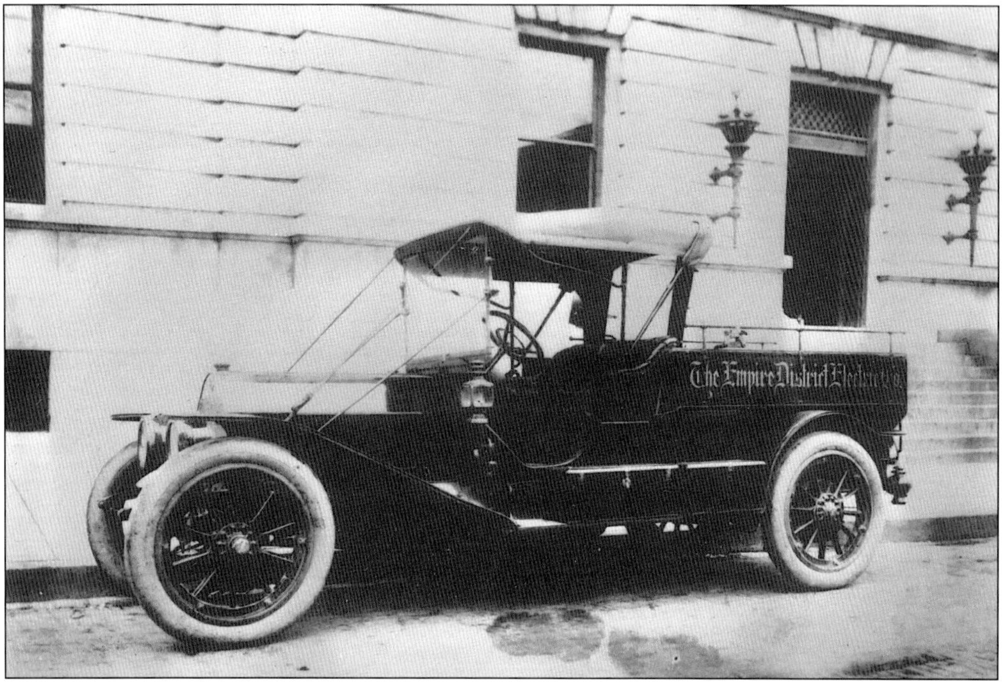

This truck with "The Empire District Electric Company" emblazoned on the side belonged to a firm that has been much involved in Joplin's history of growth. Its origin traces to 1909 when four smaller power companies—including one started by Joplin pioneers Moffett and Sergeant—were combined to light Main Street.

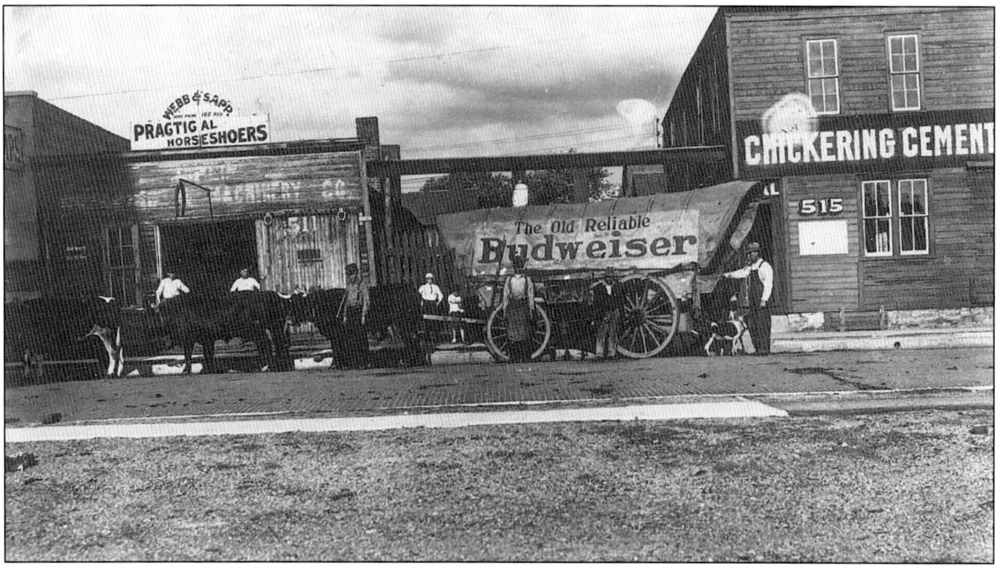

Before Clydesdales pulled the Budweiser wagon, oxen pulled the Budweiser Conestoga. They are pictured here visiting Webb & Sapp, practical horseshoers. Chickering Cement and Building Supply was founded by Henry Chickering, who moved to this area from Massachusetts in 1876. His lifelong interest was farming, and he turned his hand to it in Kansas. Unfortunately, the grasshoppers were so bad "they even ate the fence posts," so he returned to Joplin and founded the business instead. (Courtesy Helen Chickering, photo Ron Mosbaugh.)

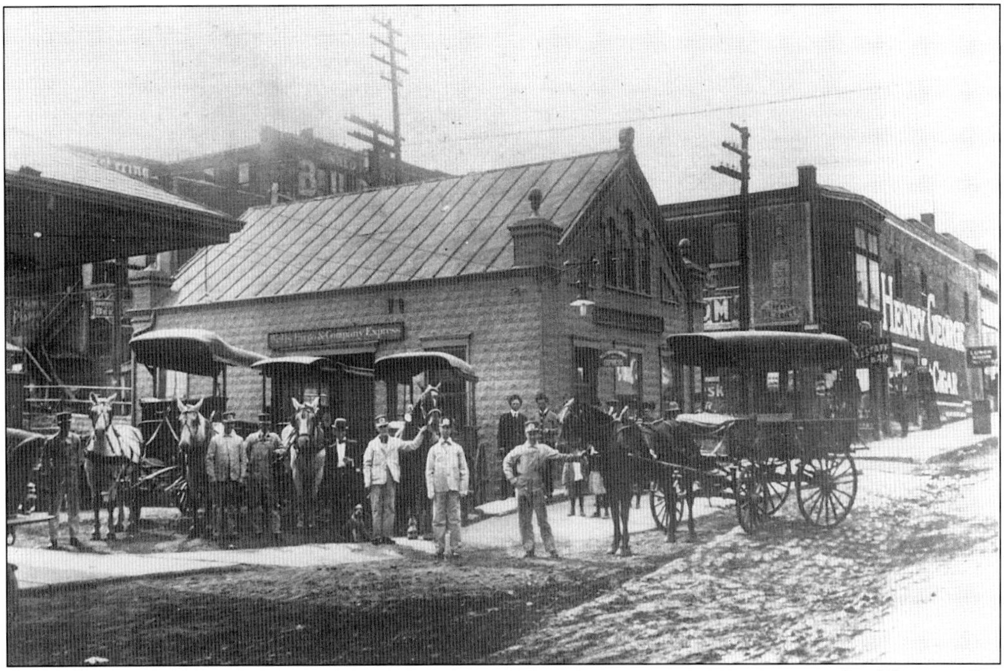

Employees with their delivery rigs pose by the Wells Fargo & Company Express building in this early photo. After a merger with two other companies in 1850 in New York, the Wells Fargo & Company was organized in 1852 to provide transcontinental express service. For a time it had a virtual monopoly in the West.

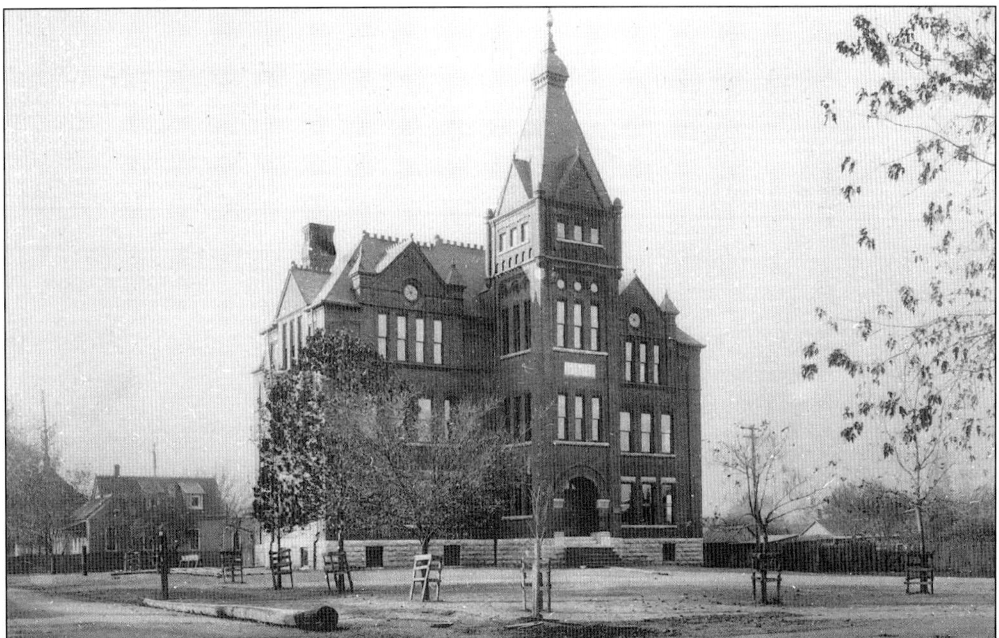

Built in 1892, the Jackson Avenue school served as Joplin's high school, as well as housing elementary classes. It was the last of the old school buildings to fall prey to modernization, being razed in 1932.

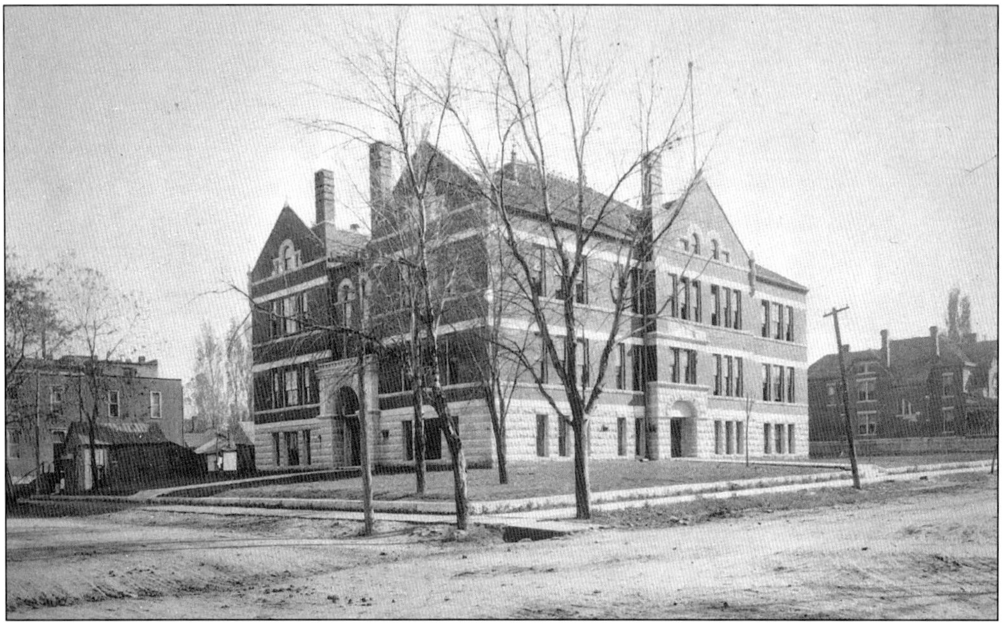

Joplin's new high school building, at the southeast corner of Fourth Street and Byers Avenue, was dedicated before a large crowd in 1897 with two of the original school board members from a quarter of a century earlier in attendance. It was an imposing edifice, but as the following photo so clearly shows, still only "half a loaf."

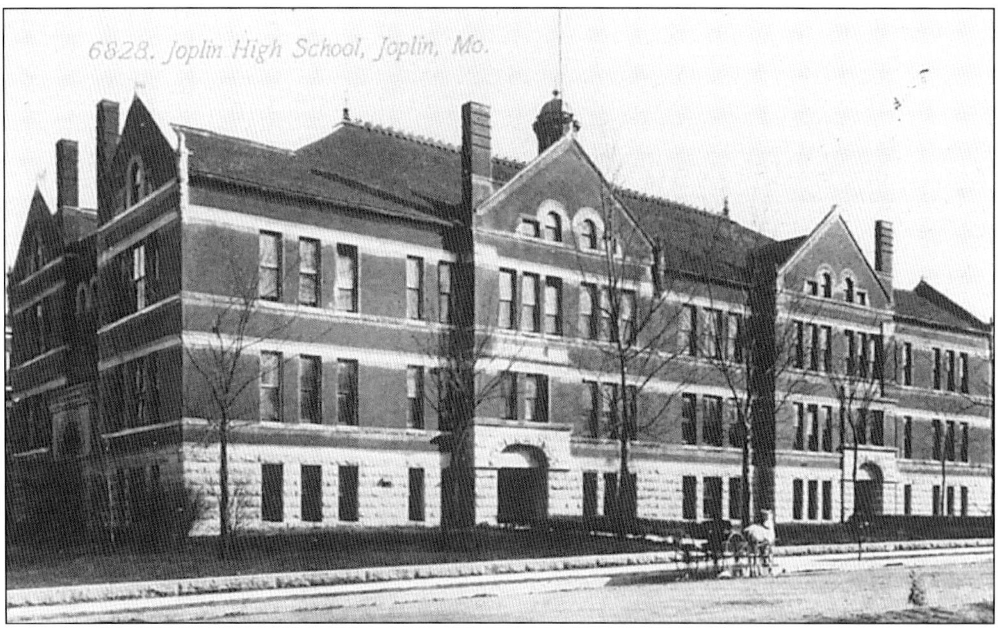

Ten years after it was built, the high school almost doubled in size, but only another decade passed before students abandoned it for a new high school at Eighth Street and Wall Avenue. Starting in 1938, the old high school building served for a time as Joplin Junior College. It was demolished in 1961.

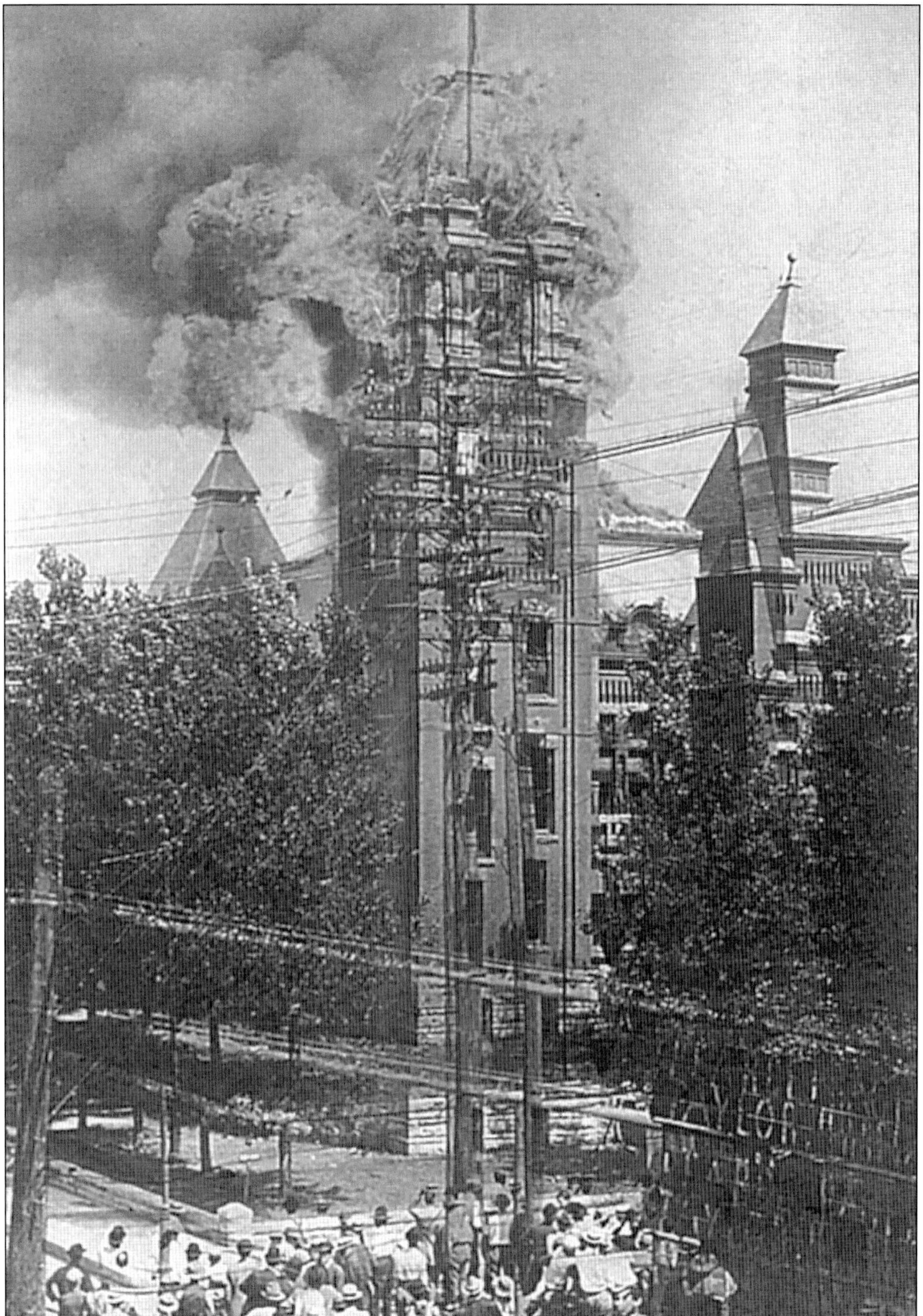

To serve the growing population and to supplement the courts at Carthage, a county courts building was erected in Joplin in 1895. Only 16 years later, on June 13, 1911, it was destroyed by fire. The cornerstone was saved and rededicated in 1994. (Photo Courtesy John Burrington.)

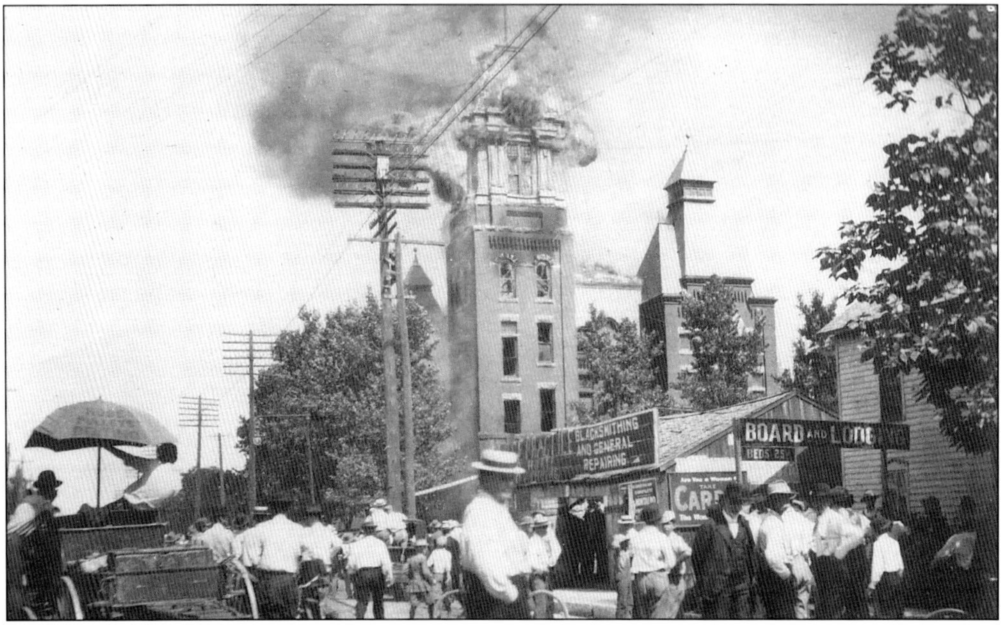

A crowd gathers to watch the burning courthouse. The lettering on the driver's umbrella spells out, "HGGE PULLEYS BELTING." The sign on the far right proclaims, "BOARD and LODGING, BEDS, 25¢."

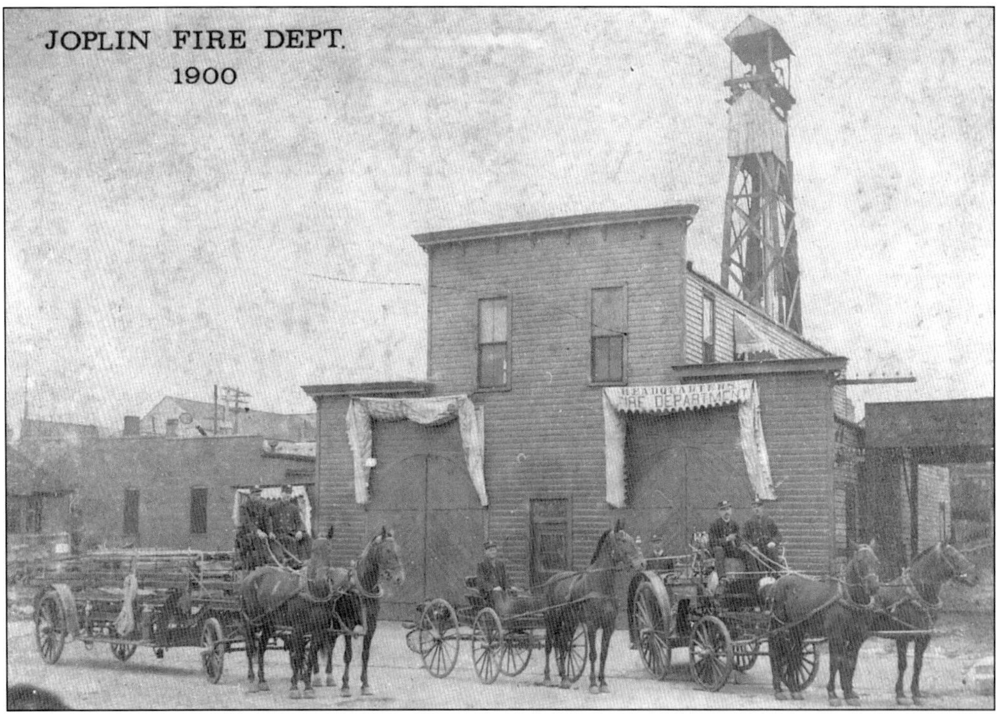

A photograph from 1900 depicts Joplin's horse-drawn fire fighting equipment. Some area residents may recall when firemen trained on the tower in the background, clambering up and down, dragging hoses and weights simulating humans being rescued.

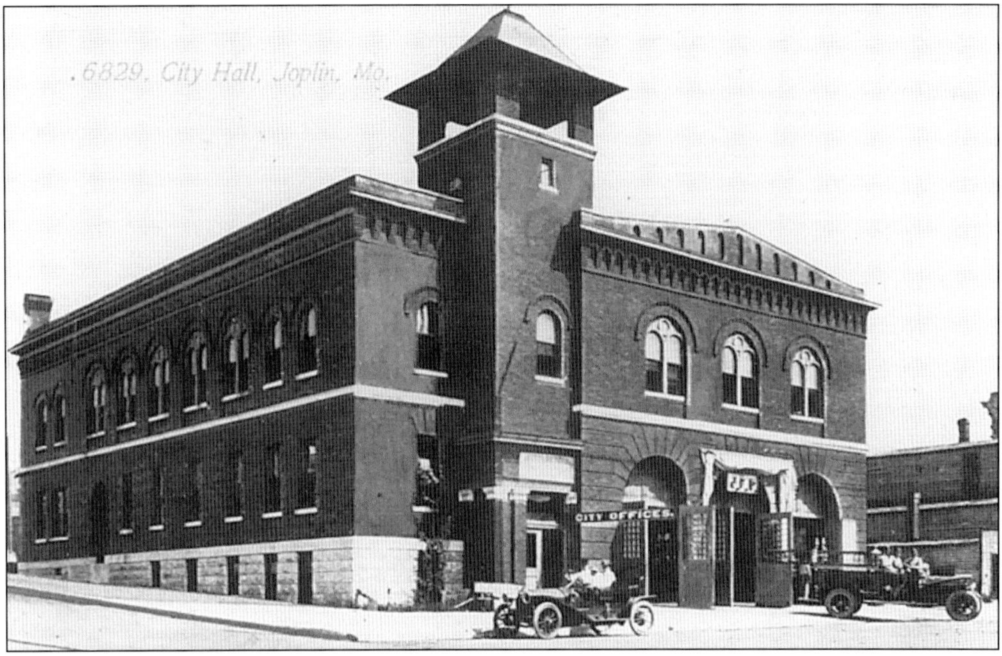

The new Joplin City Hall was opened for business in 1906. What was said to be the nation's first motorized fire-fighting vehicle, constructed on a converted taxicab chassis and later affectionately known as "the goat," was put into service one year later.

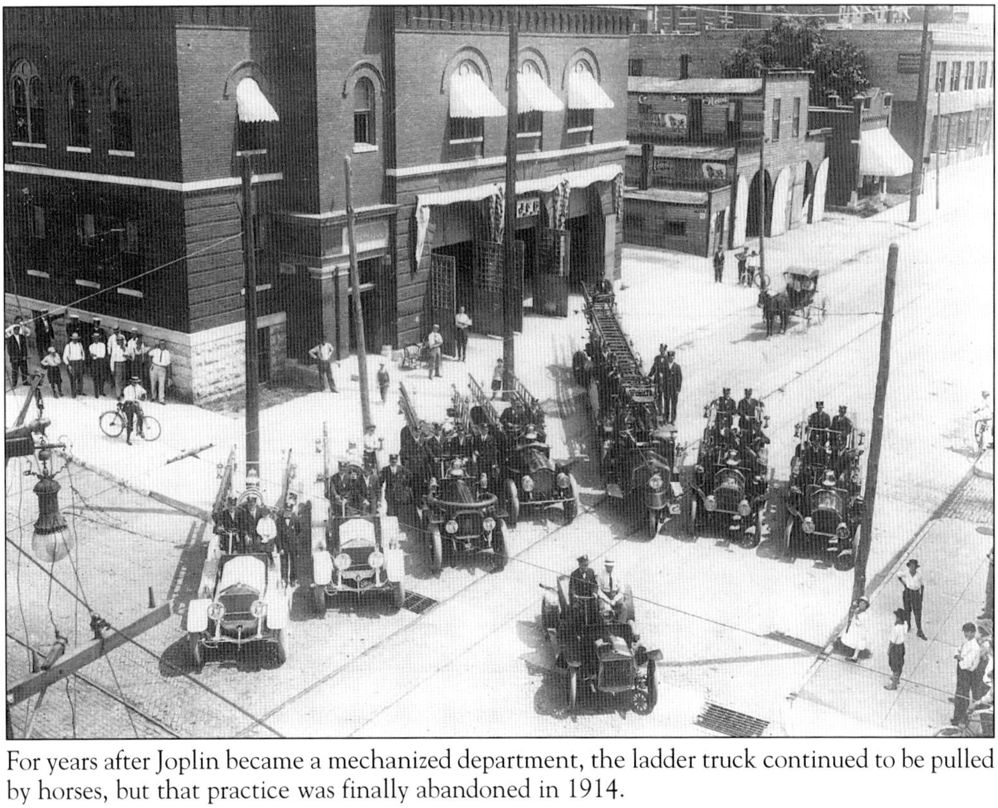

For years after Joplin became a mechanized department, the ladder truck continued to be pulled by horses, but that practice was finally abandoned in 1914.

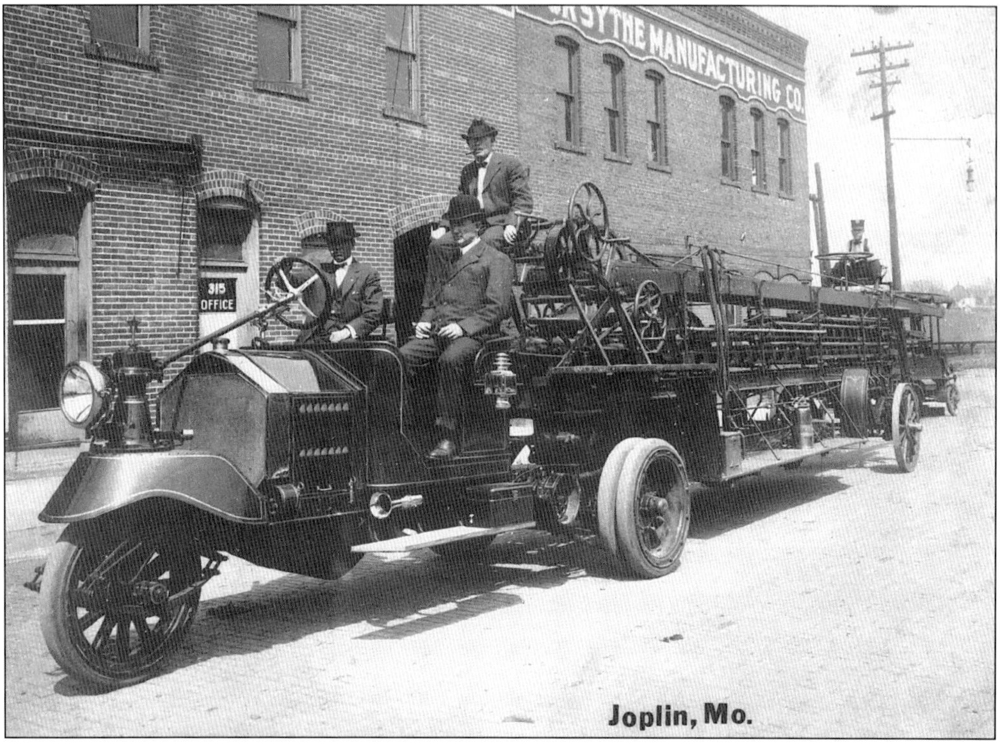

By 1910, with its motorized equipment, the Joplin Fire Department was boasting that it was the equal of the best in the nation. It didn't rest on its laurels but continued to improve. This unique "three-wheel" ladder truck was later put into service.

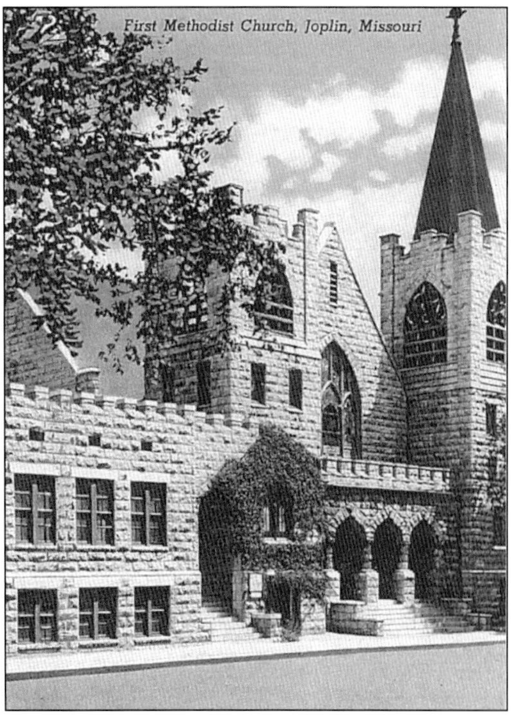

Churches and their ministers have played a prominent role in Joplin's past, with the city even deriving its name from an early Methodist pastor, Rev. Harris G. Joplin. The Methodist congregation dedicated this beautiful new church on the northwest corner of Fourth Street and Byers Avenue in June 1906. It was built of Carthage stone, accented with Gothic arches, and drew nearly two thousand people during the day of the dedication. (Photo Courtesy John Burrington)

The Catholics of St. Peter's Parish dedicated their magnificent new church on the southwest corner of Eighth Street and Pearl Avenue in November 1907. Archbishop John J. Glennon of St. Louis delivered the dedication sermon. The Gothic-style building is constructed of Carthage stone and is listed on the National Register of Historic Places. (Photo by author.)

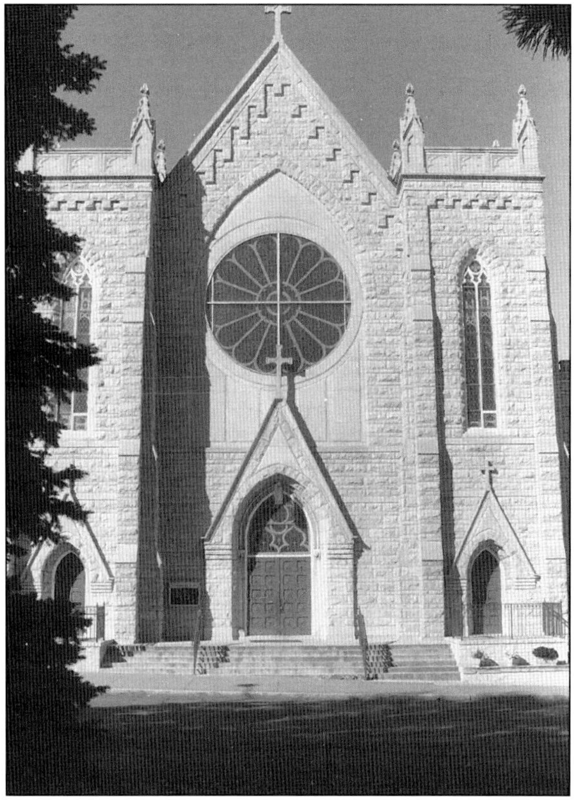

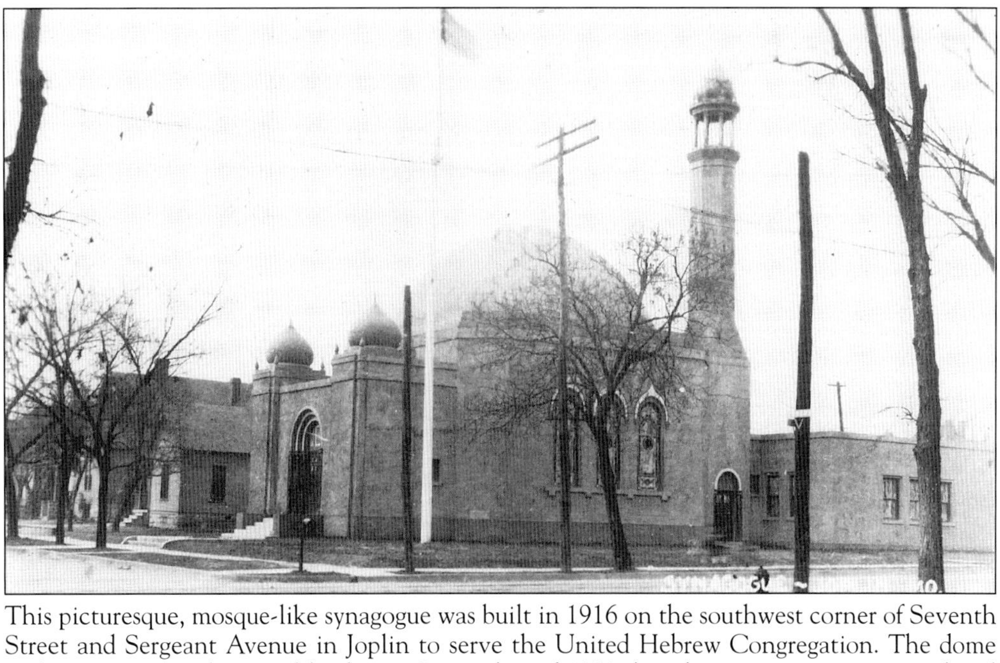

This picturesque, mosque-like synagogue was built in 1916 on the southwest corner of Seventh Street and Sergeant Avenue in Joplin to serve the United Hebrew Congregation. The dome and minaret were destroyed by fire in September of 1970, but the structure was repaired and continues in use.

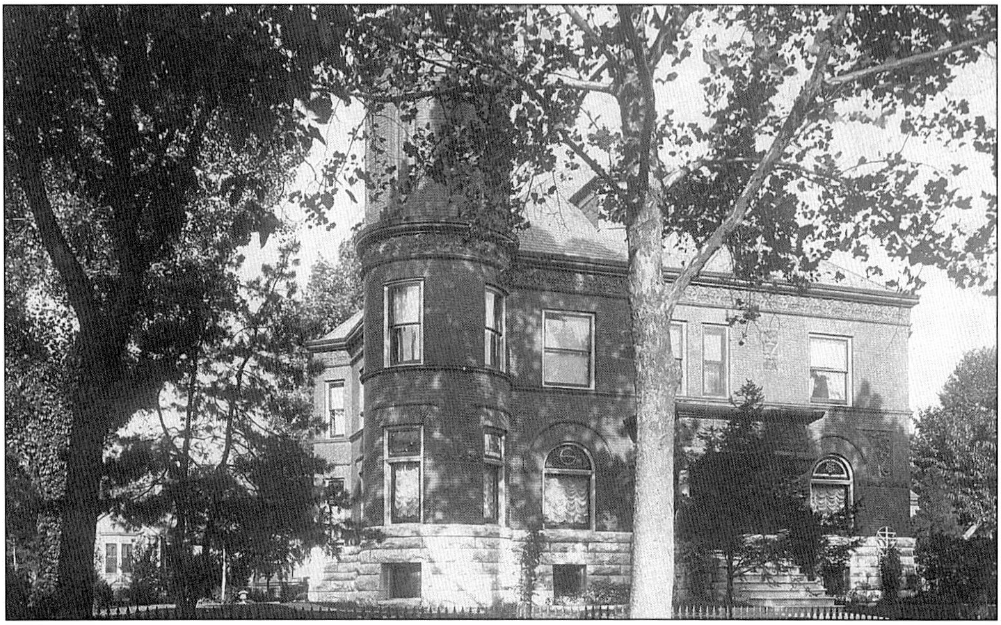

Charles Schifferdecker, whose name is one of the best remembered of the Joplin pioneers, lived in a home on the western edge of the community. The beautiful grounds were first known as Schifferdecker Gardens, then Schifferdecker Electric Park, and now just Schifferdecker Park. In an 1898 city directory, Mr. Schifferdecker lists himself as a "Capitalist." He also built this home on the northwest corner of Fifth Street and Sergeant Avenue. Mr. Schifferdecker was an immigrant from Germany. Perhaps the crenelated turret reminded him of the castles on the Rhine. (Photo Courtesy Brad Belk, Joplin Museum Complex.)

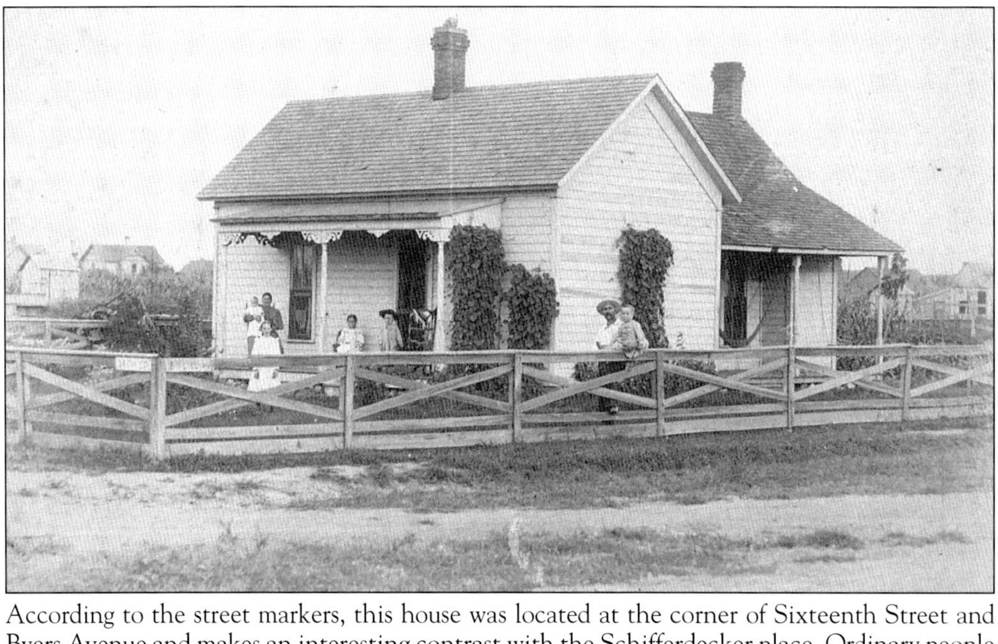

According to the street markers, this house was located at the corner of Sixteenth Street and Byers Avenue and makes an interesting contrast with the Schifferdecker place. Ordinary people lived in homes like this—and many less substantial. Notice the dearth of trees. Mining land was often prairie land. (Photo Courtesy Webb City Genealogical Society.)

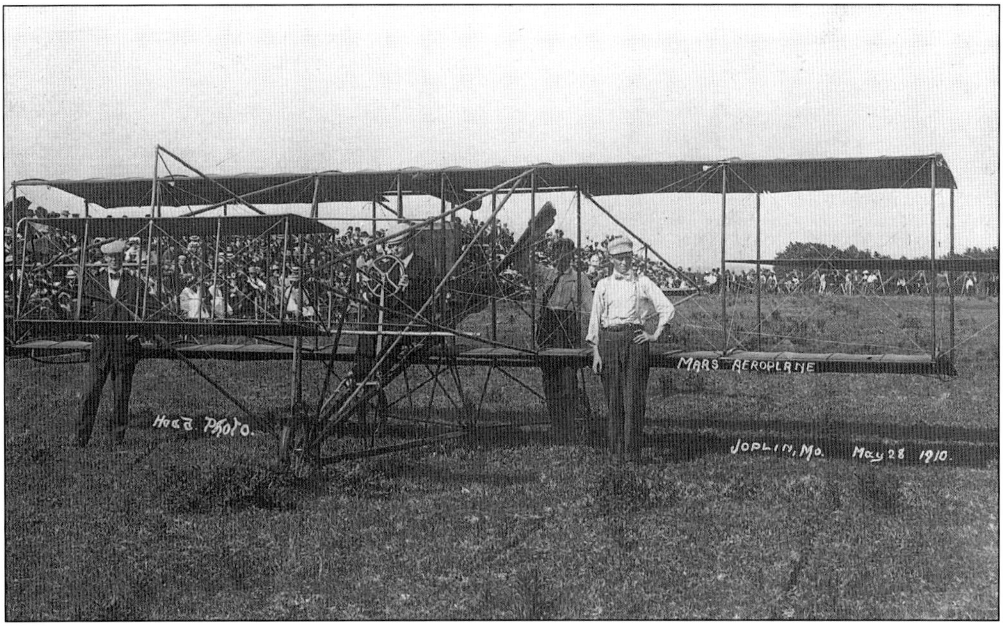

Almost six-and-a-half years after the Wright brothers made their first successful powered flight and one year after tests were completed on the two-man plane they built for the U.S. Army Signal Corps, a large crowd gathered in Joplin to watch Bud Mars make a demonstration flight in his Curtiss aeroplane. Early powered flight caught the public's imagination much as manned space flight would later in the century.

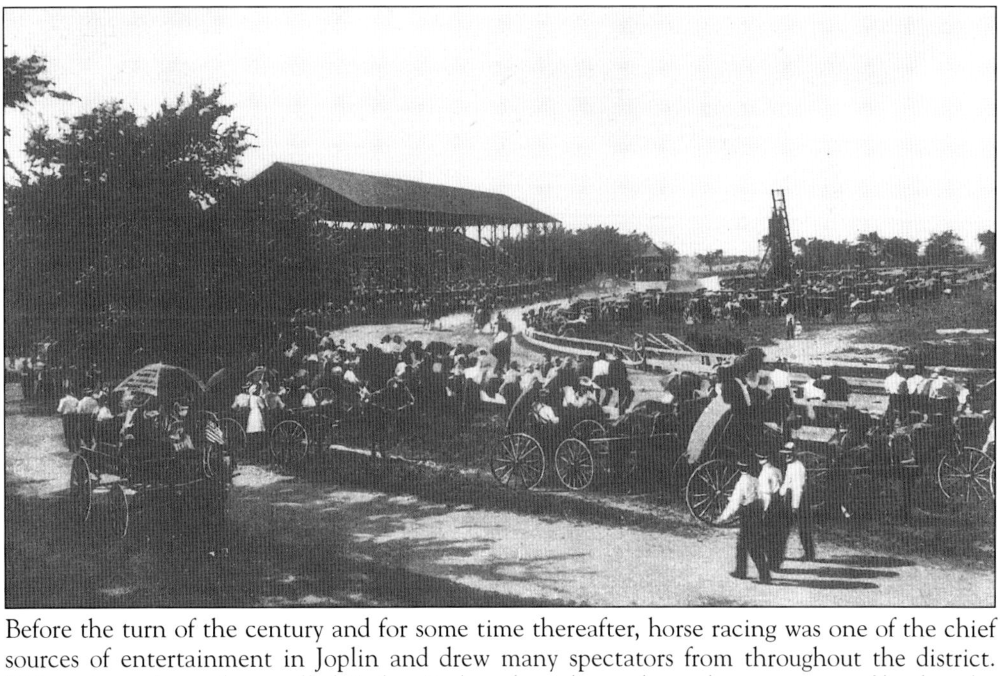

Before the turn of the century and for some time thereafter, horse racing was one of the chief sources of entertainment in Joplin and drew many spectators from throughout the district. This early track was later called Barbee Park and was located on a large expanse of land to the northeast of 20th Street and Maiden Lane. The trotting race pictured here probably occurred in the early 1890s.

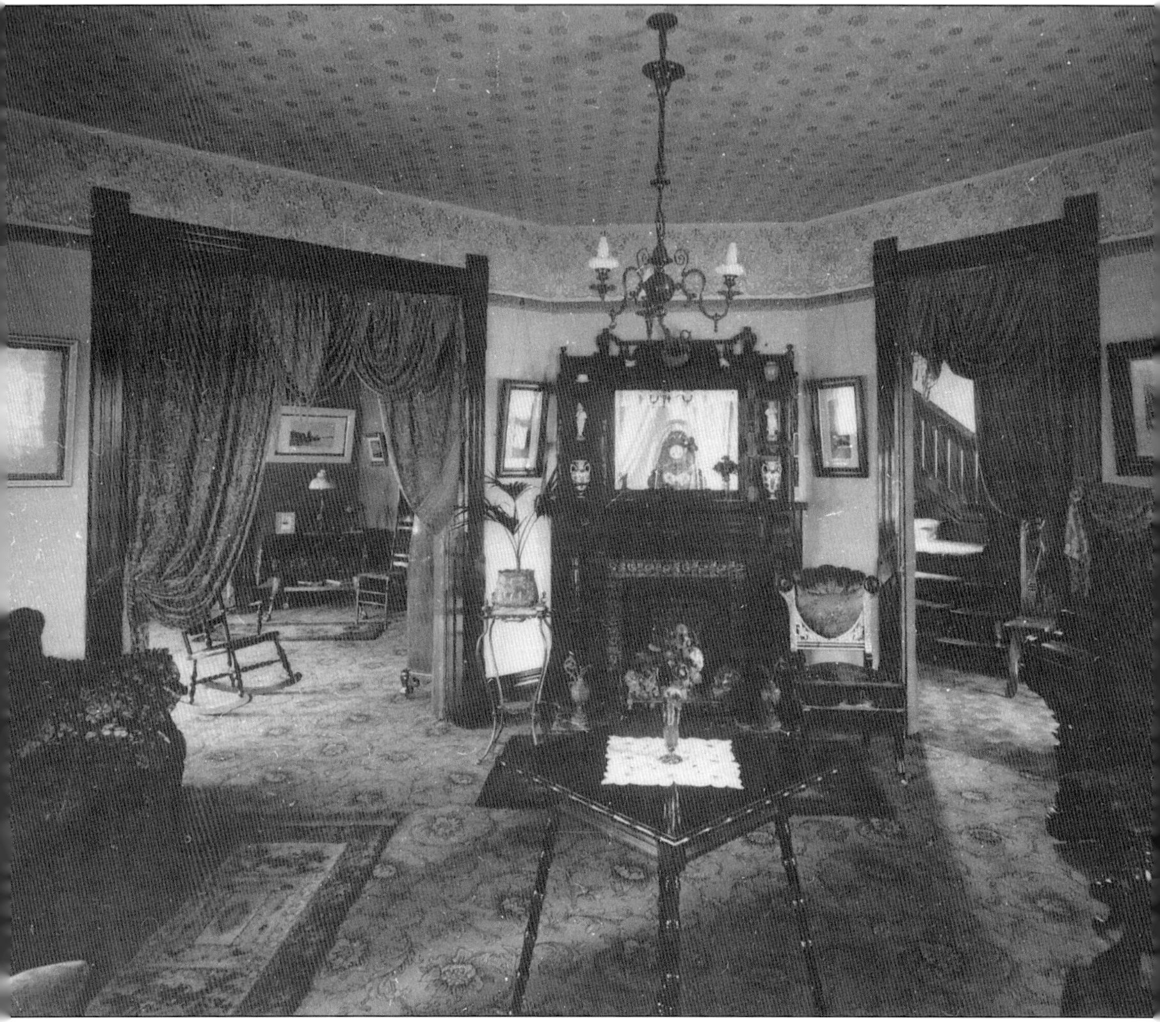

After a busy day, Joplin entrepreneurs and mine operators returned to substantial homes with inviting gaslit parlors, such as this in the Gus Weymann home at 528 Pearl Street in 1896. The Weymann name was well known in Joplin before and after the turn of the century. Henry Weymann immigrated to this country in 1868 when he was 22 years old. He became involved with the first zinc smelter west of the Mississippi in eastern Missouri. He visited southwest Missouri in 1873 and arranged for zinc ore to be shipped to the smelter from the Granby area. After joining Joplin's growing German population, he was instrumental in saving Crystal Cave, believed at the time to hold the largest crystals anywhere.

Gustavus Weymann, whose home is shown, was Henry's brother and helped manage his business enterprises.

Three

WEBB CITY

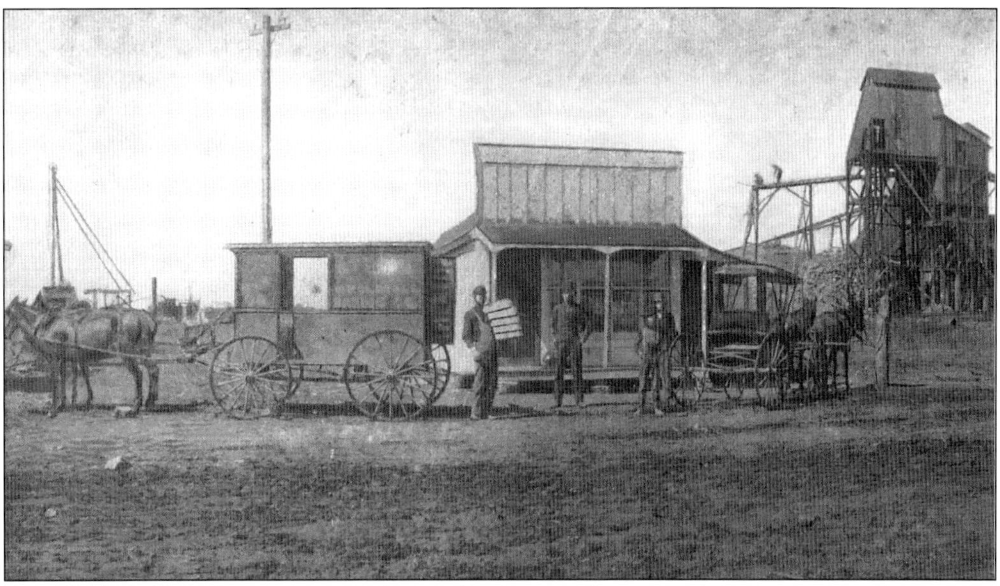

This photo taken about 1900 typifies Webb City at the turn of the century, still primitive in some ways—note the rough, rutted street—but growing and prosperous because of the booming mines. The horse-drawn panel conveyance is identified as the city's "first bakery delivery wagon, Webb City Bakery, North Main Street, Mr. Worner owner about 1900." Mr. Worner (also Wornner or Wonner) was possibly Henry Worner, a baker, who was also chief of the volunteer fire department for a number of years. Perhaps unwittingly, the photographer captured in the background two workmen extending a tramway for one of the mines, which supported the community's prosperity. (Courtesy Webb City Historical Society.)

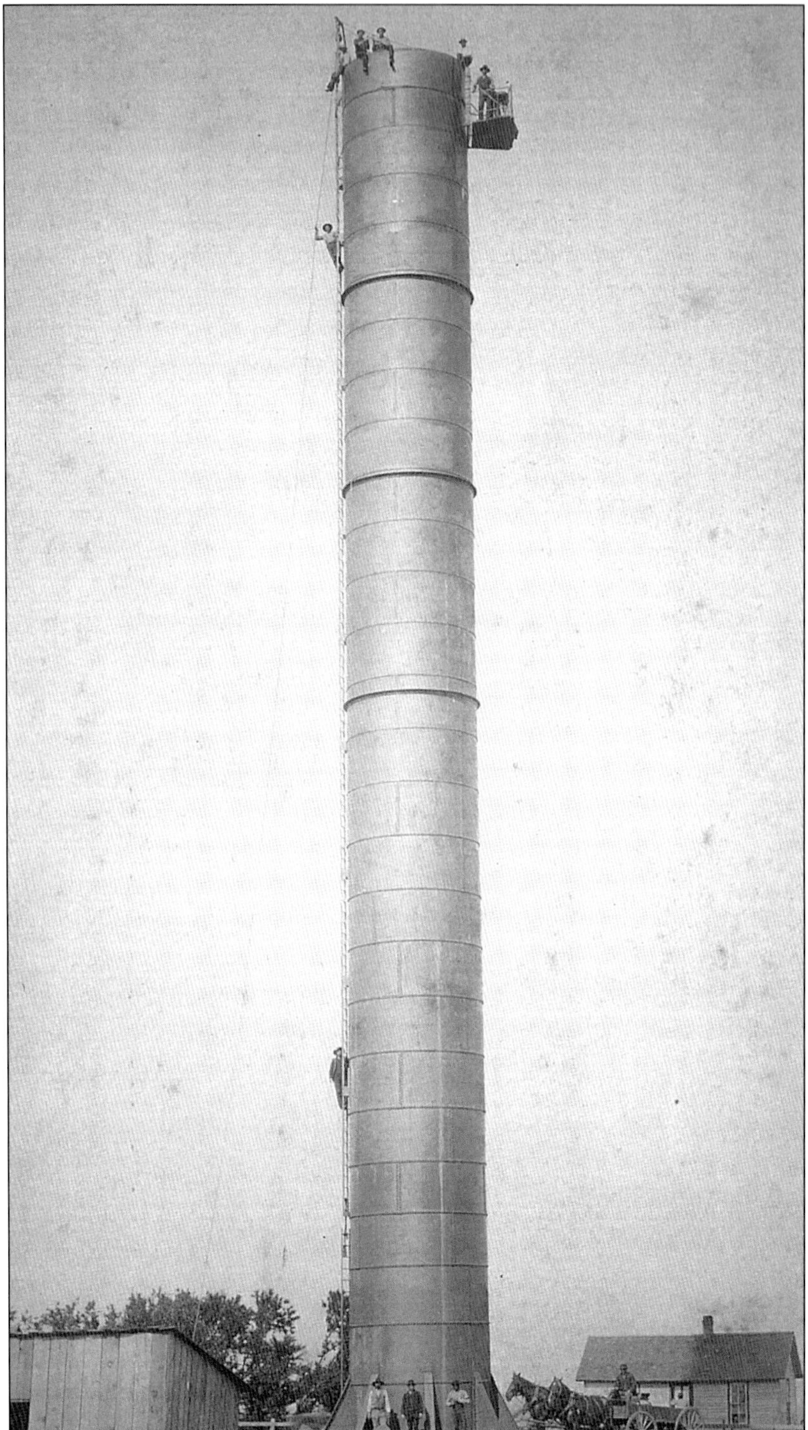

Workmen pose on the Webb City water tower, probably shortly after its completion in 1890. The photo came from Mrs. Bert Hancock, granddaughter of George Hornish, one of the men shown. (Courtesy Webb City Historical Society.)

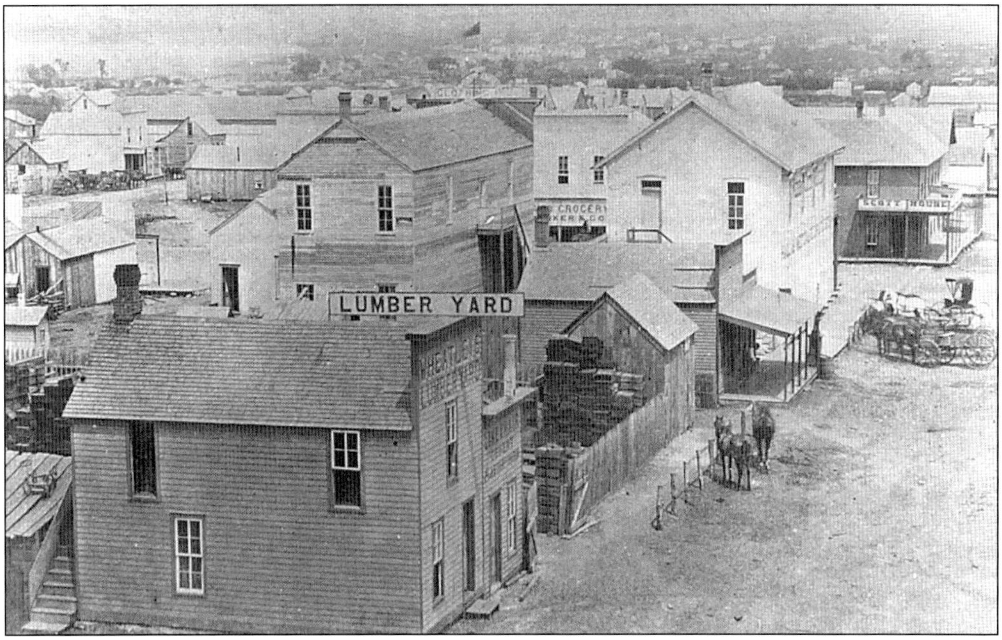

This photo was taken in the late 1880s from atop the brick home of John Webb, founder of the city. His house was located a block west of what is now Main Street where Broadway jogs south. The view is to the northeast toward Carterville. W.A. Wheatley's lumberyard, Webb City Hall (where the buggy and wagons are parked), Manker and Co. grocery store, and Scott House are identified. (Photo Courtesy Henry Hulett.)

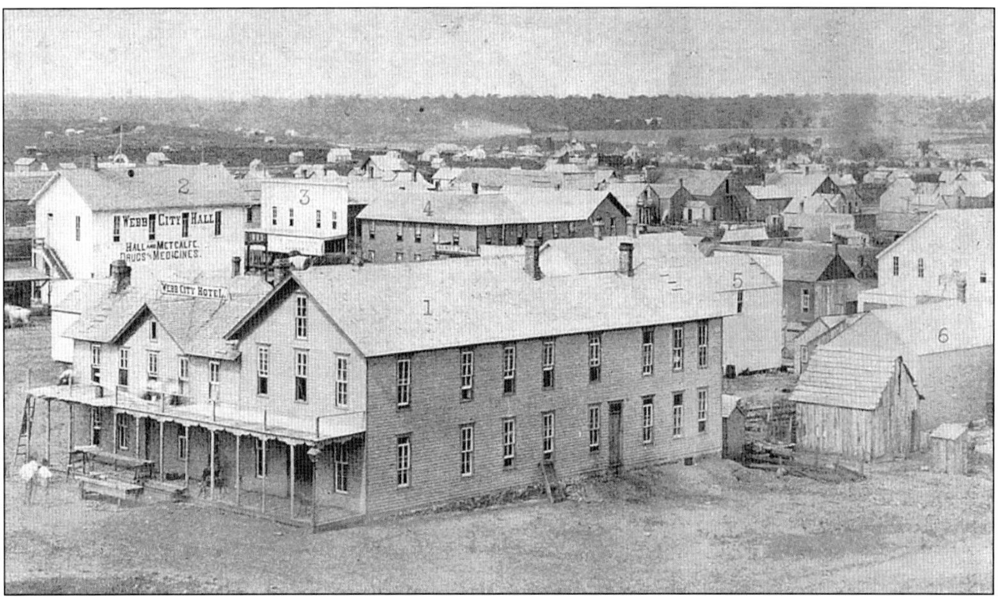

Another 1880s view of Webb City was taken from the belfry of the old Central School building located about a block south of Mr. Webb's home. The numbers identify the buildings as follows: 1. The Webb City Hotel, later the Buffalo House. 2. The Webb City Hall. 3. Manker & Co. grocery store. 4. The Scott House. 5. O.P. Sutherland & Son Groceries. 6. Alice & Sadie Rice Millinery. (Photo Courtesy Henry Hulett.)

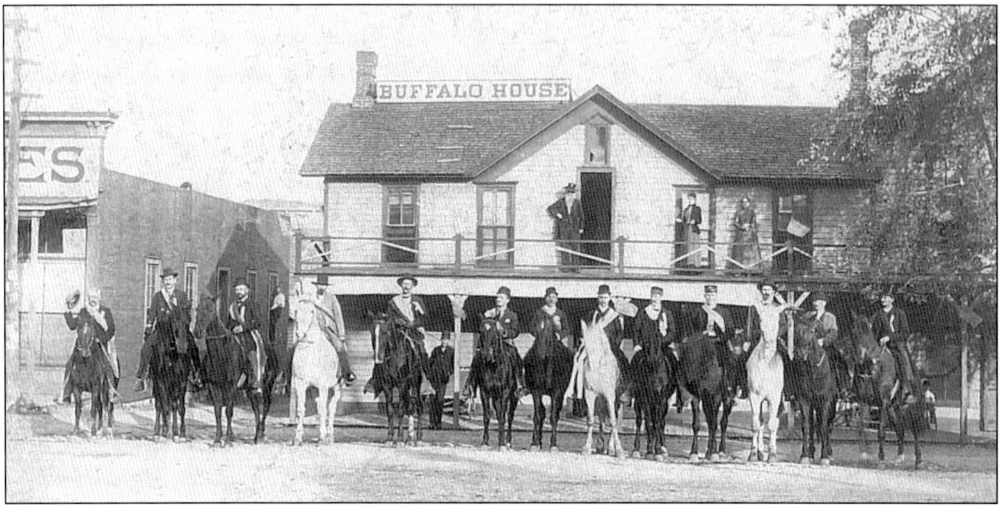

Webb City celebrated Columbus Day on October 12, 1892, with a mounted escort, here reigned in line before the Buffalo House. The riders, from left to right, are: Jos. C. Stewart, James P. Stewart, Dr. C.H. Craig, James Ball, Sam Koontz, Clarence Gaston, Grant Wilbur, Walter F. Moore, Harry H. Heylmun, Tom C. Hayden, William Hilburn, E.E. Spracklen, and Col. A.A. Hulett, Marshall.

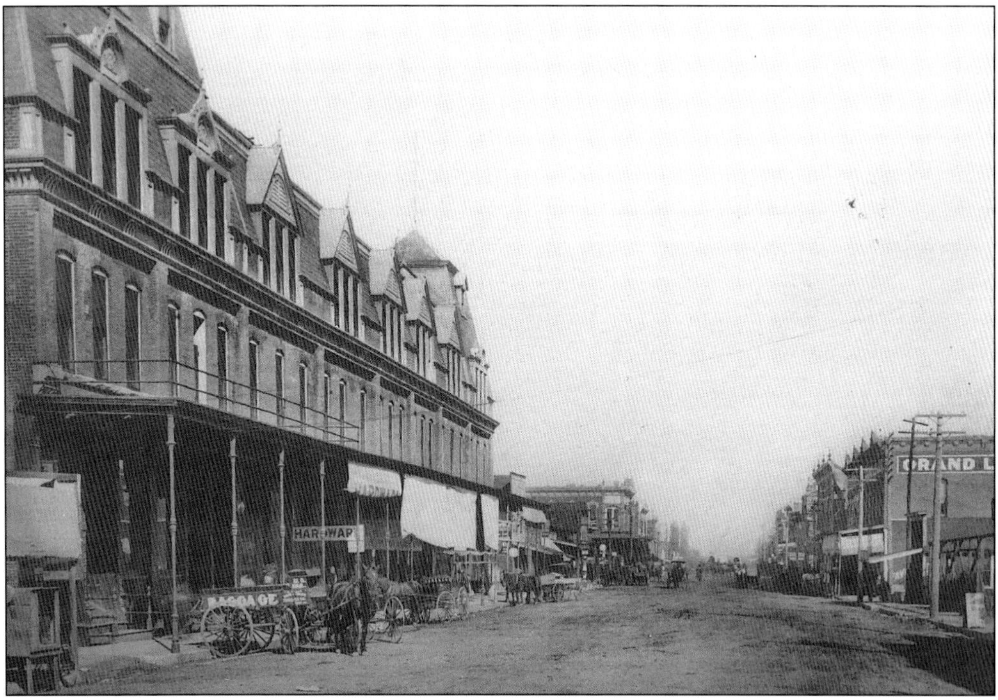

A view of Webb City's Allen Street (now Main Street) looks to the north in the 1890s. The imposing Newland Hotel, which opened with much fanfare in 1891, is in the foreground on the left. Notice the wagon with "BAGGAGE" lettered on the side. Some livery stables offered such service, meeting trains and hauling luggage to the hotel of choice. (Photo Courtesy Henry Hulett.)

An employee, Lew Bradbury, on the left, and T.C. Hayden, owner, stand in front of his harness shop at 117 East Main Street (now Broadway) about 1895. Sale and repair of harness, saddles, etc., was a thriving business at the time, but later with the advent of automobiles, income declined. Mr. Hayden's enterprise, however, changed with the times. He later operated a Ford Model T dealership in a building on present day Main Street, storing the unsold autos in crates in the basement (Photo Courtesy Henry Hulett.)

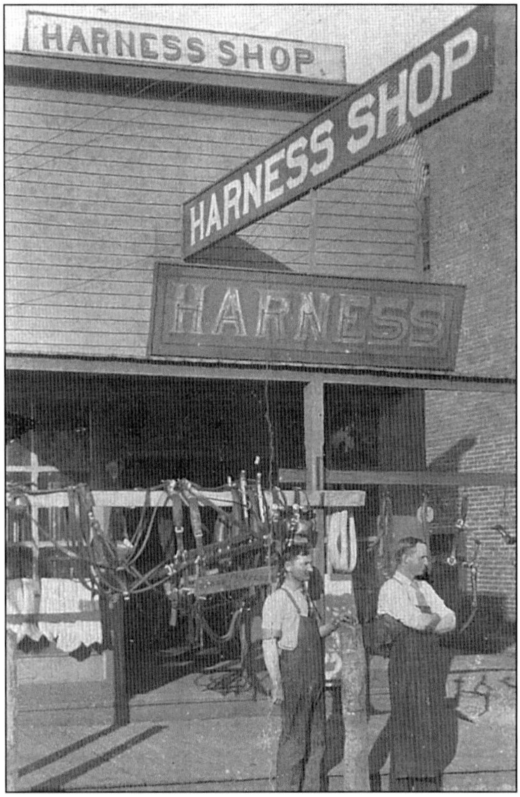

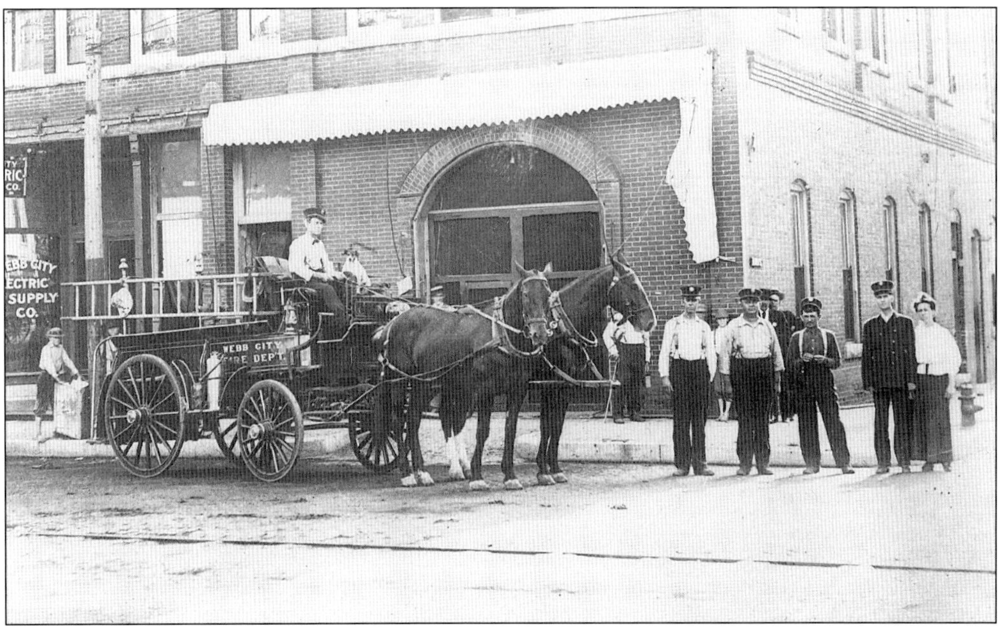

With the installation of a city water system in the late 1880s, the old horse-drawn water wagons began to disappear. Other wagons, chiefly used for transporting ladders and hoses, replaced them. Here firemen and bystanders pose with one such vehicle. (Photo Courtesy Ron Mosbaugh.)

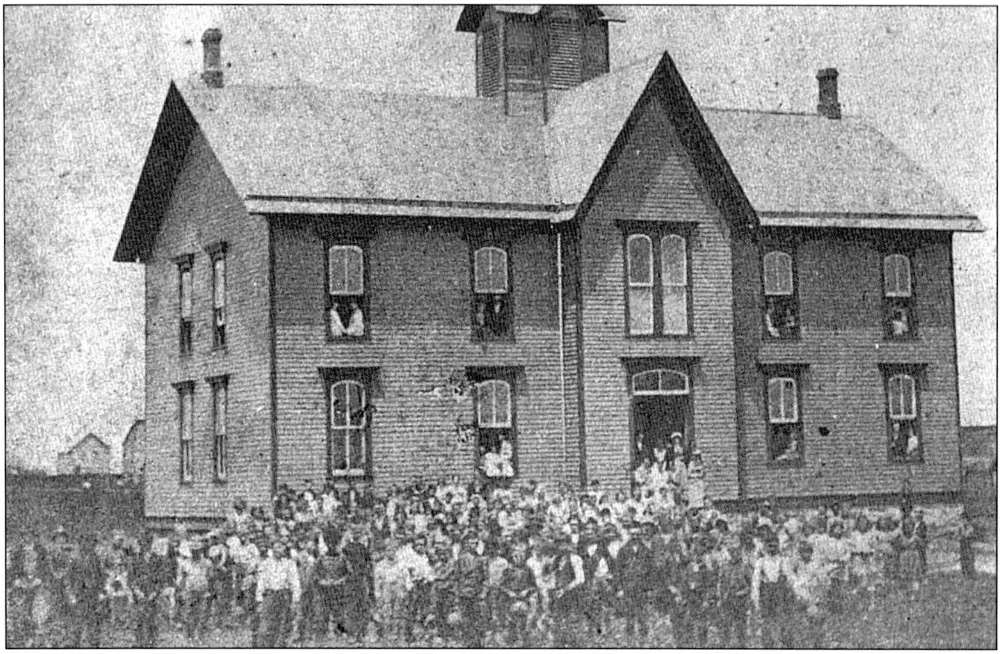

This old photo, c. 1877, shows the student body standing in front of Webb City's Central School, the first school building. It was located on Webb Street, but was later sold and moved to west Broadway where it survives to this day as a residence. (Photo Courtesy Henry Hulett.)

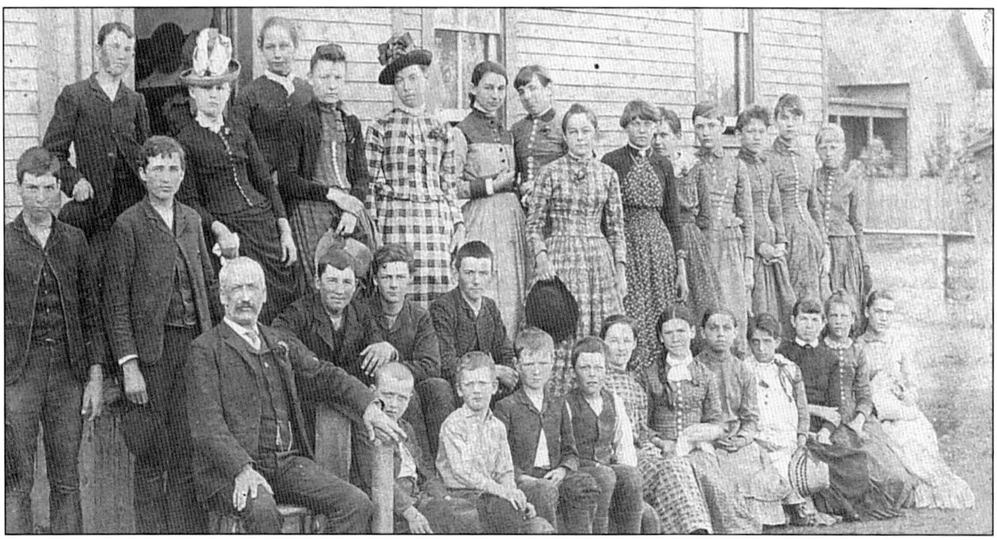

In May of 1885, the seventh and eighth grades of Webb City posed in front of the school. Pictured, from left to right, are: (first row, seated) Charlie Lewis, Fred Varner, Claude Fishburn, Frank McPherson, Julia White, ? Lewis, Leona Hatcher (later Mrs. W.H. Spain), Ida Weatherford, Emma Baker, Myrtle Lingle, and unidentified; (seated on chair) L.L. McCormick, principal; (second row, seated) George R. Sutherland, Clarence Lingle, and Harry Hulett; (standing behind principal) Ben C. Aylor and John Greenwade; (back row) Herman Baker, Hattie Carey, Beulah Robinson, Pet Johnson, unidentified, Clara Caldwell, Mamie Marvin, Rosa White, Myrtle Snodgrass, Lucy Wolfe, Maude Reed (later Mrs. Geo. R Sutherland), Cora Tholborn (later Mrs. W.F. Moore), Minnie Hancock, and Pearl Smith.

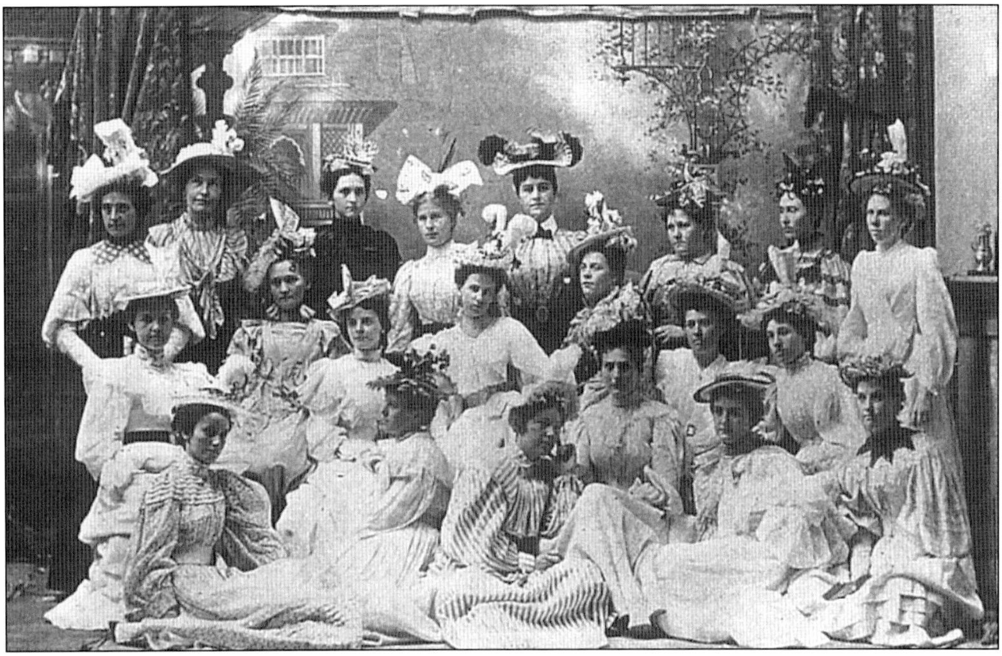

At the dawn of the century, some Webb City ladies gathered in their finery to help Laura Conklin celebrate the opening of her new millinery shop. Pictured, from left to right, are: (front row) Agnes Wolfe, Mrs. J.T. Albert, Rose Boydston, Essie Overstreet, Jessie Motley, and Lena Cooper; (middle row) Alice Hardy, Lulu Howard, Towle Hall, Laura Conklin, Carrie Overstreet, Cora Moore, and Minnie Dunlap; (back row) Ida Weatherford, Kate Hardy, Grace Frankenberger, Mrs. Frank H. Danglade, Allie Whitsett, Mollie Marrs, Miss Wiles, and Nelle Redding. Ten years later, in 1909, a brochure glowingly described the establishment as being a "most exquisite millinery store," employing "from five to ten expert trimmers, milliners and assistants." (Photo Courtesy Henry Hulett.)

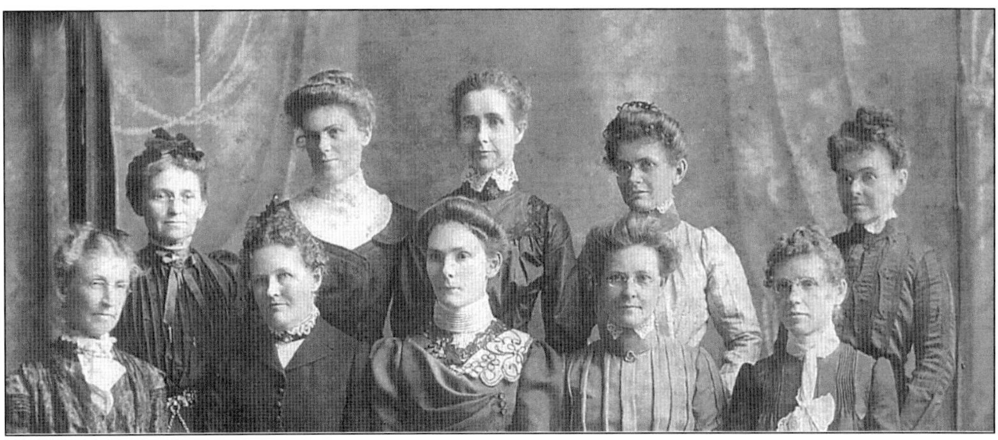

Women bearing some of the most prominent names in early Webb City belonged to the Priscilla Club. Their husbands were known as the "John Aldens." In January of 1902, the ladies posed for a formal photograph. From left to right, they are: (front row) Mrs. J.C. Stewart, Mrs. J.A. Bowman, Mrs. A.D. Hatten, Mrs. C.M. Manker, and Mrs. W.C. Stewart; (back row) Mrs. C.L. Wright, Mrs. W.A. Corl, Mrs. O.J. Raymond, Mrs. W.S. Chinn, and Mrs. A.A. Hulett. (Photo Courtesy Henry Hulett.)

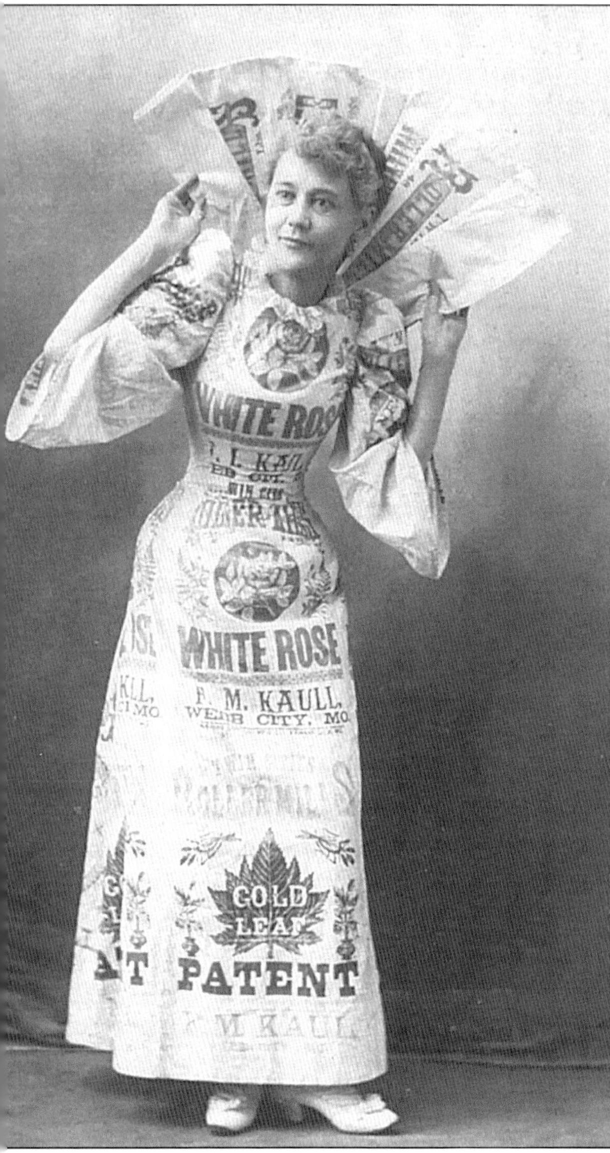

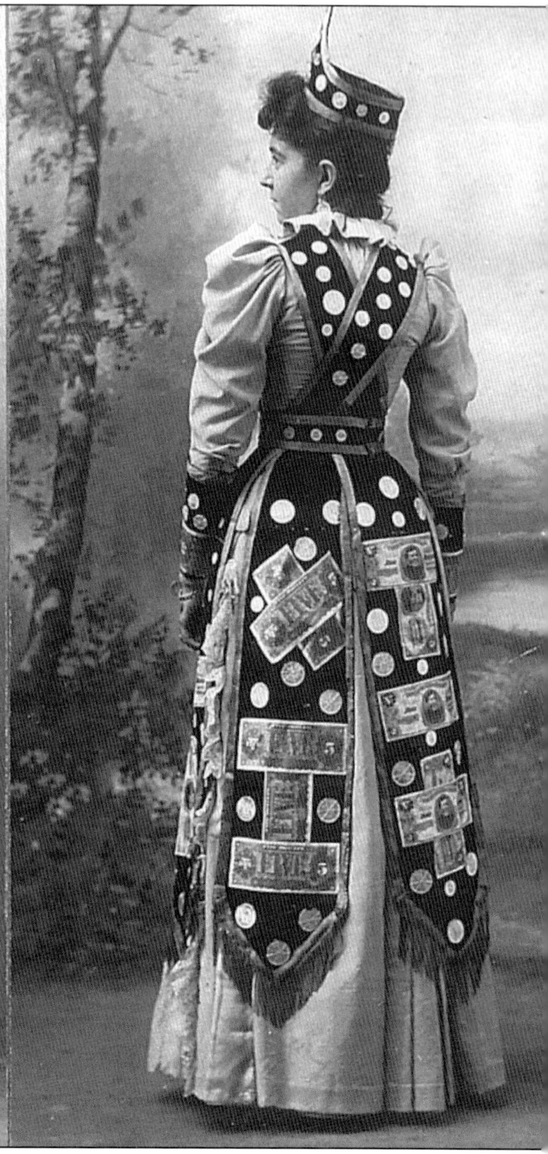

At the turn of the century and for many years thereafter, Harry Hulett, a Webb City businessman and amateur photographer, chronicled the growth of the city with photos. Some he took himself, while others he merely collected. Many had notes or comments on the back. These two are from the Webb City Merchants' Carnival of 1892, which Mr. Hulett calls "a spectacular event in Webb City in the early '90s." During the carnival, local ladies wore elaborate costumes of their own design. Mrs. W.J. Bogar poses on the left in a costume made of White Rose flour sacks. On the right Mrs. T.F. Coyne's costume is accented with $5 bills and various coins of the realm. (Photo Courtesy Henry Hulett.)

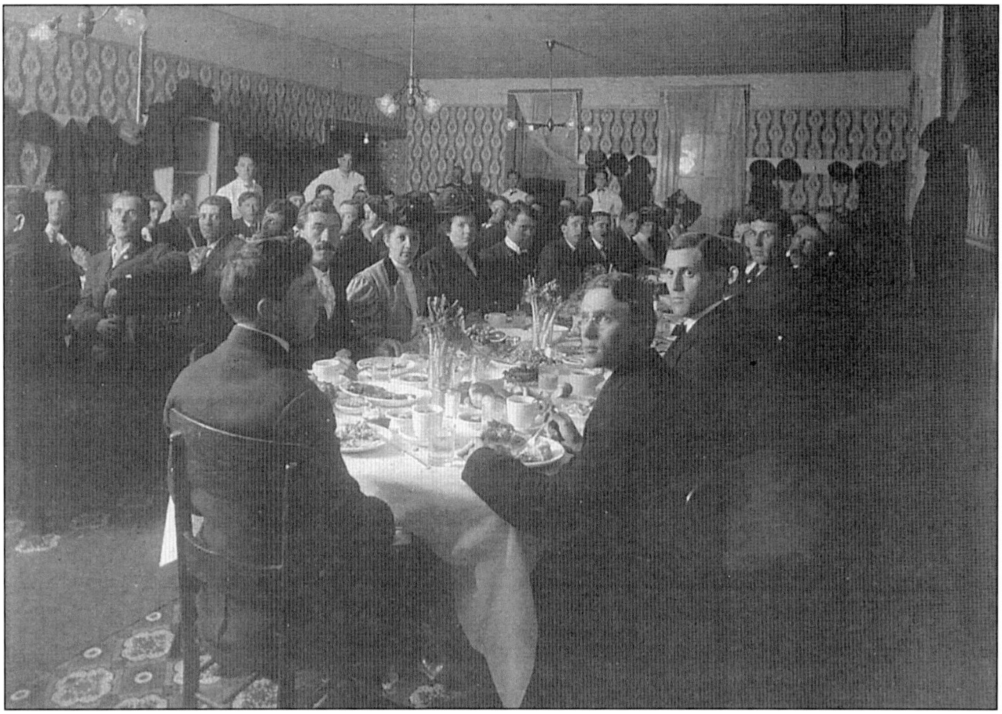

In 1898, diners at a banquet in the Newland Hotel pause for a photograph. The nearest couple on the left of the table in the foreground are Mr. and Mrs. John Etter. (Photo Courtesy Webb City Historical Society.)

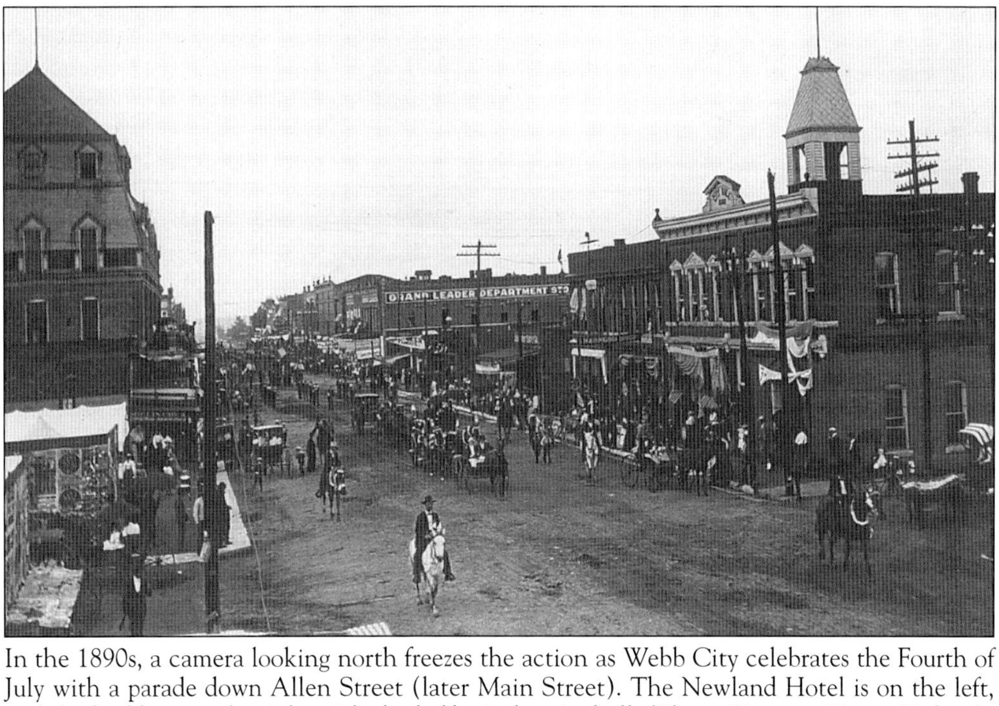

In the 1890s, a camera looking north freezes the action as Webb City celebrates the Fourth of July with a parade down Allen Street (later Main Street). The Newland Hotel is on the left, and the building on the right with the belfry is the city hall. (Photo Courtesy Henry Hulett.)

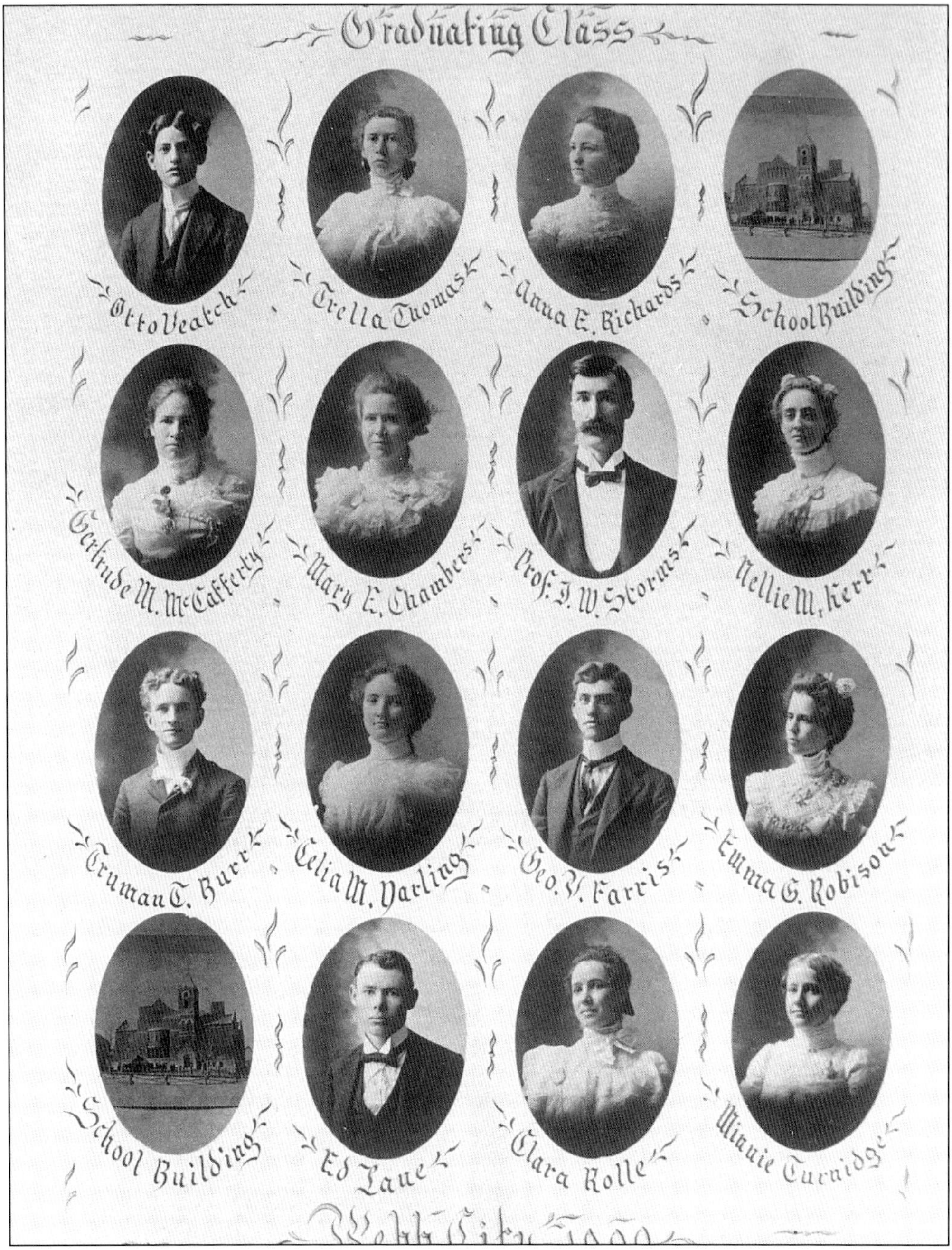

Education was not so universal a century ago in 1900, when many young people quit school early to take jobs. With the population of Webb City at approximately 10,000, the city's high school had only 13 graduates. Those who did graduate, however, presented an appearance of maturity and dedication beyond their years. (Photo Courtesy Webb City Genealogical Society.)

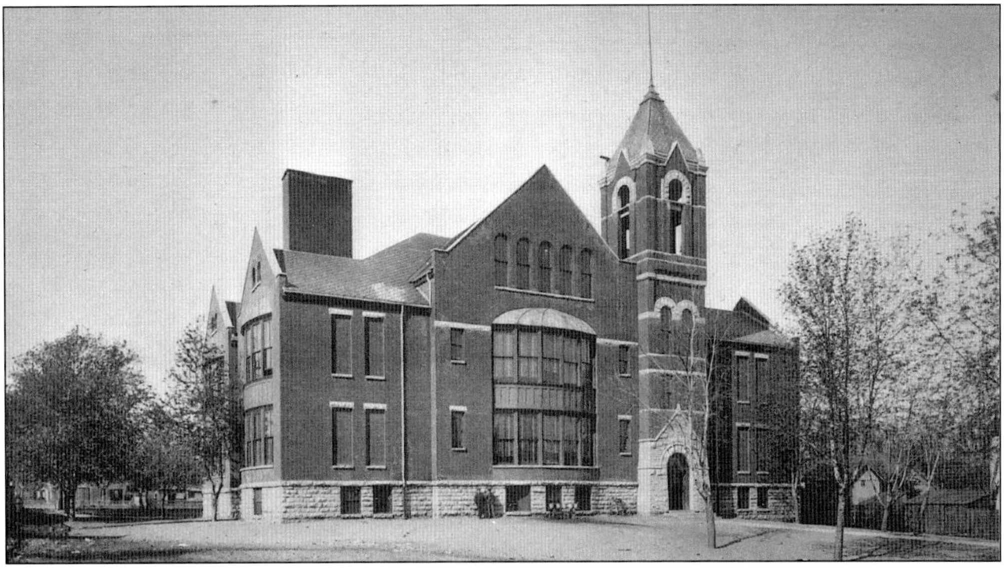

Early records indicate that students were attending a Webb City High School as early as 1888. The school building pictured here replaced the original frame structure that was located just to the east of it. It was erected in 1894 and was also known as the Central School. A new high school was built on Broadway in 1911, but this building continued in use for elementary classes until 1931 when part of it collapsed. It was later razed. (Photo Courtesy Ron Mosbaugh.)

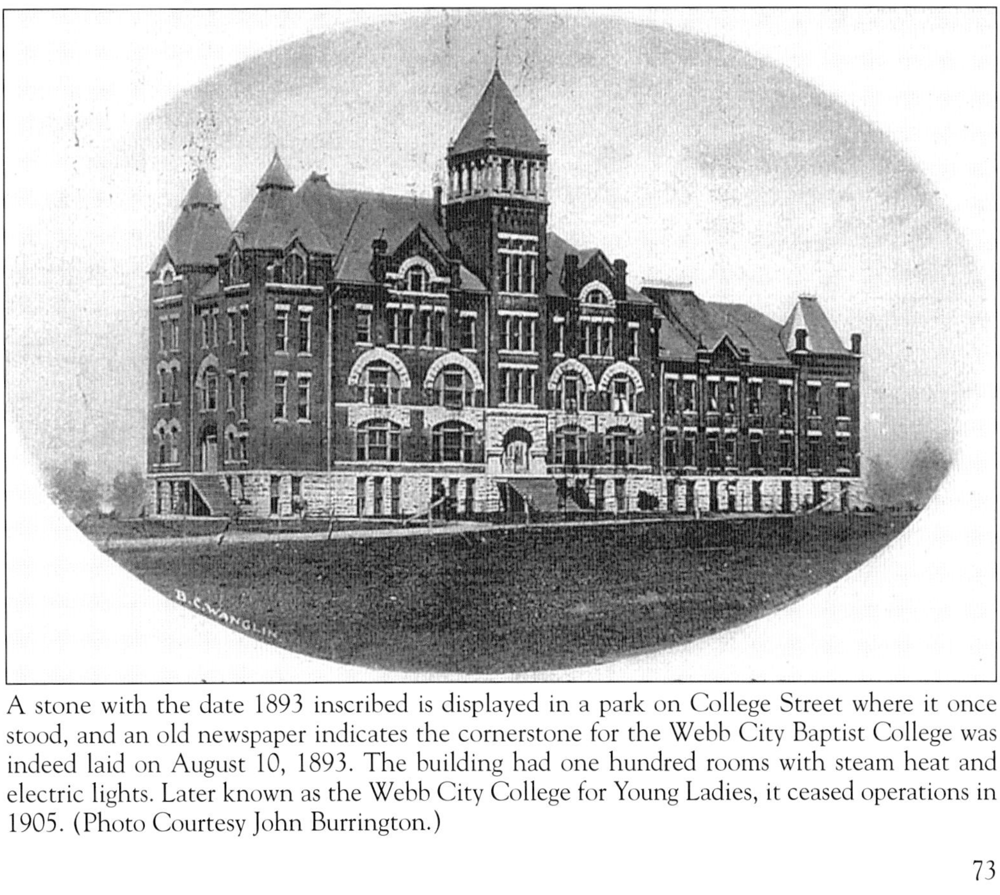

A stone with the date 1893 inscribed is displayed in a park on College Street where it once stood, and an old newspaper indicates the cornerstone for the Webb City Baptist College was indeed laid on August 10, 1893. The building had one hundred rooms with steam heat and electric lights. Later known as the Webb City College for Young Ladies, it ceased operations in 1905. (Photo Courtesy John Burrington.)

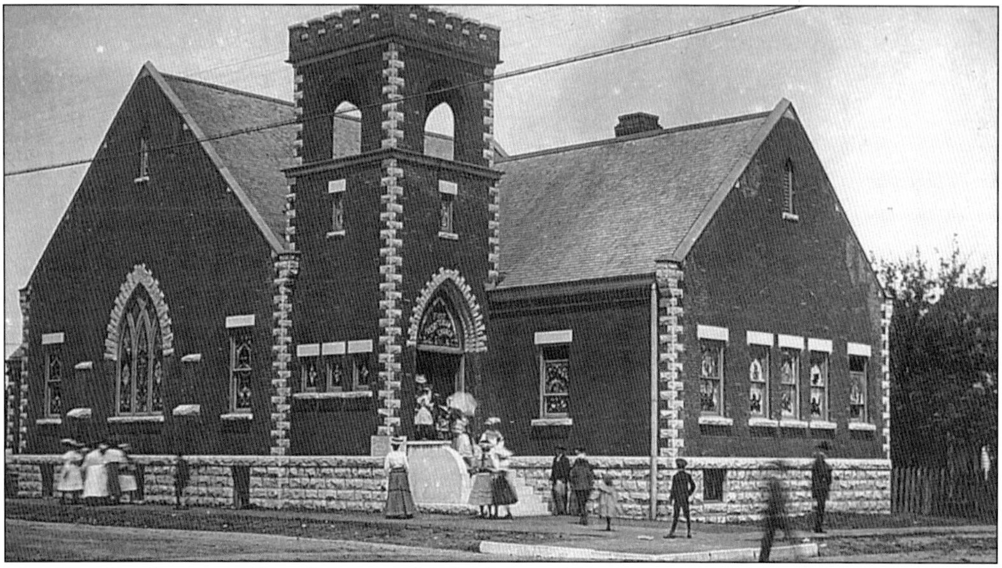

The First Presbyterian Church was established in Webb City in 1877, less than four years after the discovery of lead, and was the city's first church. The present structure (shown) was built during the tenure of Rev. Charles B. Boving (1899–1905). Webb City and Carterville were rivals. Miners from the opposing camps occasionally met at Ben's Branch for melees using bare knuckles and pick handles, but when First Presbyterian and Cumberland Presbyterian—another area church whose origins reach back to 1858—combined, they gave the old Cumberland building to Carterville for a Presbyterian Church. It was moved there with teams of horses. (Photo Courtesy Henry Hulett.)

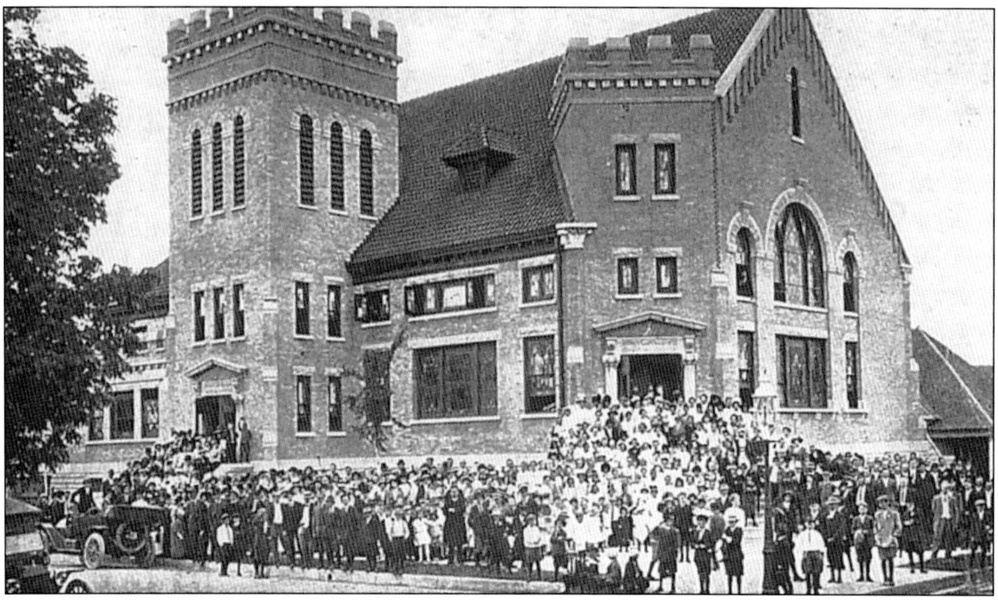

Before the Civil War, the Methodist Church in America split over the issue of slavery. On October 14, 1914, the members of the First Methodist Episcopal Church (sometimes called North) posed for a Sunday school photo in front of their church on the southwest corner of Second and Oronogo Streets. The Sacred Heart Roman Catholic congregation later occupied the building. (Photo Courtesy Central United Methodist Church.)

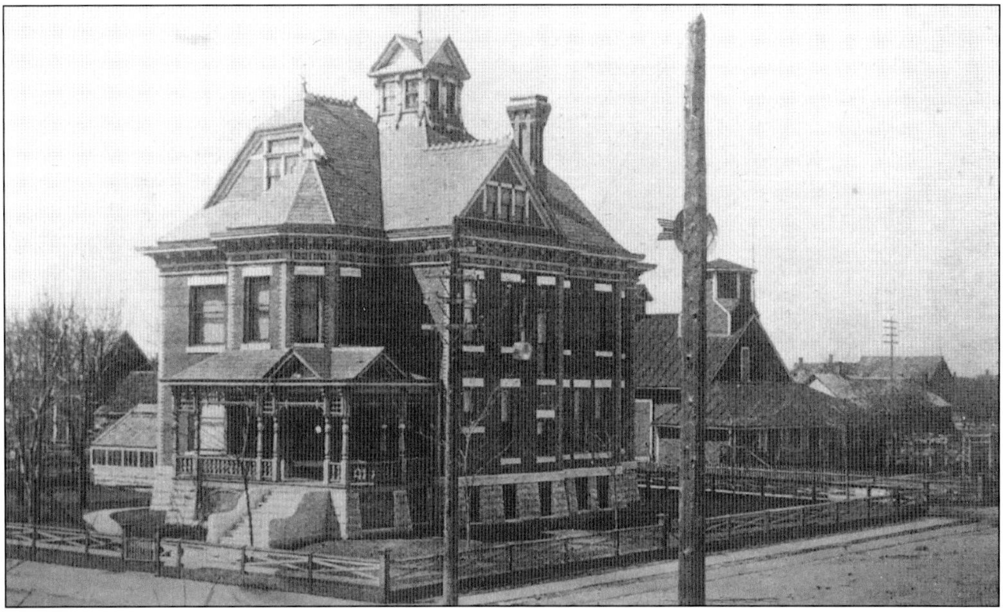

The E.T. Webb family home, with their carriage house and stable visible behind, is shown in this photograph from around the turn of the century. John Webb, E.T.'s father, had presented the Methodist Episcopal South congregation with a small brick church in 1882 on a lot he had reserved for that purpose. When they outgrew the building, a new church was constructed where E.T.'s carriage house had stood. (Photo Courtesy Ron Mosbaugh.)

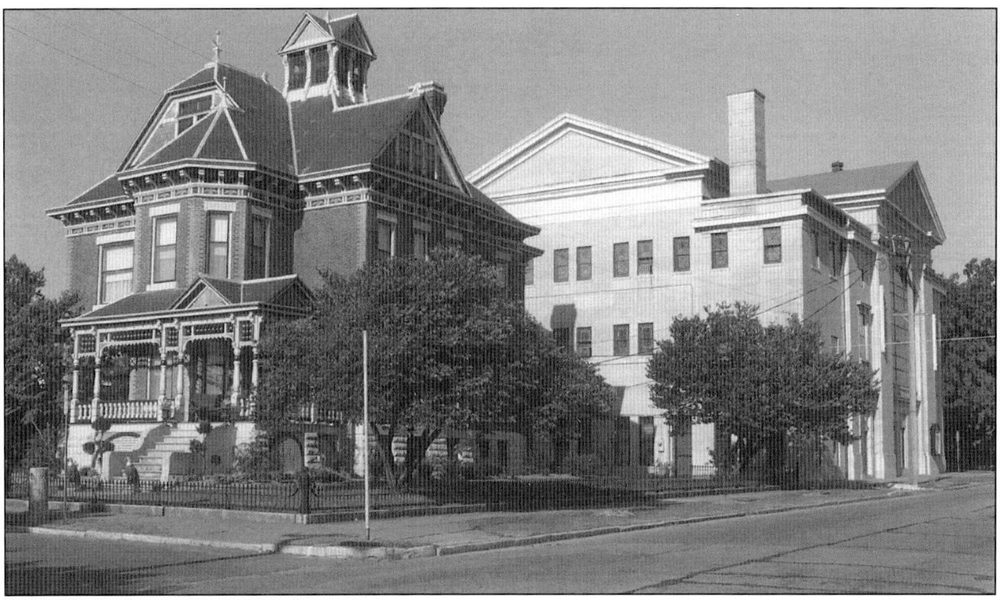

This is E.T. Webb's home as it appears today. The imposing edifice that replaced the carriage house appears behind it. The cornerstone for the new church, which stands at the southeast corner of Broadway and Pennsylvania Streets, was laid in 1908, and the building was dedicated November 19, 1909. With the reunification of the three major branches of Methodism in the United States and the addition of the Evangelical United Brethren in 1968, the church became known as Central United Methodist Church. (Photo by author.)

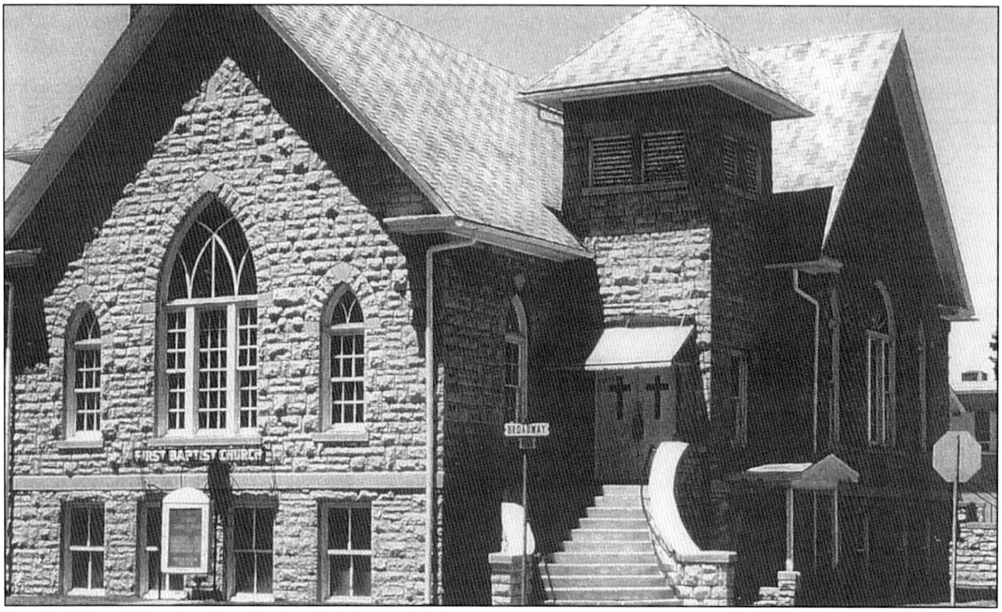

The Baptist Church began in Webb City when ten charter members started worshipping in an abandoned store building on what is now Main Street. A frame structure they later built was moved by an old-fashioned capstan, eventually arriving at Broadway and Roane Streets. It received a stone veneer in 1912. In 1935, it was heavily damaged by fire, but was again ready for use in just three months. A large, modern church and educational complex have since replaced it. (Photo Courtesy Webb City First Baptist Church.)

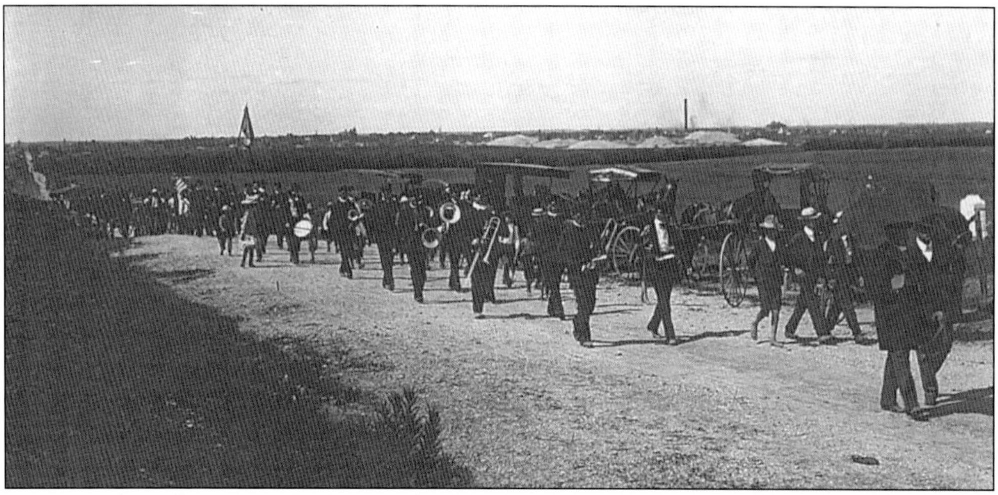

A view early in the century looking to the northeast depicts an unidentified procession led by a band approaching the Webb City Cemetery at what was then the south end of Oronogo Street. It was likely in honor of Memorial Day, or Decoration Day as it was then popularly known, and people would gather beneath the flag to hear orations eulogizing the nation's war dead, especially those of the Civil War. Oronogo Street, now a busy thoroughfare, was then hardly more than a country road. Visible in the background are the Sucker Flat mining area and the water standpipe, which had been completed in 1890. Noticeably absent are the many trees that now screen the view. (Photo Courtesy Henry Hulett.)

This frame structure occupied the southwest corner of Allen (Main) and Daugherty Streets before the Merchants and Miners Bank building replaced it. The men identified by the numbers are: Charles L. Wright (fourth from left), Will H. Wright (sixth from left), Charles' brother, Burt Wright, Charles' son (seventh from left), Wilbur Owen, an attorney (third from left), and Dr. C.H. Craig, in the window. The others are unknown. The sign on the pole at the far left reads "S.T. KLONDYKE, CLOTHING MAN, SUITS TO ORDER, $9." (Photo Courtesy Henry Hulett.)

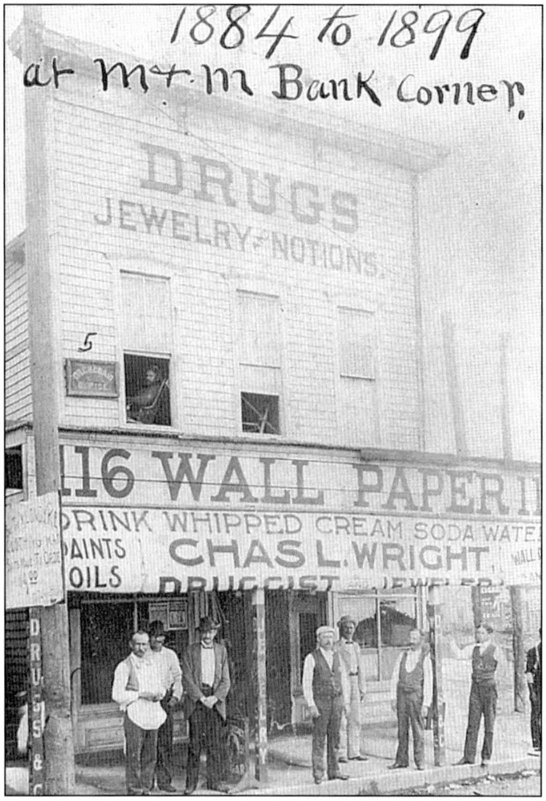

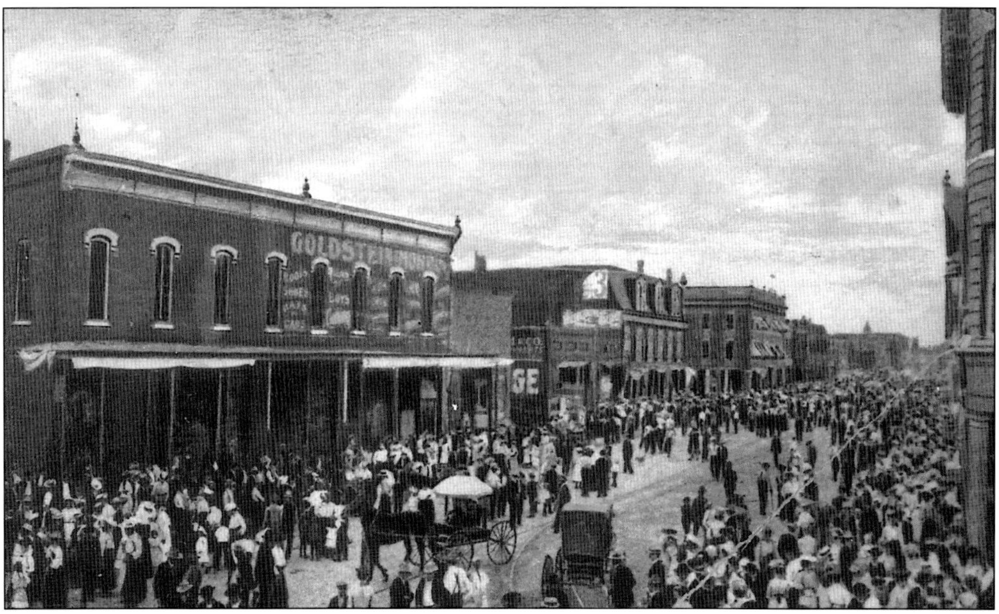

The intersection of Allen (Main) and Daugherty Streets as it later appeared. The Merchants and Miners Bank building is at the extreme right. Trolley tracks are visible on Allen Street turning onto Daugherty Streets. The throng of people had doubtless gathered for a parade. (Photo Courtesy Ron Mosbaugh.)

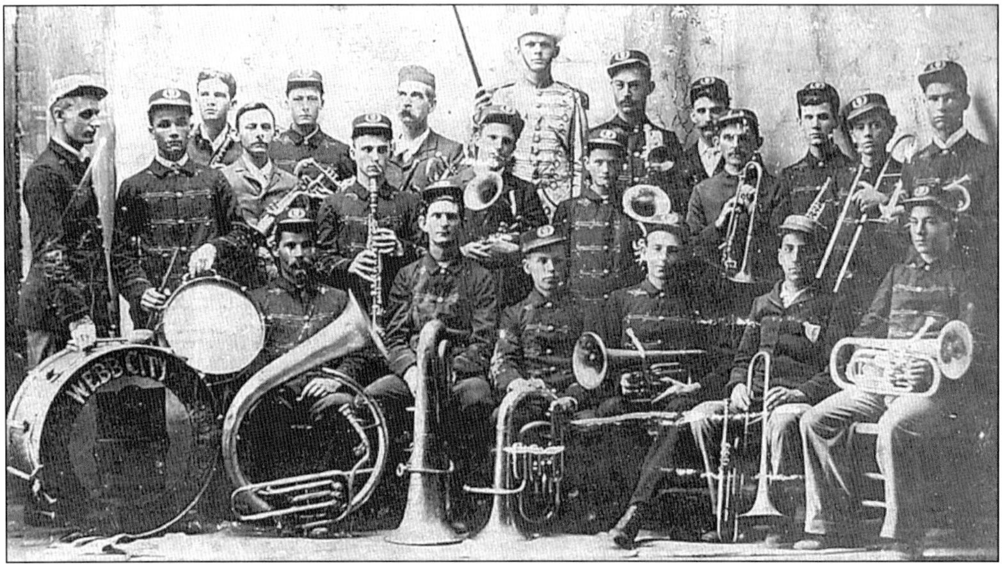

The smartly attired Webb City Band is pictured as it appeared shortly after the turn of the century. The members, from left to right, are: (front row) unidentified, Art Nye, Roy Gibbs, unidentified, Ralph Magruder, and Charlie Magruder; (middle row) Charlie Webber, Ralph Musser, Harry B. Hulett (also conductor of the Citizens Band), Frank Zimmerman, Ed Peters, Charlie Wallace, Gus Witzansky, and Frank Hoselton; (back row) unidentified, Billy Wallace, Prof. Protheroe (director), Harry Wasser, drum major, Tol Munson, Boss Hulsey, Walter Hilburn, and Roy Atchison. (Photo Courtesy Henry Hulett.)

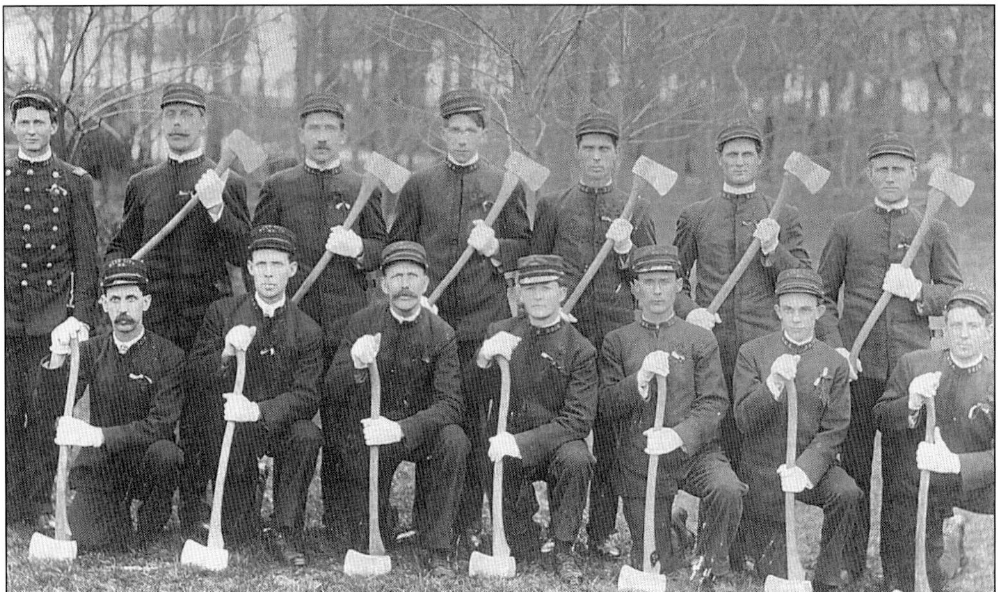

The uniform-clad members of Webb City's Modern Woodmen Lodge pose for the camera in 1892. Those identified are George Kelley, front row second from the left, and John H. Etter, back row third from the left. Lodges and clubs were an important facet of life in the mining region at the turn of the century. The Modern Woodmen of America Organization was founded January 5, 1883. (Photo Courtesy Webb City Historical Society.)

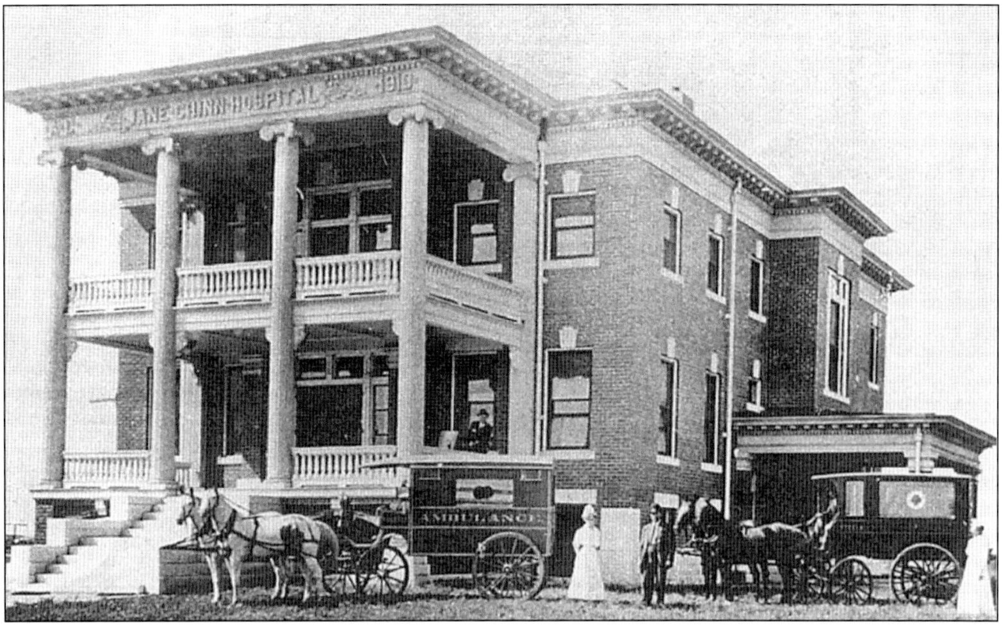

Though it was home for a short time to a Salvation Army hospital, a more permanent facility came into being in Webb City in 1910. Mr. and Mrs. Charles R. Chinn donated money to build Jane Chinn Hospital to serve the community and the surrounding area—especially the miners and their families. It continued in use through most of the century, and the building still stands on the southwest corner of Austin and Rose Streets. (Photo Courtesy Jim and Jean Horn.)

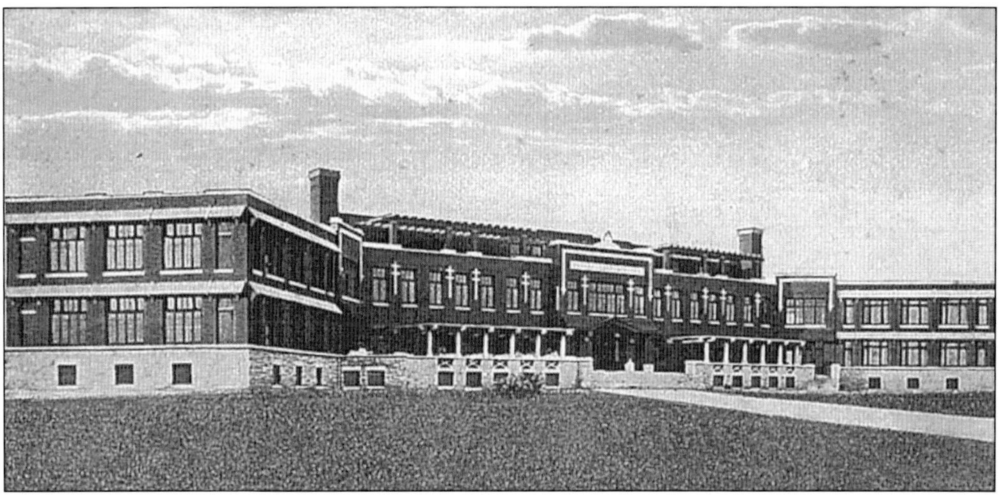

A $100,000 bond issue for the construction of a tuberculosis hospital in Jasper County was approved by voters in 1915. It was built in 1917 on a site northwest of Webb City on land furnished by J.A. Daugherty of Carterville, with the money for actual operation raised by a fund drive. It opened in 1918 with over 30 patients. As previously mentioned, a form of tuberculosis called miner's consumption or "the con" was caused by breathing dust in the mines and was widespread. In the days before medical insurance, the hospital was very beneficial to the miners and others who contracted the disease. Sam Robb, whose progress we have followed, spent the final months of his life there, dying in 1924. The building has been replaced with a modern nursing home. (Photo Courtesy John Burrington.)

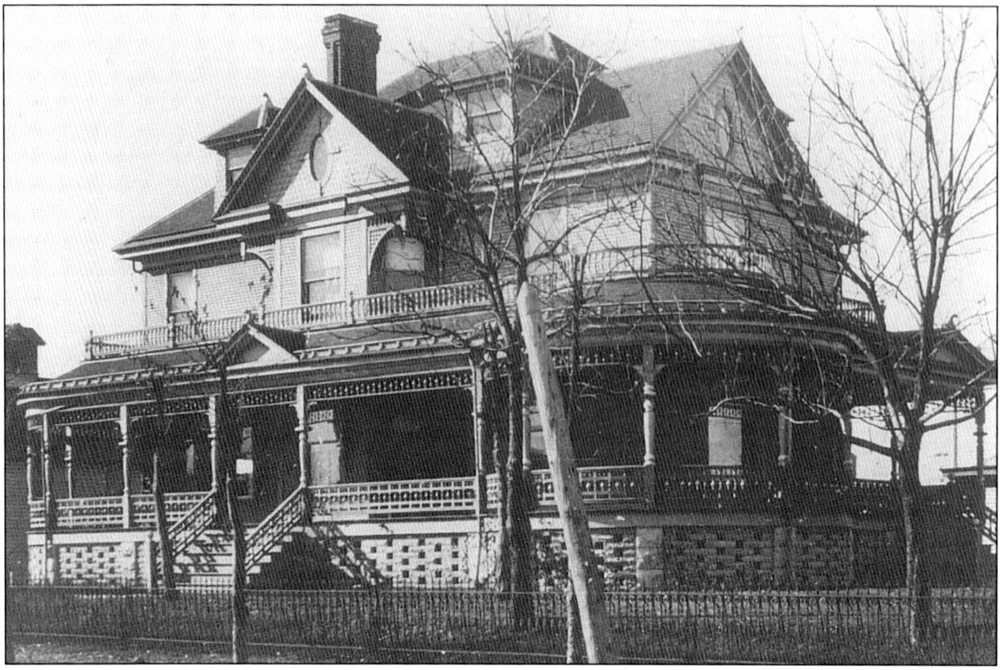

This splendid Victorian is an example of the impressive houses erected in Webb City during the heyday of the mines. It was the home of the George Ball family in 1904. Located at 804 West Daugherty Street, it has recently been restored. (Photo Courtesy Jim and Jean Horn.)

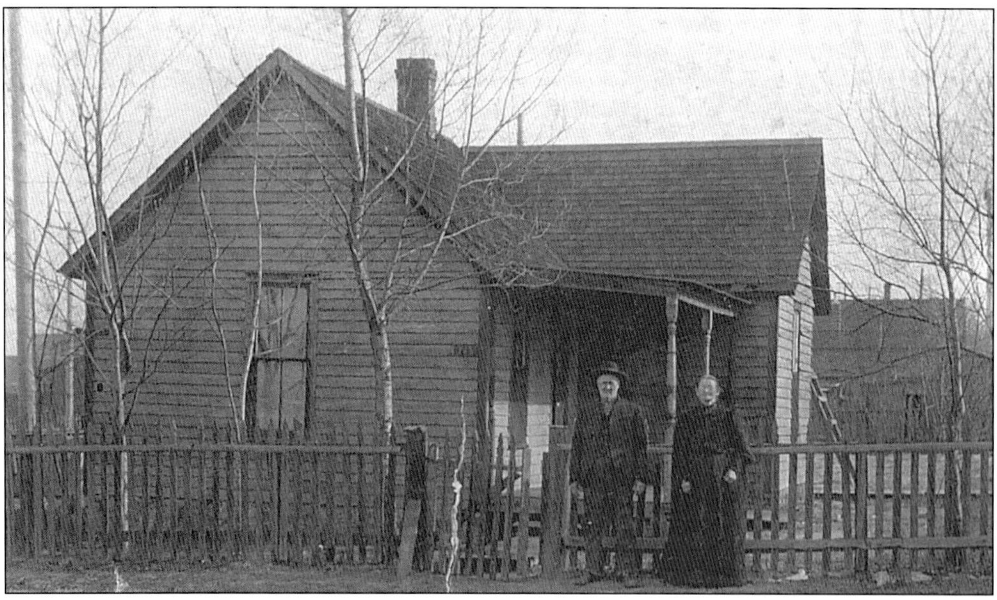

By contrast, the house of the ordinary miner was a much more modest affair, but the pride most of them felt for their homes was obvious. One man who came to the area several years after the turn of the century was impressed with the appearance of the towns—the houses painted and maintained, surrounded by neat, well-tended lawns and picket fences. (Photo Courtesy Webb City Historical Society.)

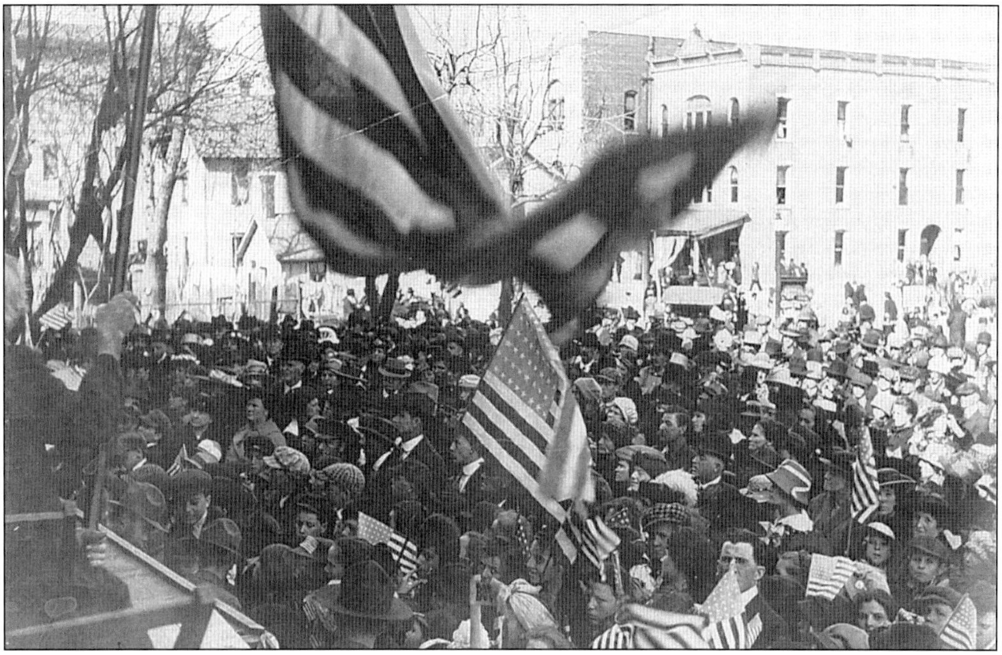

This photograph from the Webb City Historical Society archives is identified as the "celebration of the first armistice," an occasion which triggered patriotism, oration, and the assemblage of crowds. Nineteen men from Webb City lost their lives in World War I, "the war to end all wars." As of mid-February 2000, less than five thousand World War I veterans were still living.

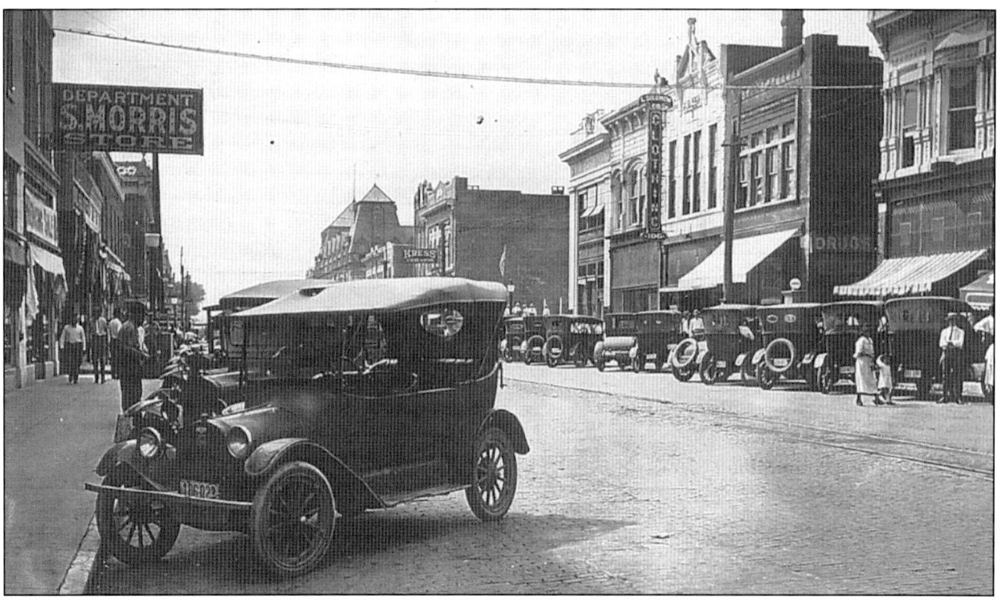

With the passage of years in the twentieth century, the character of Webb City's Allen Street changed. It gained a new name, Main Street, and was surfaced with bricks and lined with automobiles instead of buggies and wagons. However, the Newland Hotel is still visible in the distance. This view looks southwest from the southeast corner of Main and Daugherty Streets. (Photo Courtesy Henry Hulett.)

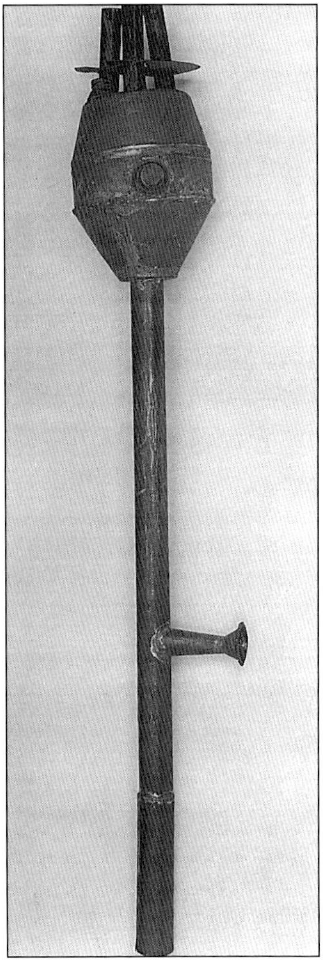

In the presidential campaign of 1896, William Jennings Bryan, a Democrat famous for the "cross of gold" speech he delivered at his party's convention, ran against Republican William McKinley. During the campaign, a group of ardent Democrats utilized the Webb City Democratic Flambeau Club with Col. A.A. Hulett as their captain. Two hundred strong, wearing uniforms of white duck trimmed in red, with knee-length red leggings and metal helmets with horsehair plumes, they were a colorful addition to any parade. But they were most famous for the device in the photo, which, with a 36-inch wooden extension (not shown), was just less than 6 feet in length. The reservoir at the top contained kerosene and was crowned with three burning wicks. Just beneath the kerosene was another chamber holding "lacapodium," a highly flammable flashlight powder. At a command from Col. Hulett relayed back along the ranks by subordinates and the sound of a bugle, each man would at once blow mightily on the mouthpiece of his torch propelling the flash powder through the burning wicks. The sight of two hundred columns of fire and smoke simultaneously blasting 15 feet into the sky with a great roar must have been impressive indeed. The Webb City Band often traveled and appeared with the flambeau club. The band had some Republican members, but apparently harmony was maintained. (Photo Courtesy Henry Hulett.)

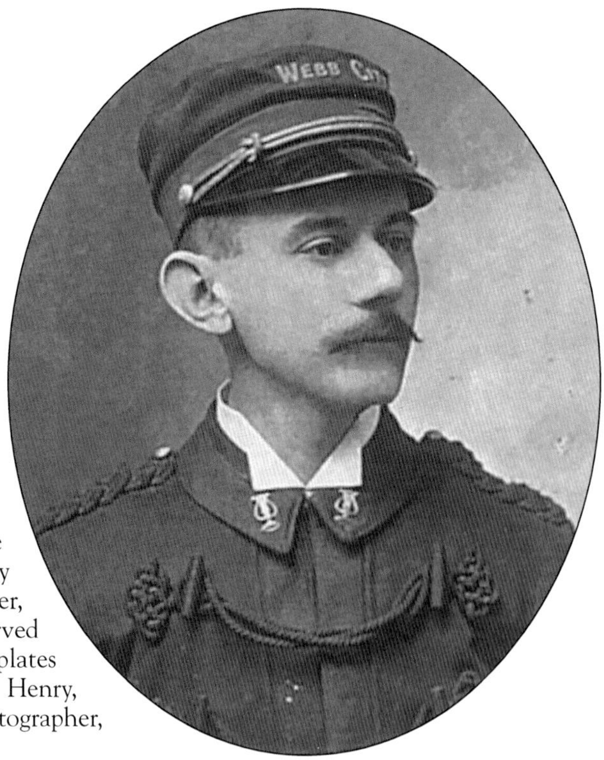

Harry Hulett (1871–1958) was the son of Col. A.A. Hulett and the father of Henry Hulett. A Webb City businessman and amateur photographer, Harry compiled, cataloged, and preserved photographs, some on rare glass plates which illustrate Webb City's history. Henry, himself for a time a professional photographer, has maintained the collection intact.

Four

CARTERVILLE

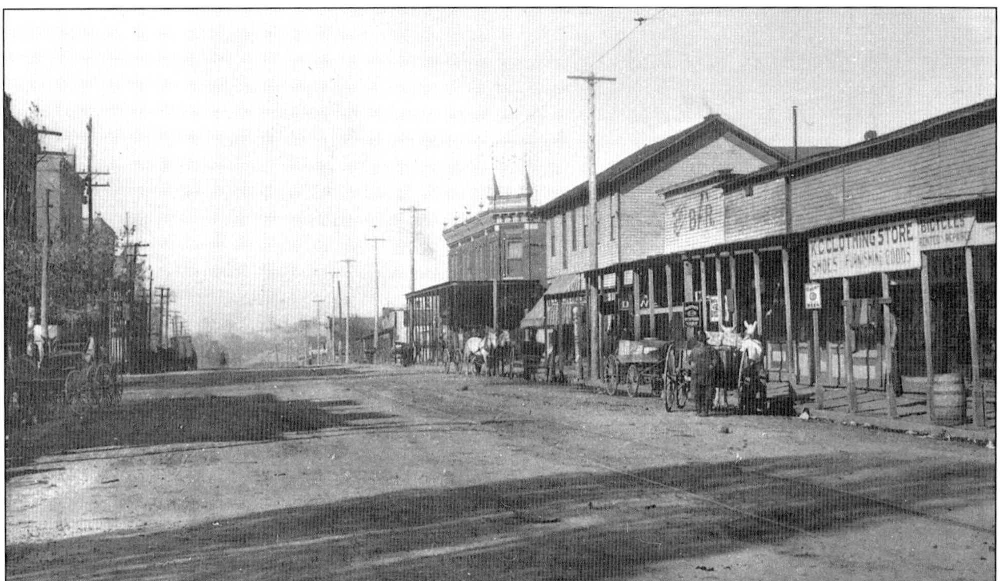

If the architectural style and building materials of the early mining camps in Southwest Missouri looked remarkably like Hollywood's version of western cow towns, there's a reason for that—they were contemporaries of those frontier towns, and the frontier was not that far away. After all, Indian territory started only a few miles to the west. This photograph of Carterville's Main Street may be the oldest in existence. It looks to the west toward Webb City and was probably taken in the late 1880s. The tracks would have been built for the early mule-drawn trolley line between Carterville and Webb City. (Photo Courtesy Ron Mosbaugh.)

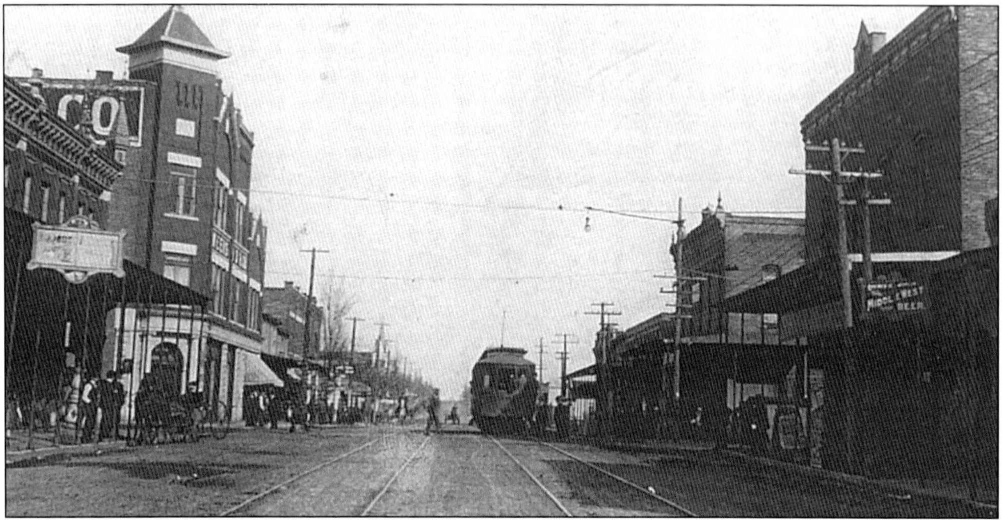

Shortly after the turn of the century, this view of Carterville's Main Street is facing to the east from west of the Fountain Street intersection. In the distance on the left, the spire of the Methodist Church is visible. There is considerable pedestrian activity. Perhaps it was Saturday—a busy shopping day. On the right, passengers board the eastbound streetcar headed for Lakeside Park and Carthage, or Prosperity and Duenweg, other thriving mining towns. People on the opposite side of the street await the westbound car to Webb City and Joplin. (Photo Courtesy Joan Johnson.)

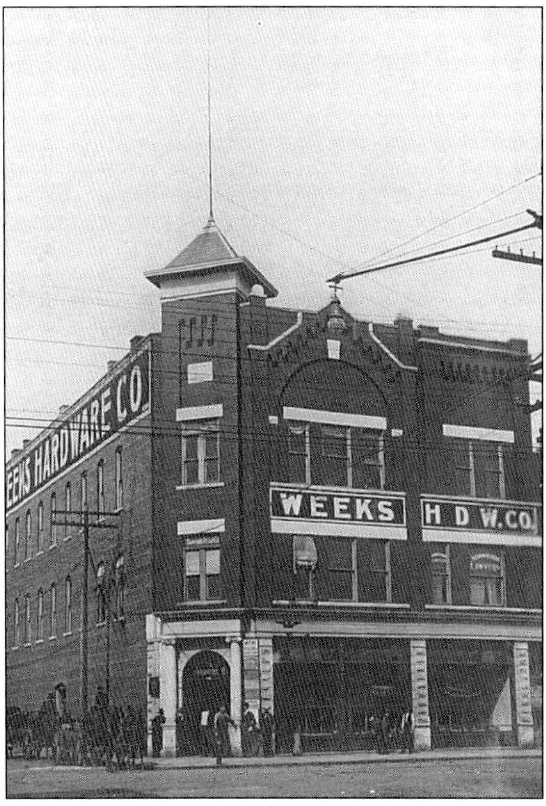

The Weeks Hardware Company was established in 1900. They manufactured and sold mining equipment and proclaimed themselves the be the "largest supply house in Southwest Missouri." Now Morton Booth Company occupies the building. They are said to be the largest manufacturer of gun cabinets in the country. (Photo Courtesy Joan Johnson.)

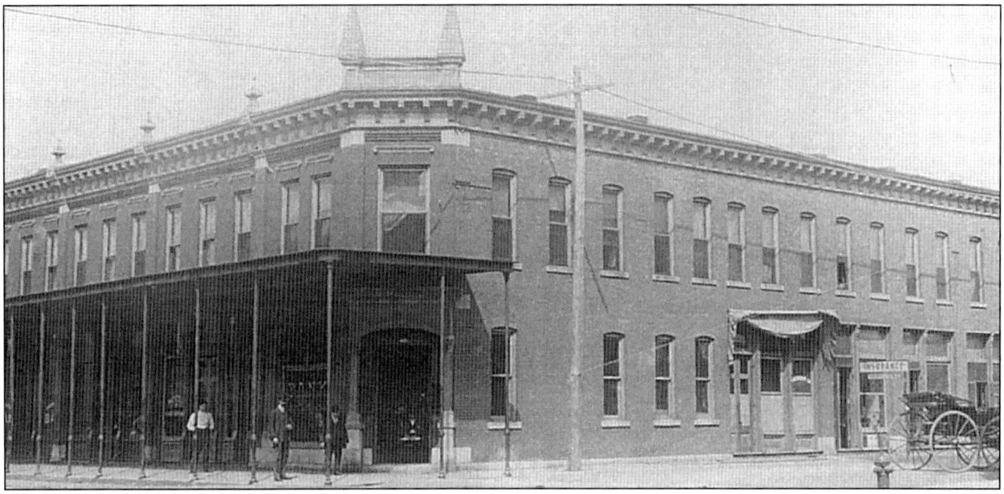

The Carterville Bank became the First National Bank (shown) in 1895 when it purchased the assets of a Webb City bank, moved them to Carterville, and assumed that name. W.A. Daugherty, founder of the city, was president and W.B. Kane was cashier. Early population figures tend to be inflated in the telling, but the 1900 census shows over 26,000 people in Joplin, 9,200 in Webb City, and nearly 4,500 in Carterville. The mining boom had put four Southwest Missouri cities among the largest in the state—Joplin, Carthage, Webb City, and Carterville. In addition, the surrounding areas were covered with shantytowns and shacks. One such notorious community between the Johnstown area of Carterville and Prosperity was known as "Monkey Town" or "Monkey Land," so named because of the pet monkey which sat tethered on a pole in one yard. (Photo Courtesy Joan Johnson.)

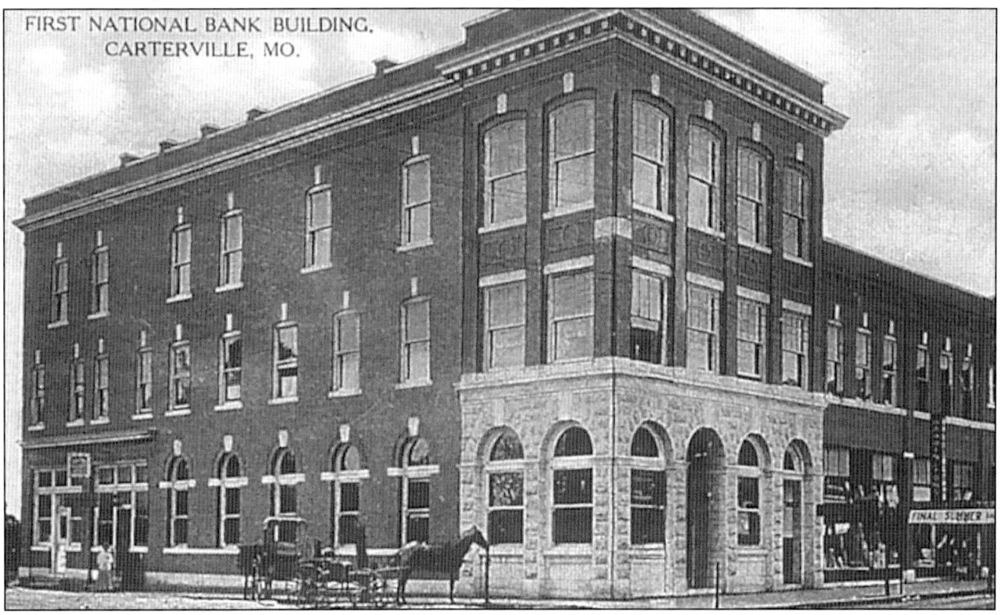

This building later became the home of the First National Bank of Carterville and also served for many years as the town's post office. The old bank building became the Carterville Hotel. The marble counters, grillwork, and solid mahogany fixtures from the bank/post office are now in the Har-Ber Village Museum on the Grand Lake of the Cherokees near Grove, Oklahoma. (Photo Courtesy Clyde Clark.)

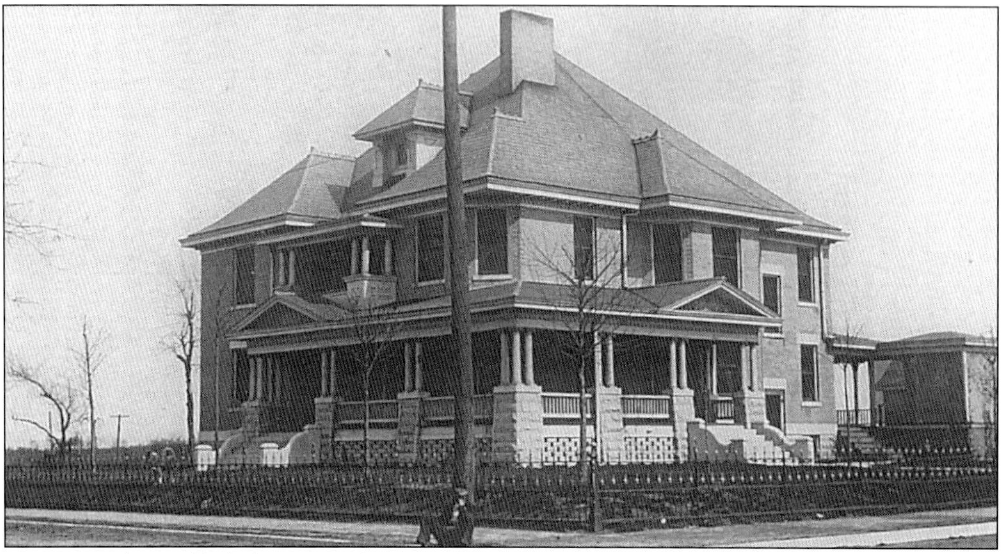

W.A. Daugherty filed the original town plat of Carterville in 1875 on land purchased from the pioneer Carter family. He built his large brick home on the southeast corner of Main and Pine Streets in 1896, and erected homes for other family members nearby. In honor of his early involvement and commitment to the development of the area, both Webb City and Carterville have streets named Daugherty. (Photo Courtesy Joan Johnson.)

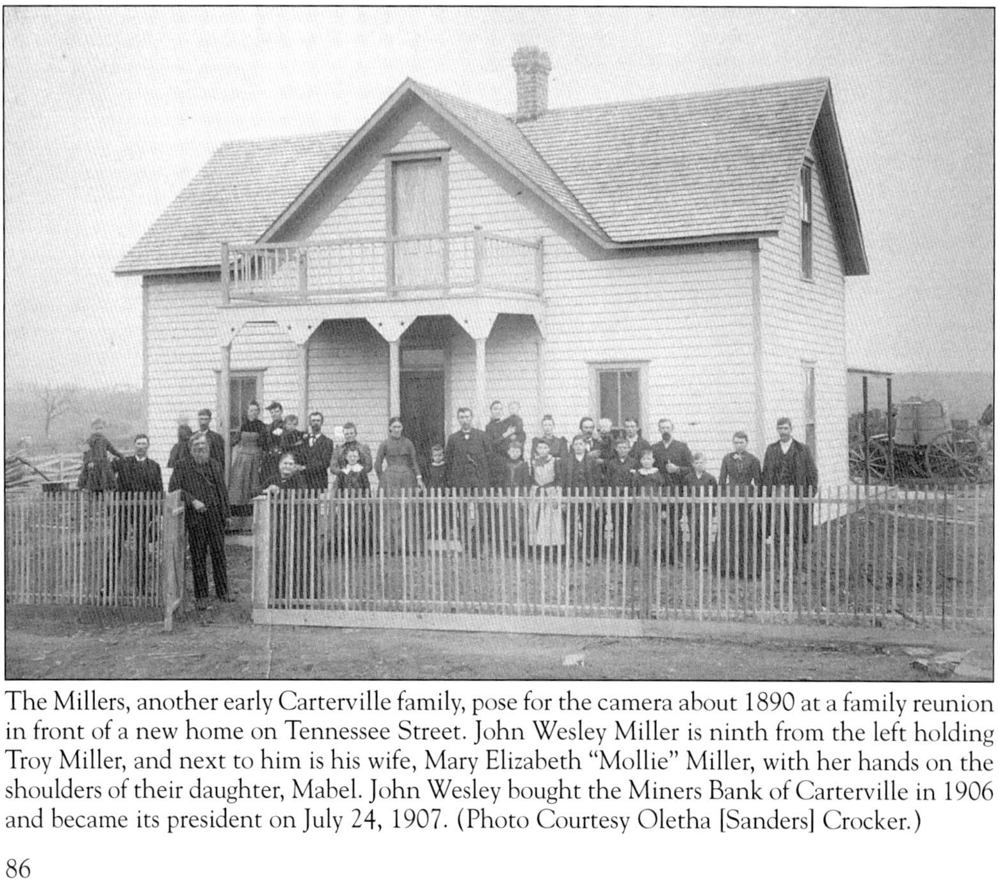

The Millers, another early Carterville family, pose for the camera about 1890 at a family reunion in front of a new home on Tennessee Street. John Wesley Miller is ninth from the left holding Troy Miller, and next to him is his wife, Mary Elizabeth "Mollie" Miller, with her hands on the shoulders of their daughter, Mabel. John Wesley bought the Miners Bank of Carterville in 1906 and became its president on July 24, 1907. (Photo Courtesy Oletha [Sanders] Crocker.)

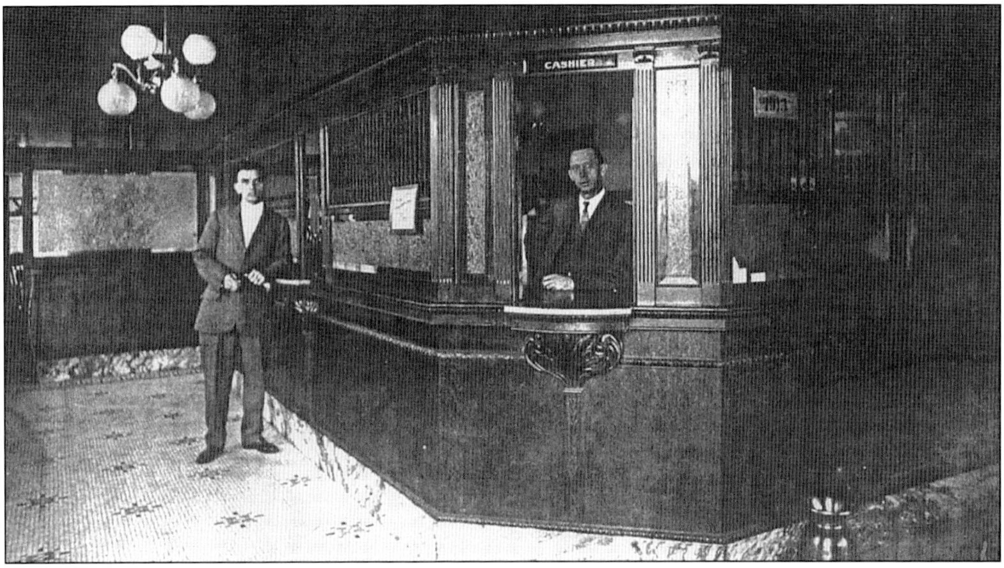

This was the lobby of Carterville's "other bank," the Miners Bank. John Wesley Miller, who had purchased the bank in 1906, contracted tuberculosis and stepped down, turning the operation over to H.B. Sanders, his son-in-law married to Mabel. Mr. Sanders is in the cashier's window. The tuberculosis killed John Wesley Miller on February 6, 1913. The disease was contagious and obviously not limited to miners. Standards of sanitation were different then. Notice the prominently located spittoon. Until recently, the Miners Bank building served as Carterville's City Hall, and the reception area in the lobby looked much the same. H.B. Sanders was instrumental from the beginning, and later served as cashier and president of the First State Bank of Joplin. (Photo Courtesy Oletha [Sanders] Crocker.)

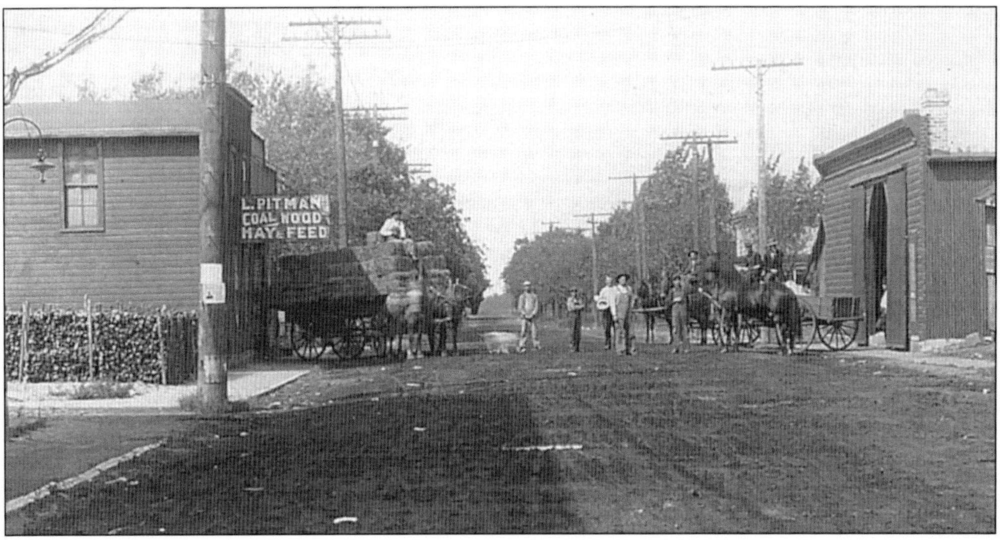

Pitman was another name well known in early Carterville. In this view from 1912, the camera looks to the north showing buildings on both sides of Fountain Street at the Hall Street intersection. Leander Pitman brought his family to town in a covered wagon and soon started a business selling and delivering coal, wood, hay, and feed. The town was built on what was originally crop land and prairie. The shade trees—often quick-growing elms, catalpas, and soft maples—were planted after the streets were constructed. (Photo from the Hammer Collection.)

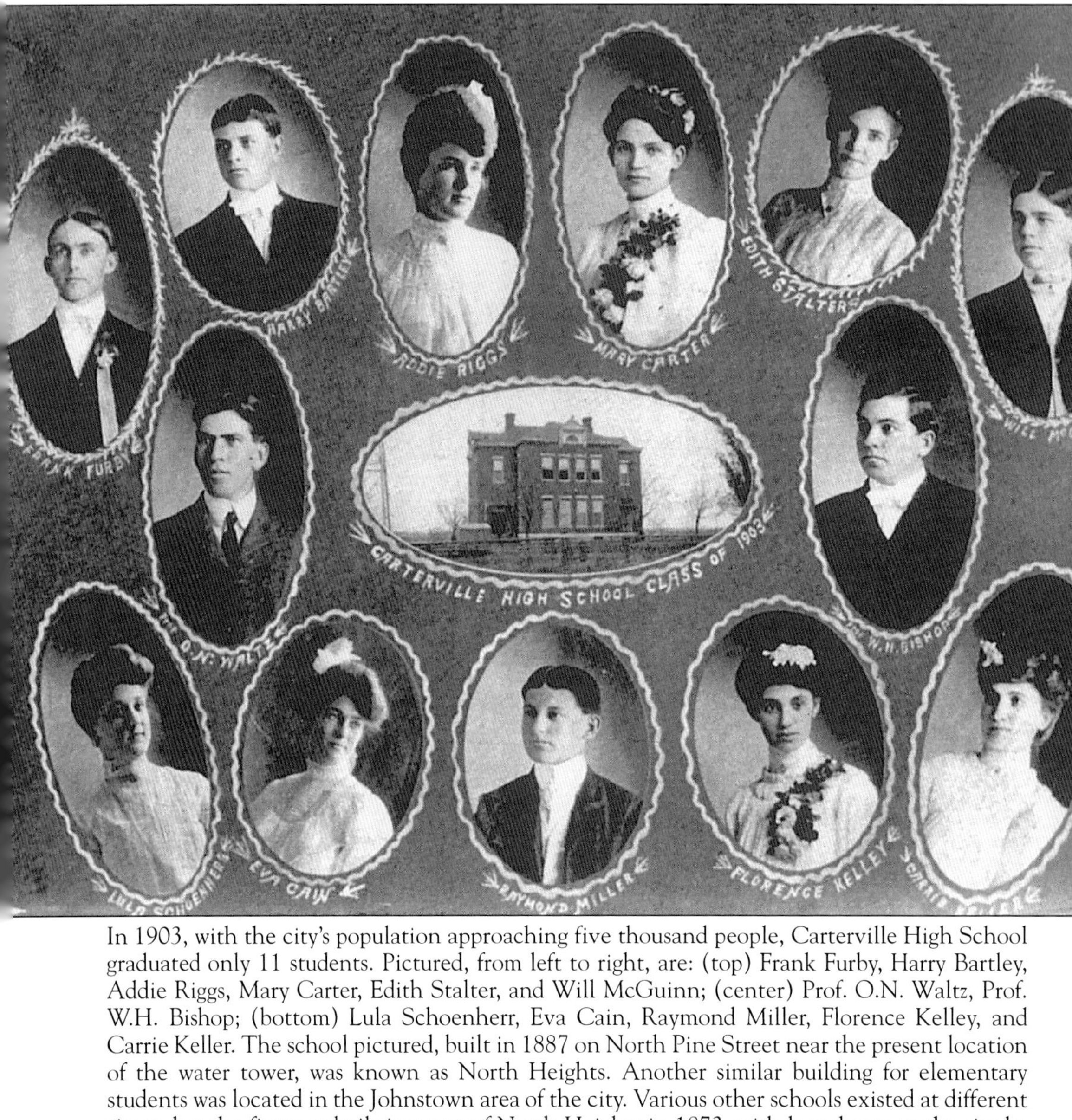

In 1903, with the city's population approaching five thousand people, Carterville High School graduated only 11 students. Pictured, from left to right, are: (top) Frank Furby, Harry Bartley, Addie Riggs, Mary Carter, Edith Stalter, and Will McGuinn; (center) Prof. O.N. Waltz, Prof. W.H. Bishop; (bottom) Lula Schoenherr, Eva Cain, Raymond Miller, Florence Kelley, and Carrie Keller. The school pictured, built in 1887 on North Pine Street near the present location of the water tower, was known as North Heights. Another similar building for elementary students was located in the Johnstown area of the city. Various other schools existed at different times, but the first was built just east of North Heights in 1873, with logs drawn to the site by oxen. The first teacher was John Walker. Jenny Crenshaw, a longtime teacher who set down a history of the city, wrote that a smallpox epidemic hit Carterville in 1882. A "pest house," where the afflicted were forcibly quarantined, was built north of North Heights near the railroad. One "beautiful woman" was said to have lost her eye and part of her nose and mouth to the disease. Those who died were buried nearby in unmarked graves. The Schoenherr family changed their name to Shaner at the onset of World War I. One member, Dolph Shaner, later wrote a history of Joplin, as cited by Marvin VanGilder, a respected current local historian from Carthage. (Courtesy Clyde Clark and Karl Shaner.)

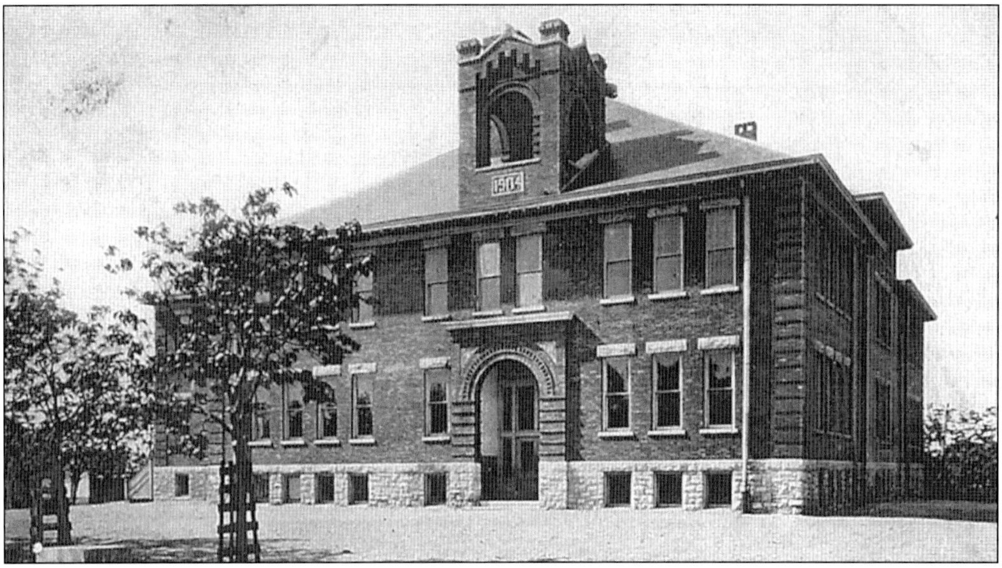

Centrally located, this school building was erected on Hall Street in 1904, but North Heights continued to serve as the high school. Mrs. Fay Furry, who taught in Carterville schools for many years, recalled that her graduating class in 1921 was the last at North Heights before the high school moved to the central location. An elementary school and the gymnasium were located just to the east. The large stone block in front originally held the water barrel. (Photo Courtesy John Burrington.)

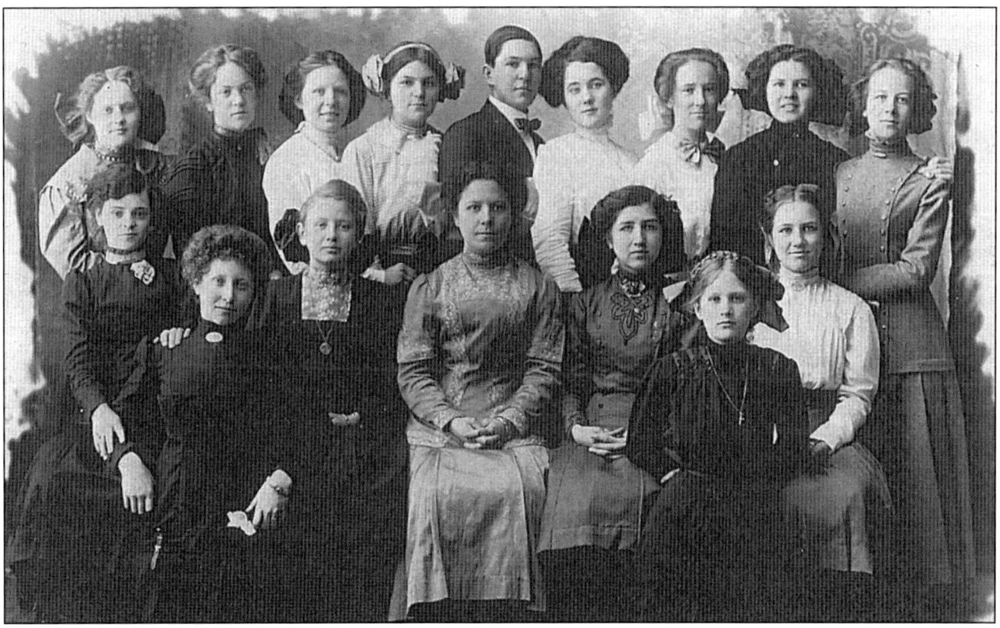

Ella Mae Robb (far left, seated), fifth child of Samuel and Ella Mae (Silverthorn) Robb, poses with her Sunday school class, c. 1908, at the Carterville St. John's M.E. Church, South. Another Robb daughter, Marguerite, told her eldest son of listening to the mine whistles blow and watching the ambulances go out to bring the bodies back to town. One somber day at a funeral in the same church, the mourners watched aghast as the catafalque collapsed, dropping the casket and spilling the body onto the floor. (Photo from the Hammer Collection.)

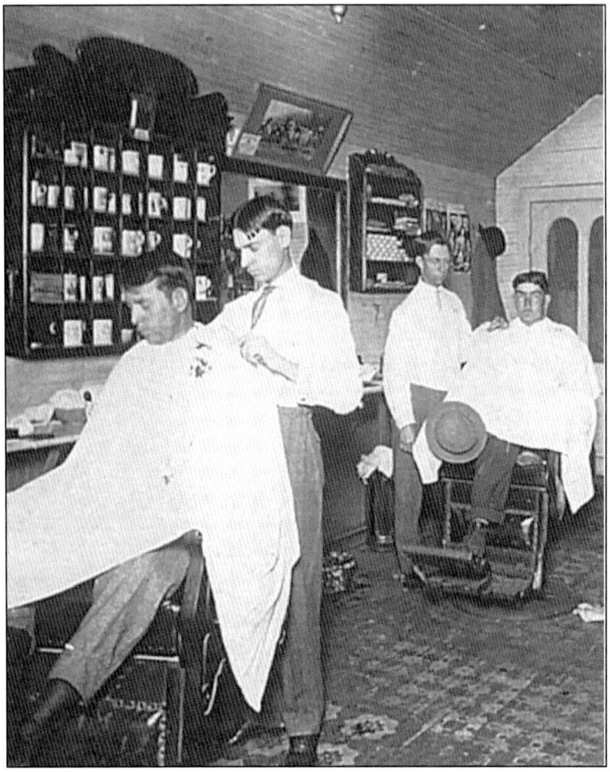

Personalized shaving mugs await their owners in this early Carterville barber shop. That is thought to be Jess Stewart working at the front chair. Jess owned and operated the shop until forced by age and ill health to retire. Displayed above the back chair is Rosa Bonheur's "The Horse Fair." Even after the shop was later remodeled, the reproduction remained, bringing a bit of fine art to a generation of patrons. (Author's Collection.)

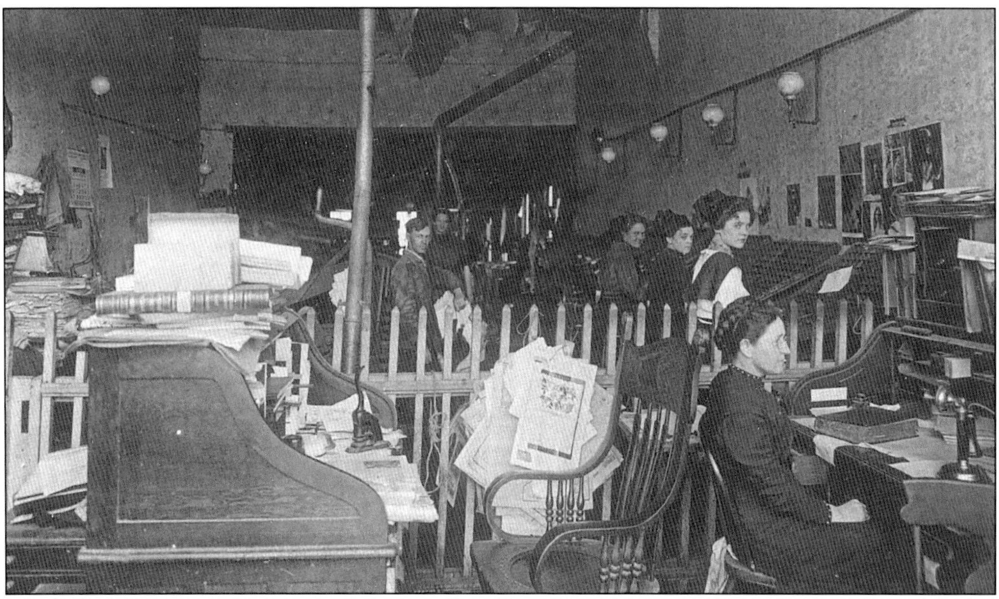

In a cluttered and gas-lit building at 217 West Main in Carterville, six people labor in 1911 to produce a newspaper. H.E. Moody was editor, and his wife, Jessie E. Moody (at the desk), was city editor. From left to right, the young women setting lead type by hand, are Gertie Hess, Ella Robb, and Ethel Sherman. A fragment of one of the old papers in the remaining portion of the "Lodge Directory" column has a partial listing of 12 lodges. The Rebekahs Lodge was said elsewhere to be the oldest in the state. (From the Hammer Collection.)

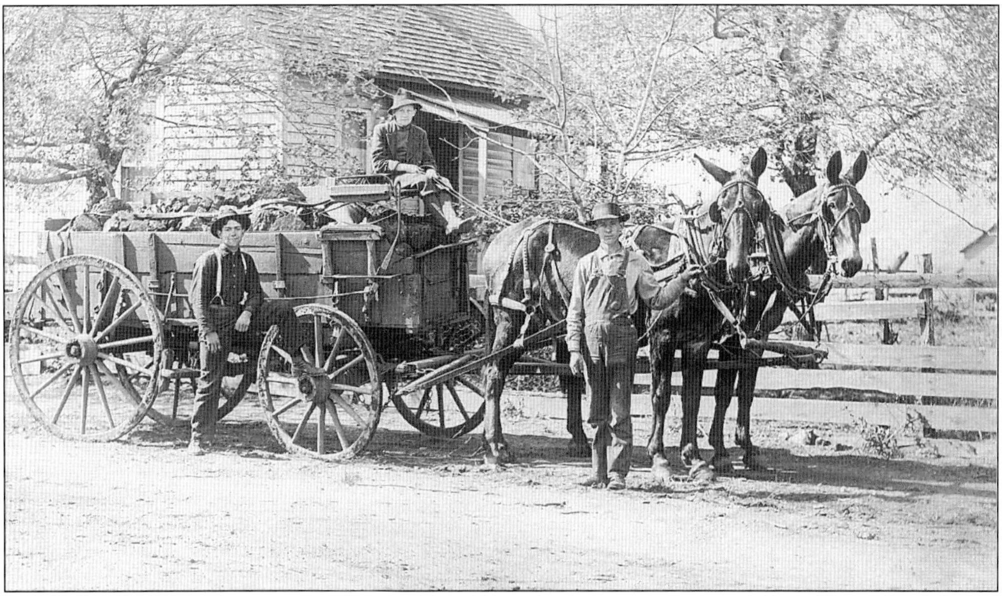

One of the Pitman mule-drawn rigs loaded with coal is ready for a delivery. Mr. Pitman had a horse named "Old Tom" that is still remembered by some early Carterville residents. That is Floyd Hammer with his foot on the hub. (Photo from the Hammer Collection.)

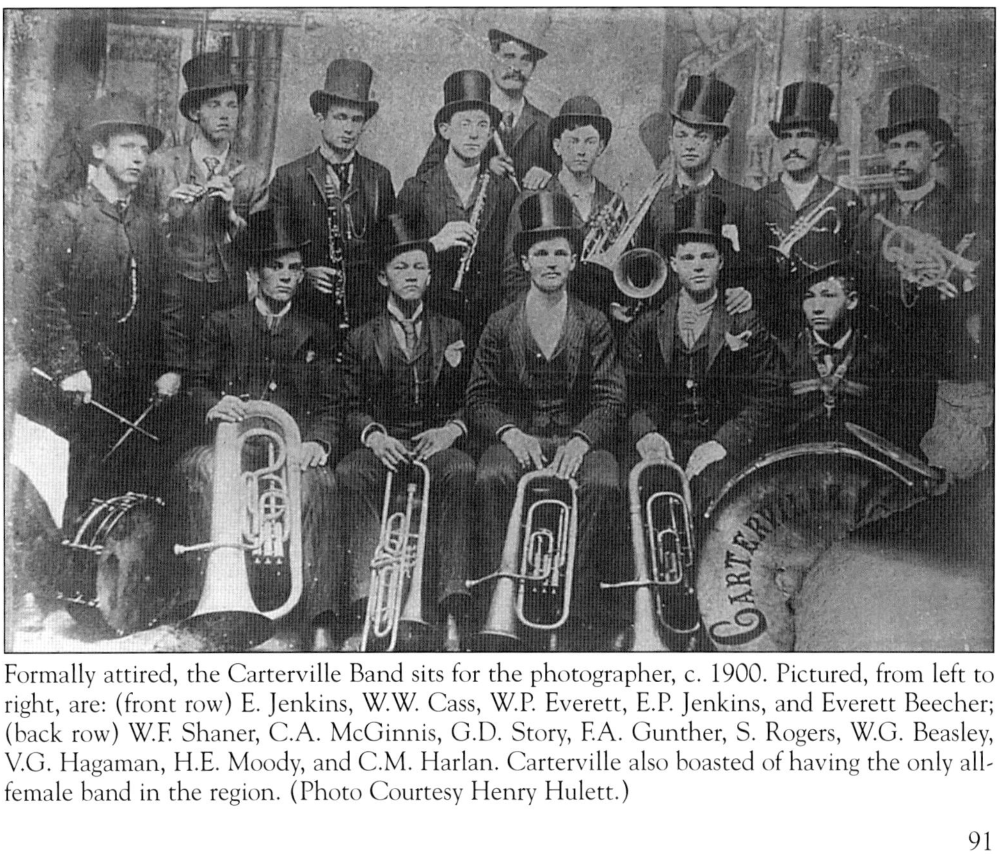

Formally attired, the Carterville Band sits for the photographer, c. 1900. Pictured, from left to right, are: (front row) E. Jenkins, W.W. Cass, W.P. Everett, E.P. Jenkins, and Everett Beecher; (back row) W.F. Shaner, C.A. McGinnis, G.D. Story, F.A. Gunther, S. Rogers, W.G. Beasley, V.G. Hagaman, H.E. Moody, and C.M. Harlan. Carterville also boasted of having the only all-female band in the region. (Photo Courtesy Henry Hulett.)

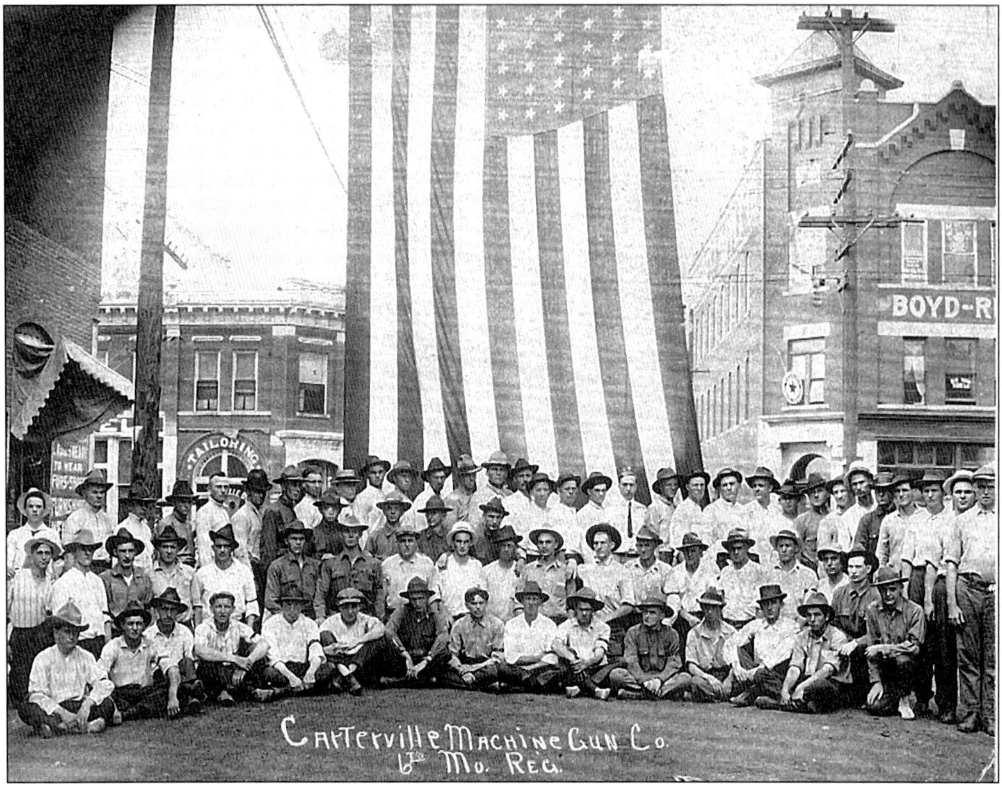

Before heading overseas in World War I, the Carterville Machine Gun Company gathered beneath a giant American flag at the south Fountain Street intersection with Main Street. The story, perhaps apocryphal, is told that the Army liked miners to serve in machine gun companies since they were used to the loud clatter of machinery in the mines. Some area communities organized such companies bearing their town's name. Webb City's company was the 130th, Carterville's was the 140th. Siblings would sometimes deliberately join different companies so the same battle might not claim them both. Many fading memories lie hidden in a group photo such as this. Perhaps an exploration of a few would be of interest—thirteenth from the left in the back row is Joe Bly. Joe carried the mail in Carterville for 25 years, retiring because of ill health in 1949. Third from the left, of the four soldiers seated together in uniform (third row), is Arch Baird. His son, Ralph, became a lawyer and politician and served for a time as Carterville's city attorney. At the extreme right of the photo is Chet Ropp who ran the Carterville Plumbing Shop for many years. While overseas, Chet's specialty gained him the use of a horse which he would sometimes lend to Joe to relieve him of the hated marching. In the second row, kneeling sixth from the left is Lawrence DeWitt, killed in bitter fighting in the Argonne. Mr. DeWitt was identified in the photo by Bob Basnett, a cousin still living in Carterville. Bob's father, John Basnett, an early Carterville city marshal, was famous for ending a pool hall fight by shooting a cue ball from a man's hand. Joe Bly's brother, Ernest, and a friend, Bob Hill, elected to join the Webb City company. Eleven servicemen from Carterville lost their lives during World War I. (Photo Courtesy Webb City Historical Society.)

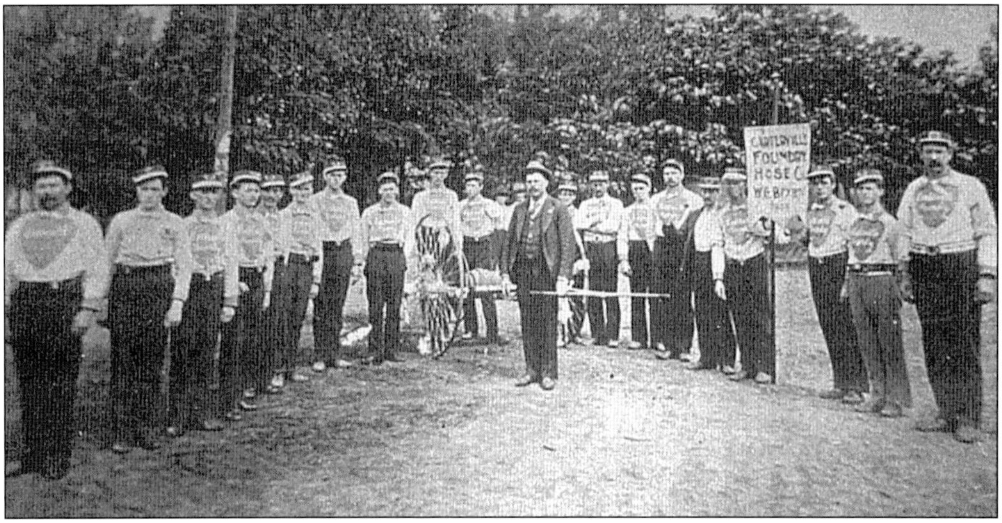

Installation of city water systems gave birth to a new sport—matches between the different fire companies. With considerable success in local contests to spur them on, the Carterville Foundry Hose Company (shown here with a competition reel) entered and, perhaps a bit to their own surprise, tied for second place out of four hundred entries at the St. Louis World's Fair in 1904. The men are unidentified, but three known to have been in the company are Noble Caughenbaugh, Finis Wheeler (also a Trolley League baseball player of some note), and Virgil Green. (Photo Courtesy John and Cherry Furry.)

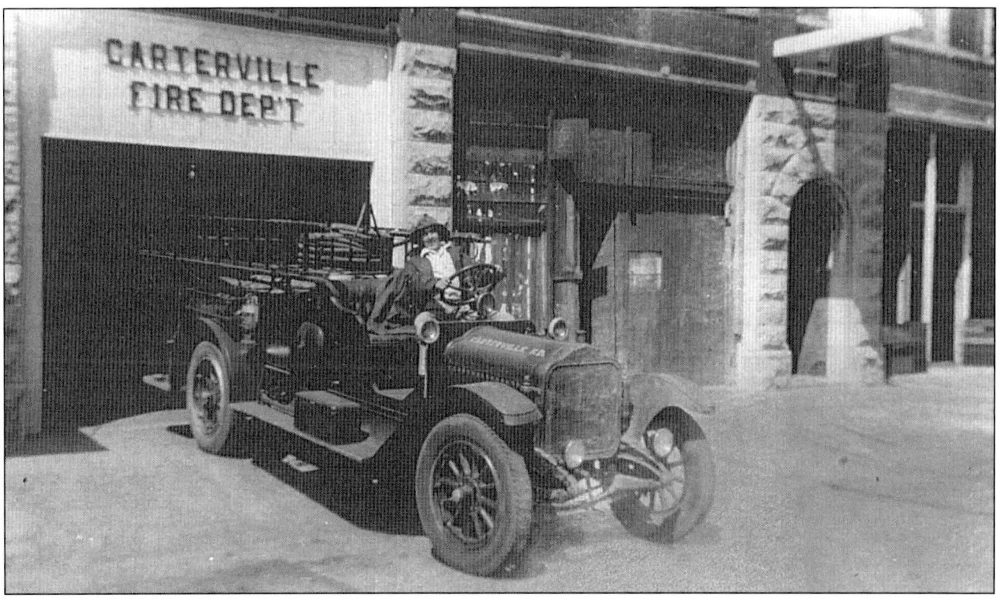

Sometime in the early 1940s, Eddie Snyder backs the old 1918 White fire truck into its home at Sample's Body Shop. With the demise of the mines, Carterville's population declined. The fire department was volunteer, and for many years this truck was the city's only protection. It had to be cranked by hand to start, but it had a compression release, and one volunteer recalled that it "usually started." The volunteer fire department has come a long way since then with a large new building housing nine fire and emergency vehicles in five pull-through bays. (Photo Courtesy Eddie and Julia Snyder.)

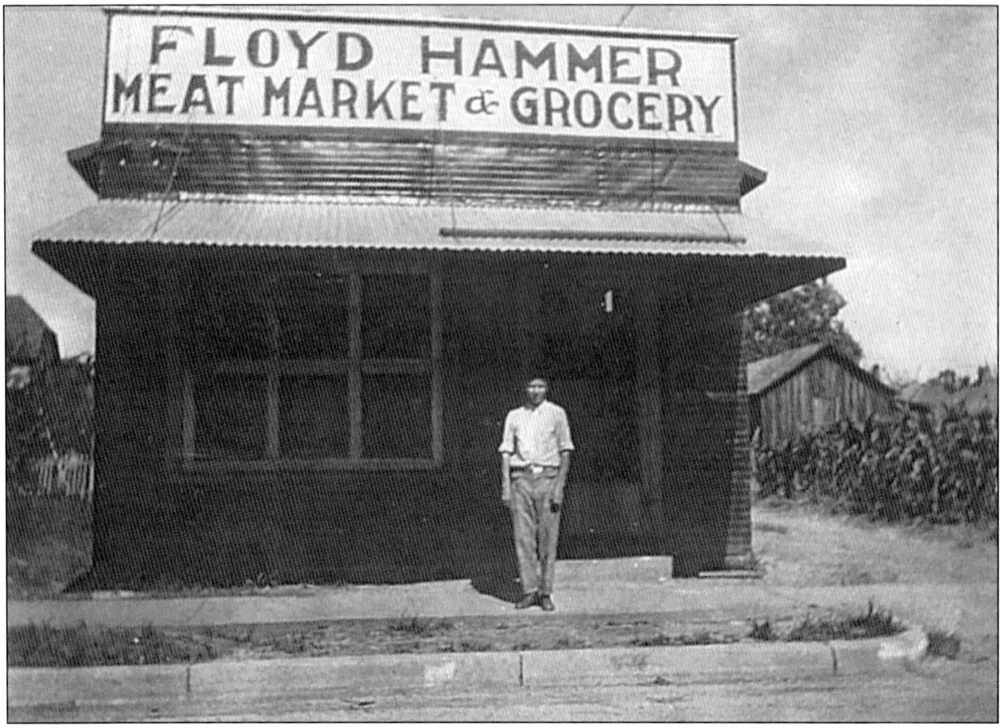

Floyd Hammer never worked in the mines. He was employed for a time by the Pitman feed and fuel company, then managed the Sanitary Grocery in Webb City. He married Ella Robb, and in 1922 they opened this small grocery on a 20-foot strip of land on Washington Street in Carterville. Floyd had a recipe for a particularly tasty pork sausage he made in the store that drew customers from throughout the area. Small stores like this were sprinkled across the town. Another, operated by Paul Pearson, was located only a block north. Both did well. It is estimated that as many as 15 were in operation at one time. Some of the other early grocers' names were Terry, Cardwell, Hunt, Carter, Stalter Bros., Hall, Ault, Sanders, Morse, Breeden, Cooper, Herrod, and Baldwin. The Baldwin store on West Main had a hitching rail and water trough in front long after the customers quit using horses.

Mom-and-pop stores such as these were a microcosm of small town life. At the Hammers' store, orders called in during the morning hours were filled and delivered as far away as Webb City and Lakeside. The customer didn't have to be home—in those days most doors were left unlocked. Also occasionally found in that store was an old one-eyed man who would sharpen pocketknives for the neighborhood boys. He'd spit on the stone, make some circular strokes with the blade, and then examine it critically with his one good eye. "Don't tetch her now, hoss. She'll bite ye." Of course they checked to see if that was true. It was!

Sometimes an unsuspecting lad would stroll in, maybe for kerosene to be filled in a container whose spout was stopped with a potato rammed on the end. Floyd had devised a paper wad shooter from a hollow elderberry branch with an elm wood plunger, and he would cackle with glee as the boy howled in indignation when a slap of the plunger propelled the wad upside his head.

In Jack Foltz's barn, visible behind the cornfield, Floyd and Jack once staged a show for the neighborhood kids, assuring a special prize for anyone who could sit through the entire performance. Most bailed out when a shower of water descended from the loft. (Photo from the Hammer Collection.)

Even with city water lines in place, some private wells and storage tanks remained. This frigid scene at the northwest corner of Hall and Pine Streets was captured by a man named Walton on February 20, 1918. The house survives, but not the tank. Another well existed on South Pine Street at the ice plant. Before city water, homes had water barrels, and private tank wagons made deliveries twice a week. The token shown was good for one barrel of water which was issued by L.W. Clark's water company. The diameter of the token is slightly in excess of one inch, and it sold at first for a nickel, and then later for a dime. It is said that Mr. Clark had three sons who were doctors (one in Carterville) and another who played in a circus band. (Photo Courtesy Clyde and Helen Clark.)

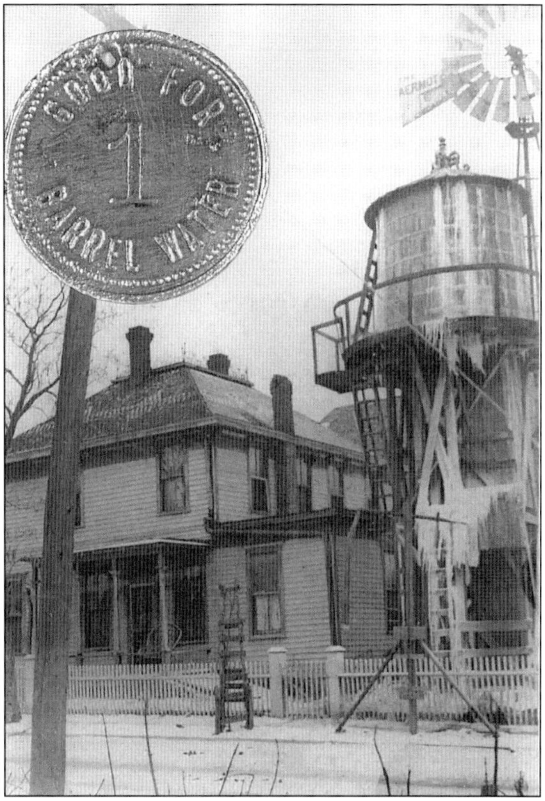

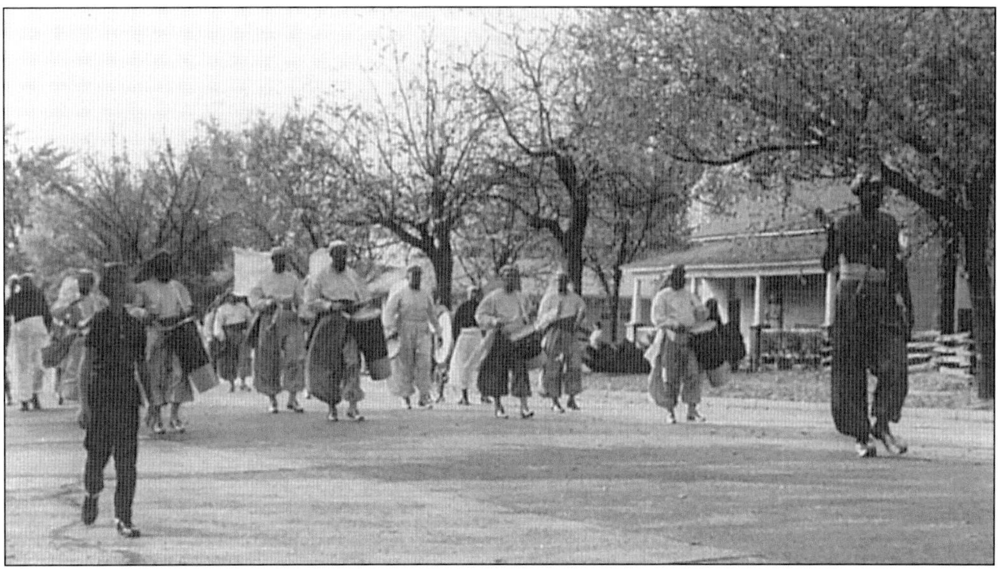

Waving a large scimitar, Dr. Paul Swanson leads the Joplin Shriners' Abou Ben Adhem Oriental Band along Carterville's Main Street in 1963. The immense stature of Dr. Swanson impressed youngsters, who thought he was truly a giant. The occasion marked the laying of the cornerstone for the town's new high school, which survived only until 1968 before becoming an elementary school in the reorganization of school districts. (Photo by author.)

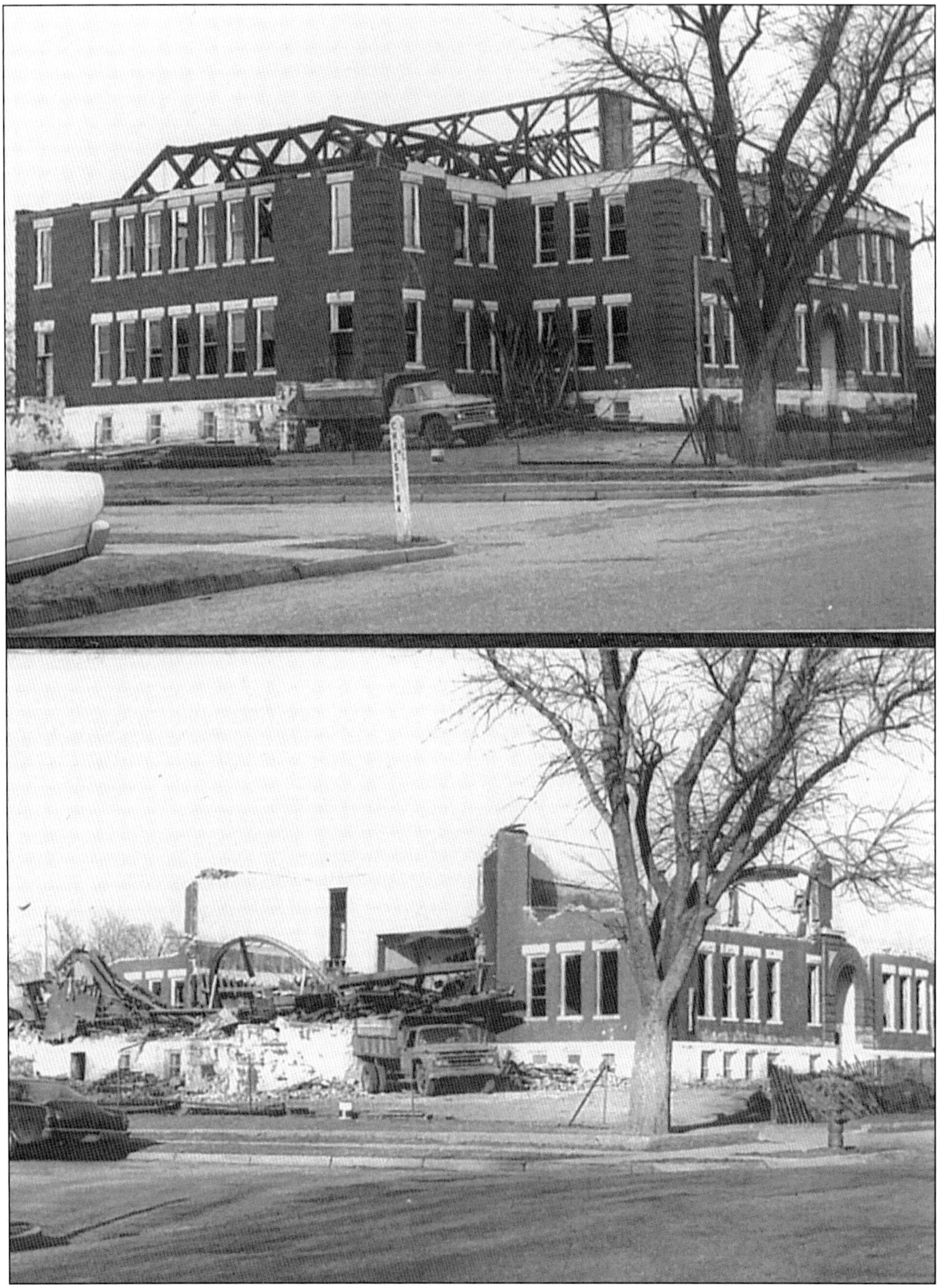

Going, going...just as the towns of Prosperity and Waco had many years before, small cities such as Alba, Duenweg, and Carterville, fearing diminished identity, resisted to no avail the loss of their high schools. Carterville's old high school building was razed in 1971. The new high school just to the east became an elementary school and was incorporated into the Webb City district, as was Alba's school. Duenweg became part of the Joplin district. (Photo by author.)

Five

Other Area Cities and Towns

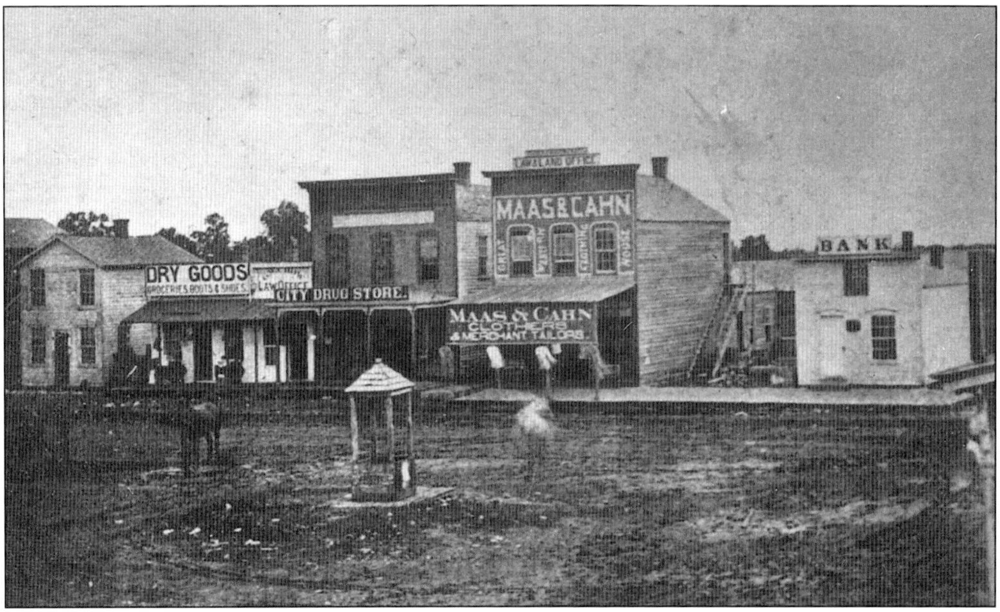

Carthage became the county seat in 1842, but the courthouse was burned in 1863 during the Civil War. Although a temporary frame structure located on the outside of the square served as a courthouse, the center of the square was devoid of buildings for over three decades. For a time, the weed-choked wreckage of the old building remained. Here a horse drinks from a puddle with the north side of the square in the background. Maas and Kahn was one of the early stores established after the war. The First National Bank later occupied the corner where the small frame bank stood. (Photo Courtesy Ron Mosbaugh.)

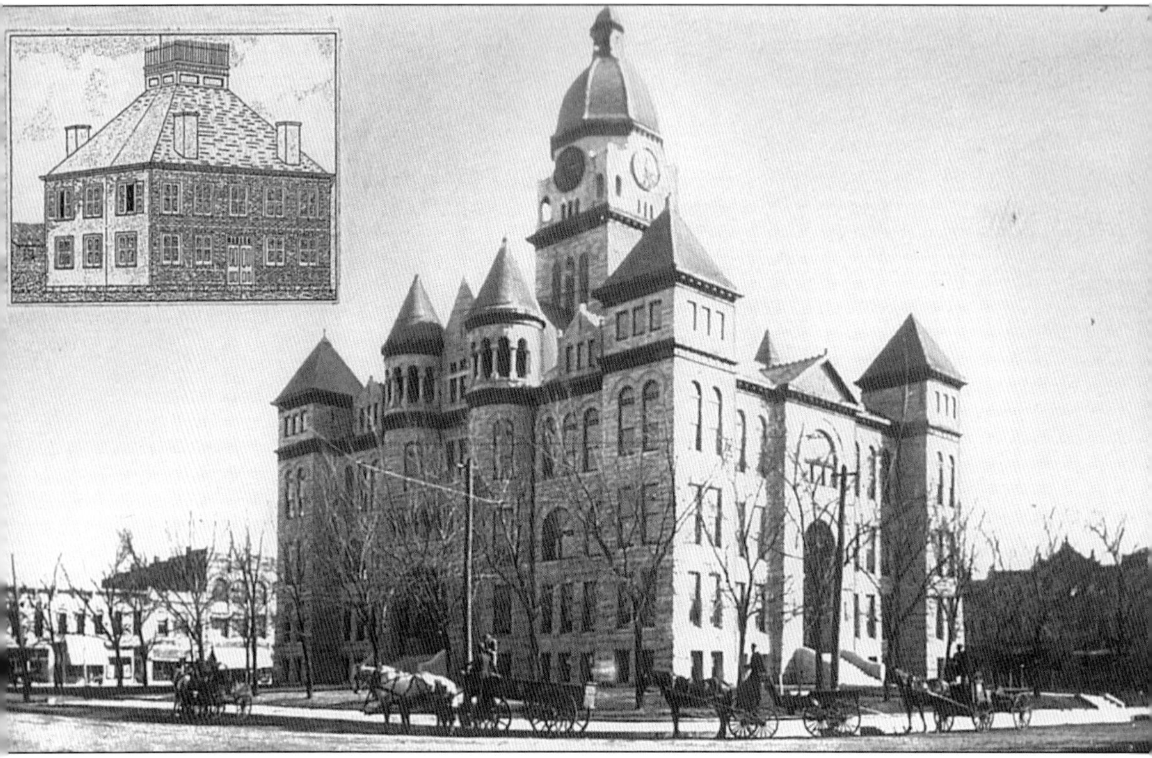

Jasper County finally replaced the courthouse destroyed during the Civil War (inset) with this remarkable structure dedicated in 1895. As its architectural fame spread, some mystics even sought to read esoteric symbolism into its unique design. After the war, some of the earliest settlements such as Sarcoxie survived as permanent towns. Others like Galesburg, Georgia City, and Medoc remained as scattered homes, and still others such as Sherwood, once one of the larger towns in the county, virtually disappeared from the map. A few small towns, notably Prosperity, Asbury, Waco, and the "Tri-Cities" of Alba, Purcell, and Neck City (Hell's Neck), which had boomed with the mines, later declined and remained as pleasant rural communities. Duenweg has had some renewal more recently with the location of large truck terminals there. Carl Junction (widely famous before the dawn of the twentieth century for the discovery of three mastodons in one of its mines) and, to a lesser degree, Oronogo, have enjoyed a dramatic resurgence as bedroom communities. (Photo Courtesy Ron Mosbaugh, inset State Historical Society of Missouri, Columbia.)

Area pioneers soon learned that the local limestone made excellent building material. However, it was buildings such as the First National Bank located on the northeast corner of the Carthage square, built in the mid-1880s, and the courthouse, built about ten years later, which really made it popular. After that, it was used on buildings and homes throughout the region. When a method was devised for polishing the stone for interior use, it became "Carthage marble," and—cut, shaped, and polished by a number of companies which sprang up northwest of town near the quarries—was used on buildings throughout the area. (Photo Courtesy Ron Mosbaugh.)

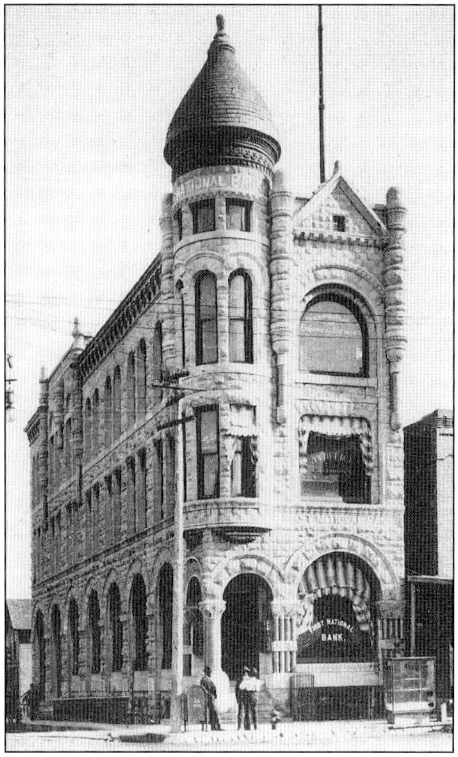

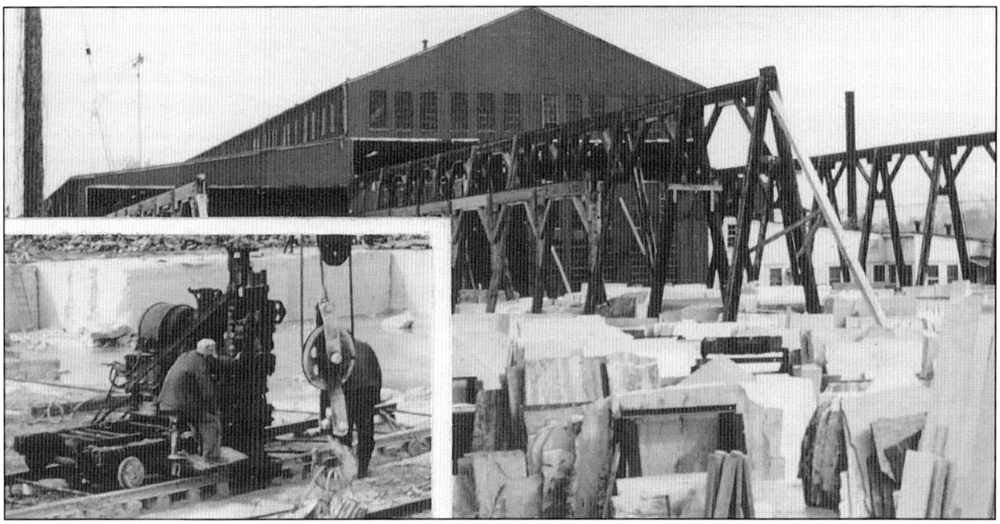

Carthage Marble Corporation became the largest producer of the shaped and polished local marble for use on buildings nationwide. In the mid-1950s, a channeling machine (inset) cuts blocks of the stone in the quarry. The company also imported foreign marble from throughout the world to meet architectural demand. Outside the "big red shed" (No. 8), the marble can be seen stored for later use. "Travelers" on overhead tracks carried the heavy slabs inside to be sawed, shaped, and polished. Albert Hinz, a world renowned stone carver (sculptor), was employed there. He made statuary and exquisite carvings for use throughout the country. Frank Pryor, who had married Marguerite Robb, ran carborundum planers and lathes there for most of his working life. (Photos by author.)

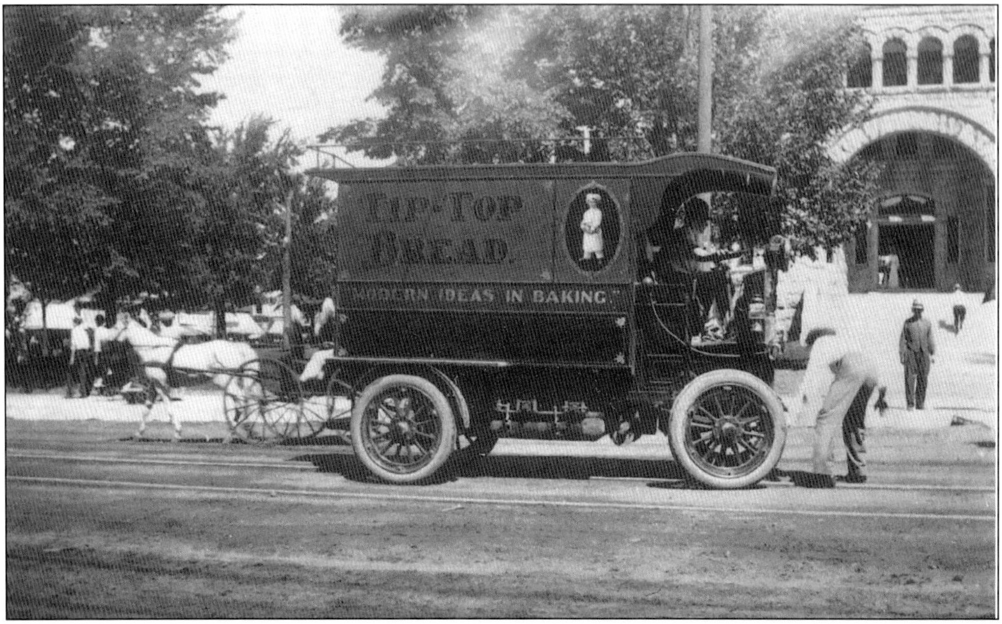

Probably about the time of World War I, a driver is pictured here parked in front of the courthouse receiving some assistance in starting his delivery van. Cranking the early vehicles could be hazardous, resulting in broken arms and dislocated shoulders should the engine backfire. The bakery proclaims "modern ideas in baking." (Photo Courtesy Ron Mosbaugh.)

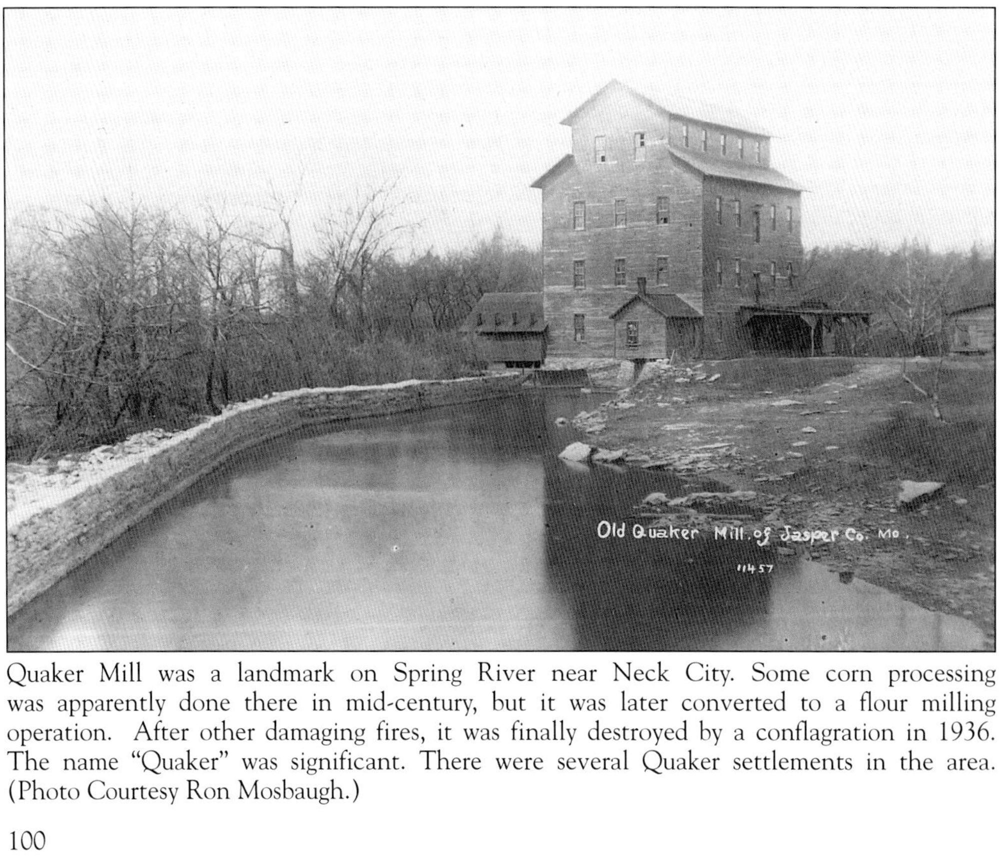

Quaker Mill was a landmark on Spring River near Neck City. Some corn processing was apparently done there in mid-century, but it was later converted to a flour milling operation. After other damaging fires, it was finally destroyed by a conflagration in 1936. The name "Quaker" was significant. There were several Quaker settlements in the area. (Photo Courtesy Ron Mosbaugh.)

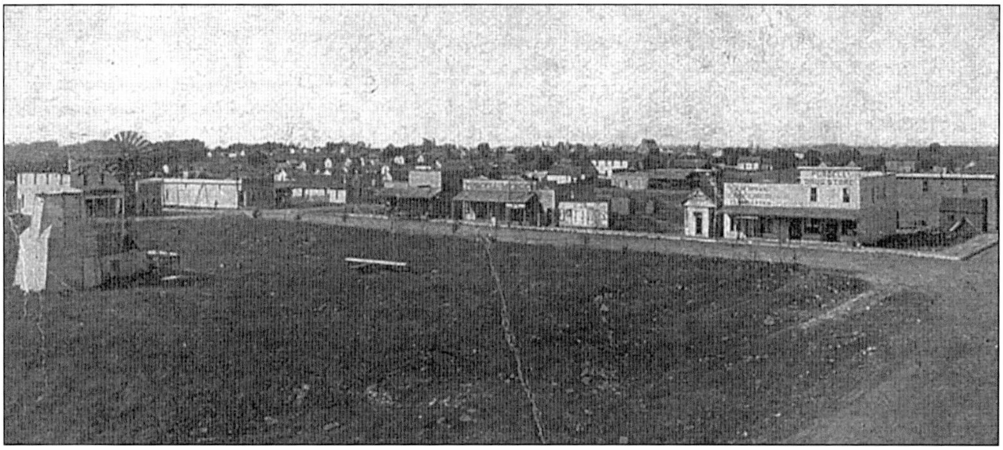

Purcell was the center of the three adjacent towns known as the "Tri-Cities" of Jasper County. This early view looks from the square to the northwest. Purcell and Neck City, located just to the west, were mining towns. Alba, a mile or so to the east, had some mines in the vicinity, but it was more of a farming community. It was at Neck City that one miner was trapped all night by rising water in a drift. After he was rescued, he quipped that he had picked rocks from the ceiling and breathed the air out of their sockets. (Photo Courtesy John Burrington.)

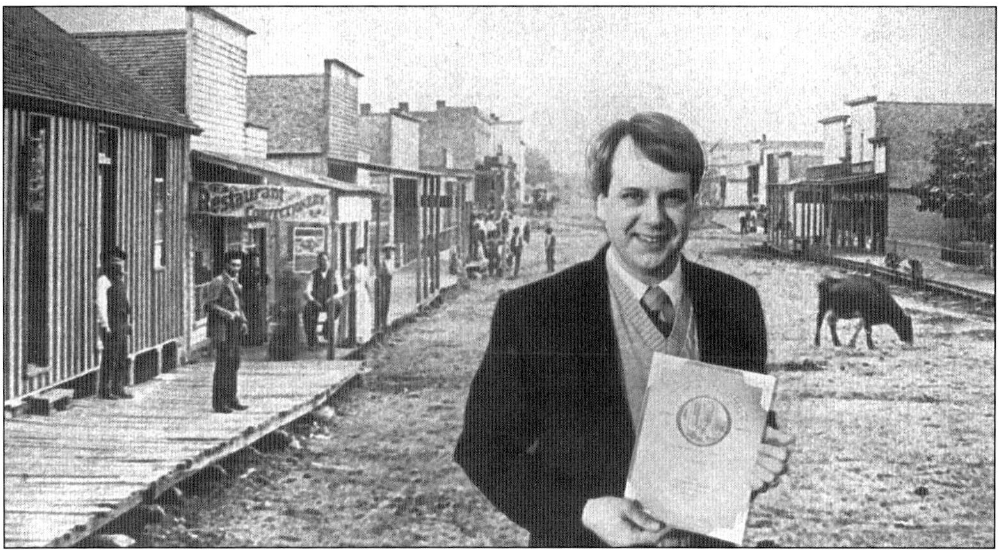

Gary D. Roney bears the name of one of the seven founding families of the Lehigh/Carl Junction community. His great, great grandmother, a widow, moved to the area in the early 1880s with three sons. One of those sons, W.T., along with a partner, established the first general store at Lehigh, later moving his business to Carl Junction. To help commemorate the city's centennial in 1984, Gary—here superimposed on an early view of Carl Junction's Main Street—wrote a history of the area. He worked on the book during several summer vacations from his profession as an educator, hence the title, *Summerstory*. The unique illustration was created by Bob Foos of the *Webb City Sentinel* for an article about the book's publication.

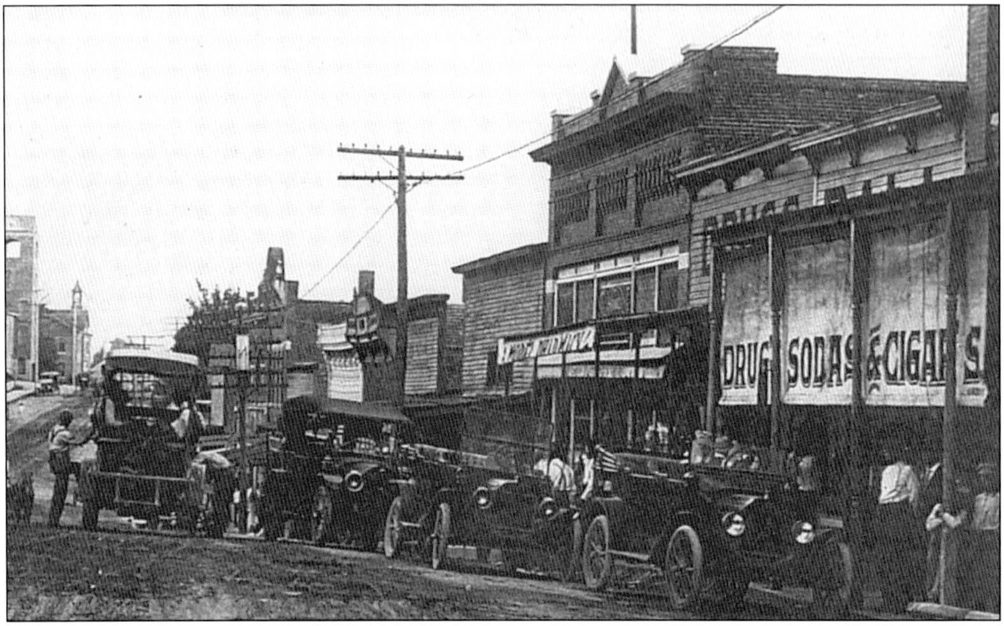

Granby is generally considered to be the oldest mining town in Southwest Missouri. Madison Vickery discovered lead there in 1850 while digging a well, and possession of the resulting mines was contested during the Civil War for the lead they produced. After the war, the town grew and prospered, and, as this photograph looking south on Main Street depicts, was a busy place in the 1920s. According to Arrell M. Gibson in *Wilderness Bonanza*, the residents resented the rapid growth of Joplin. They refused to call it a town, but continued referring to it as a "camp." (Photo Courtesy Ray Savage/Granby Miners Museum.)

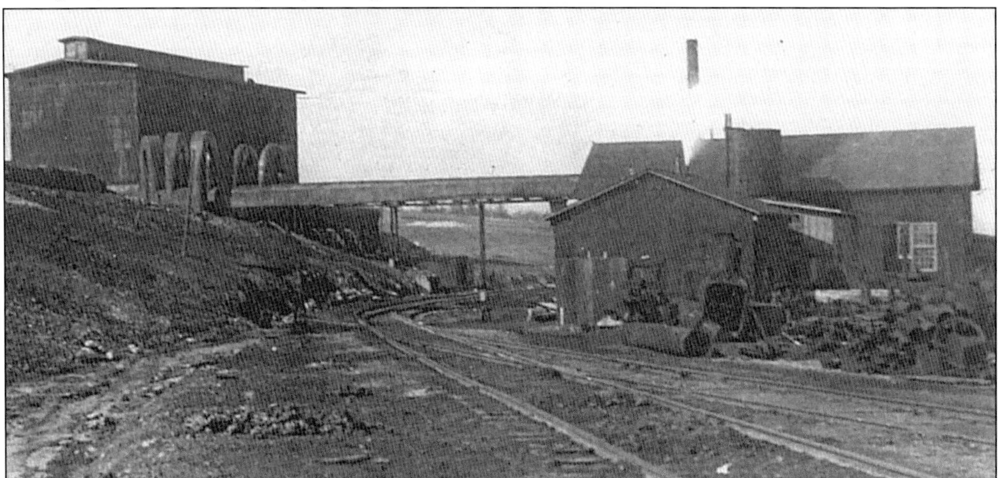

Primitive smelters existed in the vicinity well before the Civil War to melt and separate lead, but a more commercial venture was started in the mid-1850s by Henry T. Blow (said to be the discoverer of Marvel Cave of Silver Dollar City fame) and a Mr. Kennett. The Granby Mining and Smelting Company was organized in 1865. This view shows the later evolution of the smelter. A Missouri Pacific rail line bisected the operation. The tree-covered knob, now without buildings, is still known locally as Smelter Hill. (Photo Courtesy Ray Savage/Granby Miners Museum.)

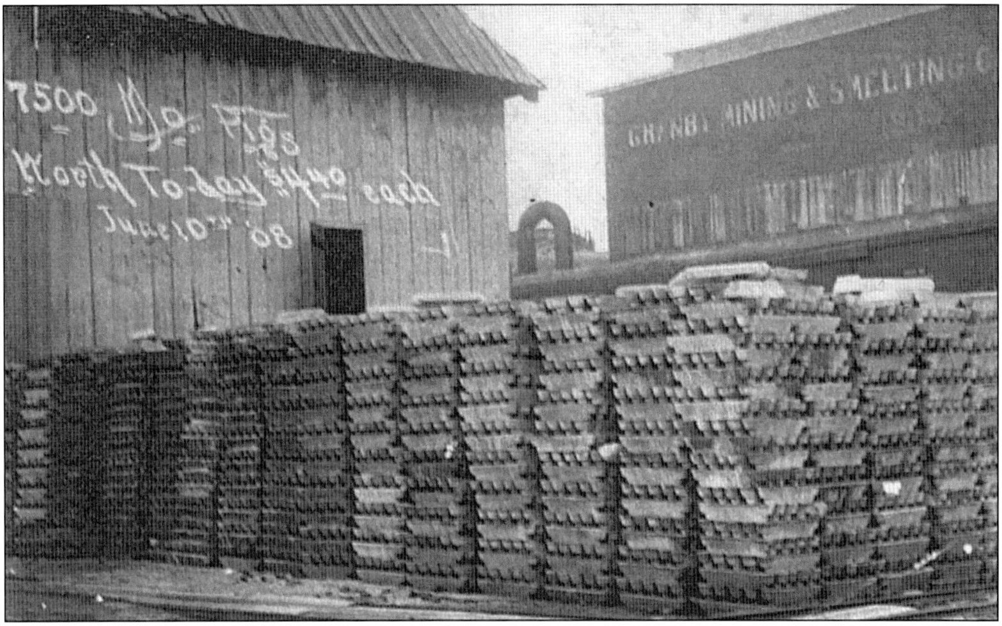

On June 10, 1908, these 7,500 "pigs" of lead produced by the Granby Mining and Smelting Company were stacked for shipment. A pig of lead weighed 100 pounds. The name derived from the term "pig iron" as used in the steel mills back east. (Photo Courtesy Ray Savage/ Granby Miners Museum.)

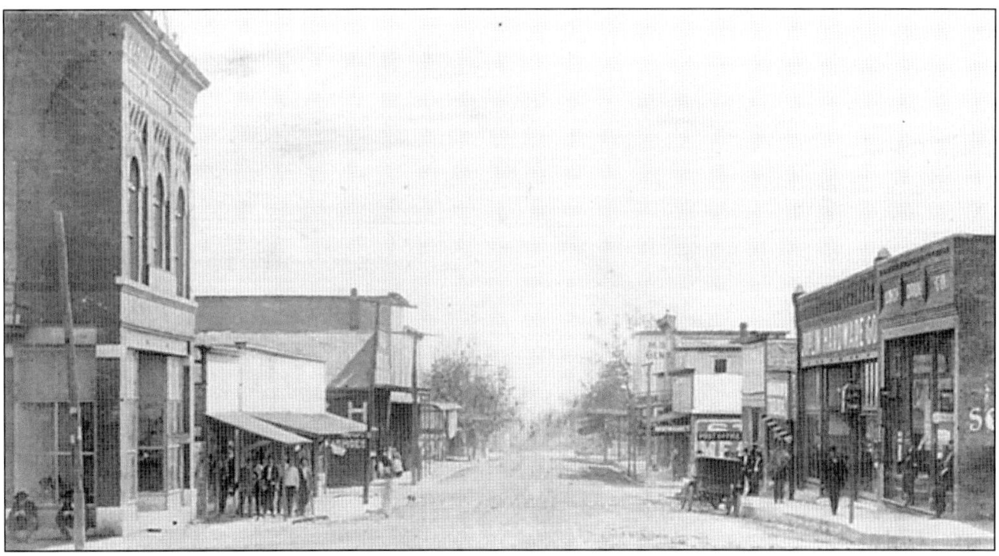

Some lead mines had existed at Minersville (which would become Oronogo) before the Civil War, and the Granby Lead and Zinc Company was instrumental in their restoration after the fighting ended. Because another community with the name Minersville existed in Missouri, it was necessary to choose a new name for the one in Jasper County. Legend has it that at a meeting called for that purpose a miner, perhaps inebriated, rose to his feet and said, "Boys, it's ore or no go," thereby gaining instant approval for the town's new name. This early view looks south on Second Street. (Photo Courtesy Ron Mosbaugh.)

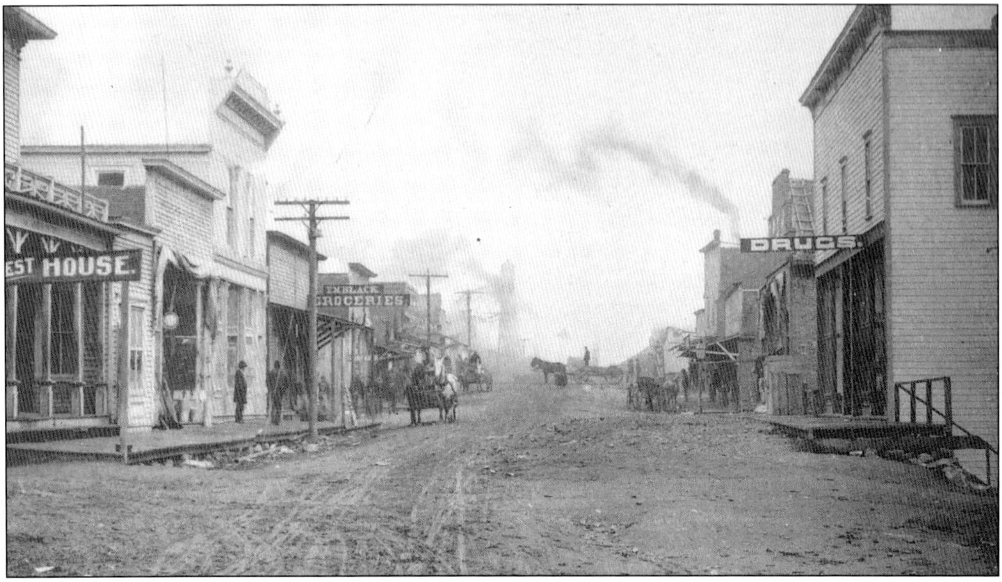

A Jasper County map printed in 1874 continues to show the name Minersville instead of Oronogo, but locates it about a mile to the west. In May of 1883, Oronogo was virtually destroyed by a deadly tornado, commonly called a cyclone at that time, but the demolished homes and businesses were removed and the town was rebuilt. (Photo Courtesy Ron Mosbaugh.)

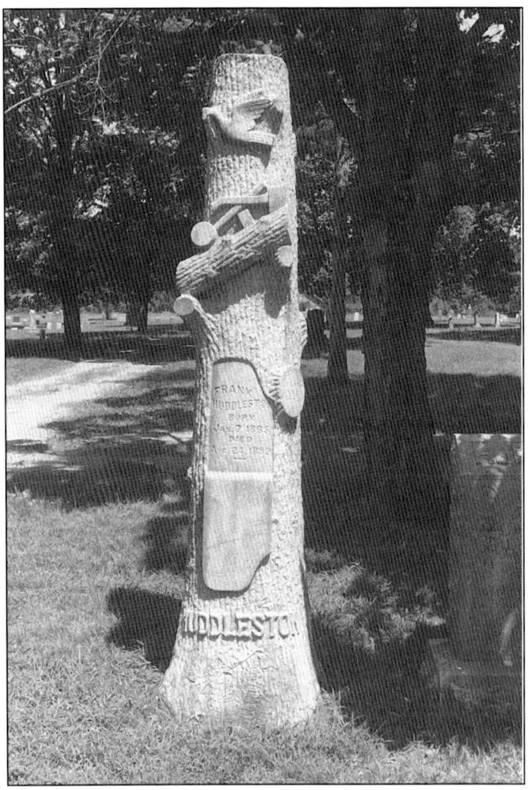

People from elsewhere sometimes look with wonder at the many stone tree trunks scattered throughout the local cemeteries. This one standing 8 feet tall in the Oronogo Cemetery is a good example. Woodmen of the World lodges were popular in the mining region. Men could obtain insurance and burial monuments through their lodge. This one was furnished by the Carterville Camp No. 9, the low number indicating it was one of the early lodges in the state. Nearby is a marker for C.E. Elliott inscribed, "1833–1918, born in Morwenstow, Cornwall, England," and "...the last survivor who laid out the city of Joplin." (Photo by author.)

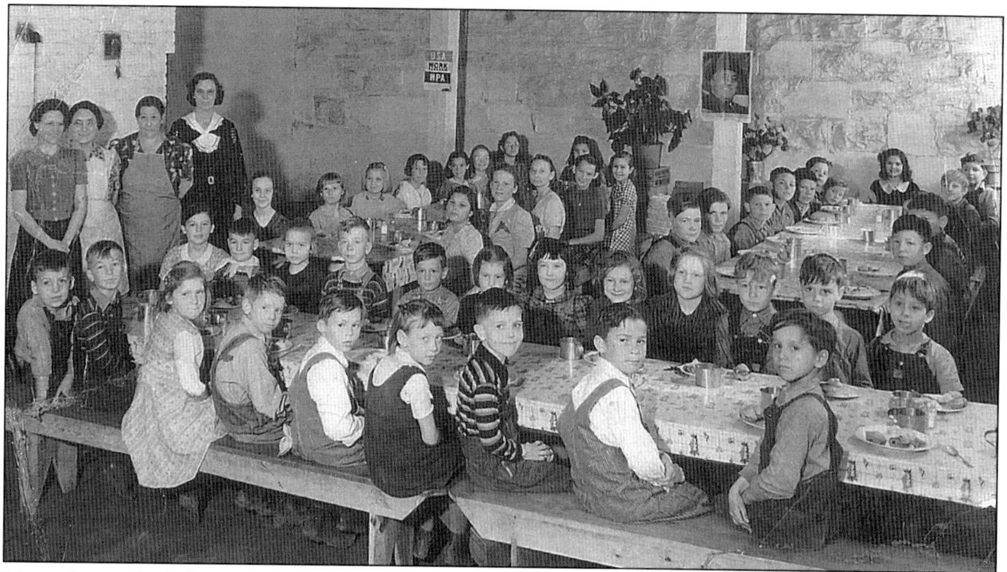

Near the end of the Depression, c. 1940, these students, participating in a program sponsored by the WPA, gathered in the basement of their Prosperity school for a hot lunch. The adults, pictured from left to right, are: Mrs. Gerkey, teacher; Mrs. Whitten, cook; Mrs. Valentine, cook; and Mrs. Hance, teacher. Posted on a column behind, a smiling portrait of FDR beams approval. Once overwhelmed with an enrollment in excess of one thousand, which forced half-day sessions, Prosperity schools saw the numbers decline sharply after the mines closed. Students are now bused to schools in Webb City. (Photo Courtesy Leeman Dorsey, information Charles and Melvin Misner.)

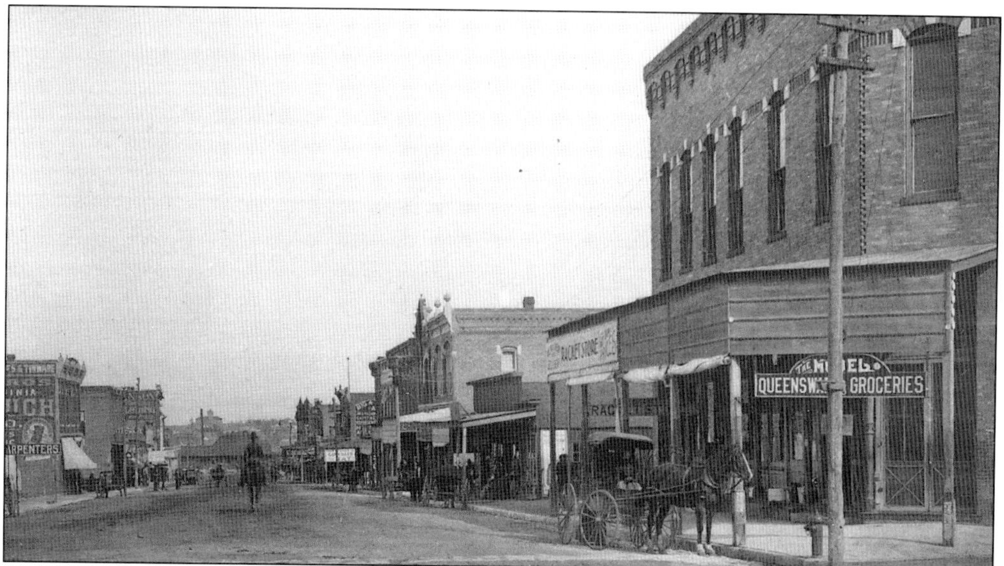

The first settlers arrived in the Aurora Township region of Lawrence County in 1834. John McNatt, on whose land Aurora would be platted on May 9, 1870, came in 1837. John was the area's first merchant, and Carr, his son, would later become the city's first mayor in 1886 and the president of the first bank in 1887. By 1872, the Frisco Railroad depot was completed. This is Madison Street looking north toward the depot before the turn of the century. (Photo Courtesy Ron Mosbaugh.)

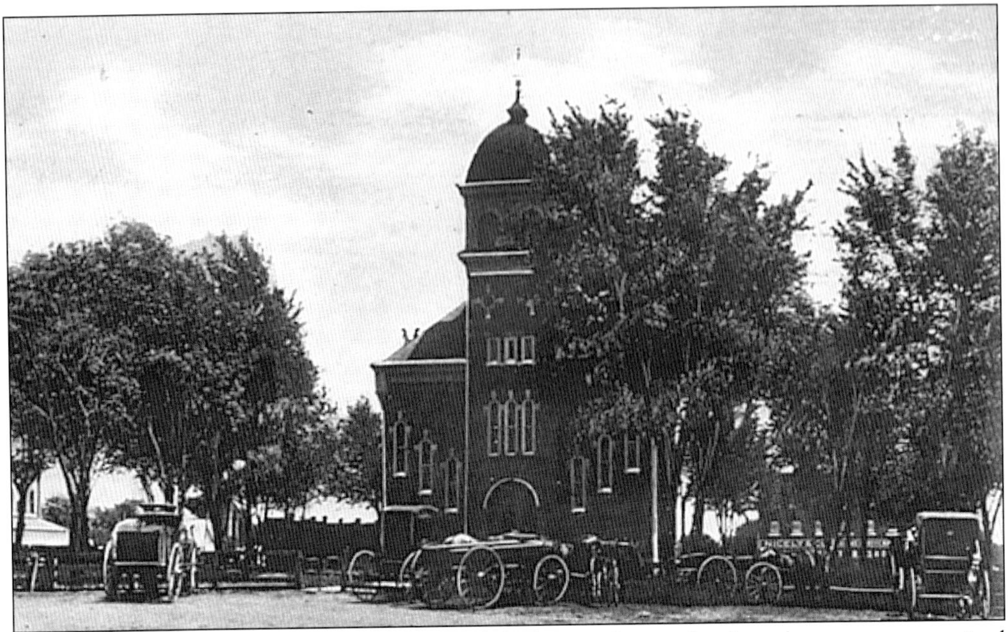

To entice the railroad to come to Aurora, half the lots in the newly-platted town were promised to defray the cost of the new depot. By 1872, the depot had been built and the railroad was in operation. In 1885, workmen digging a well uncovered a rich lead ore deposit (sound familiar?), and the mining boom began, with people living in wagons, tents, and shanties. The first city hall (pictured) was built in the early 1890s, and burned on March 30, 1913. The present city hall replaced it in 1915. (Photo Courtesy Ron Mosbaugh.)

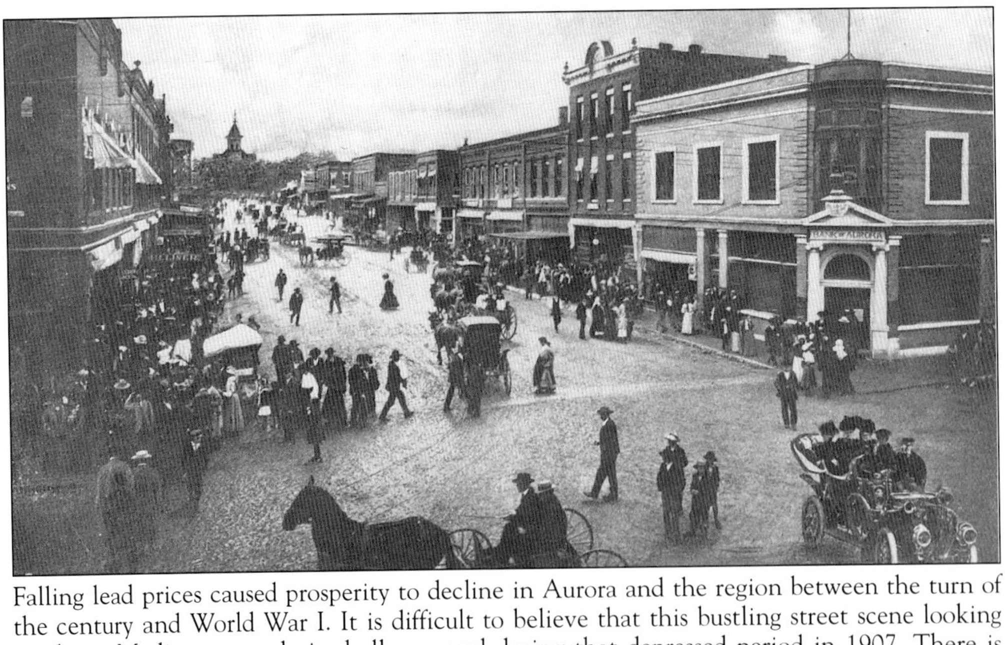

Falling lead prices caused prosperity to decline in Aurora and the region between the turn of the century and World War I. It is difficult to believe that this bustling street scene looking south on Madison toward city hall occurred during that depressed period in 1907. There is an interesting melding here of the old and the new, with horse-drawn conveyances and a newfangled horseless carriage sharing the street. (Photo Courtesy Ron Mosbaugh.)

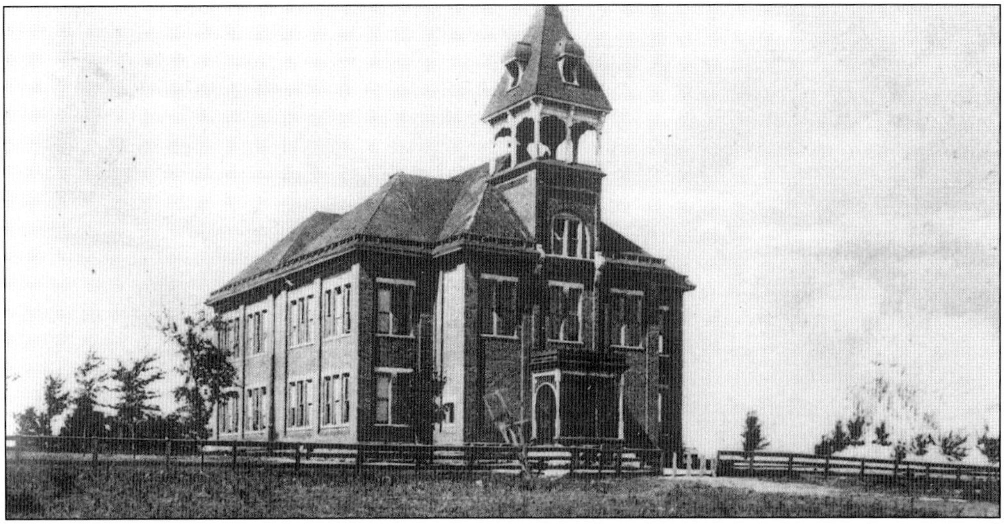

In its boomtown incarnation, Aurora was typical of such communities—wide open and generally lawless. There were eleven saloons, and, legend has it, seven fist fights occurring simultaneously in one block. Cattle often roamed freely in town, and the saloons were not regulated until the mid-1890s when the gamblers were booted out. But there was always another side to Aurora. There were churches with frequent revivals, newspapers, bands, clubs, fraternal organizations, and schools. The old Central School built in the mid-1880s (shown) at one time held all classes, including the two high school grades. It was utilized until 1910. (Photo Courtesy Ron Mosbaugh.)

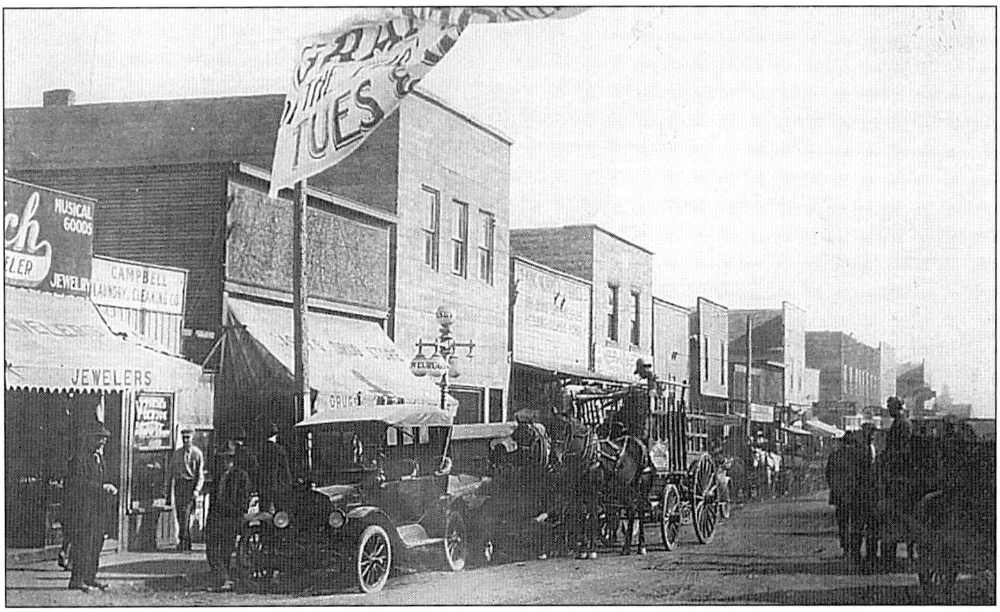

The mining region later became known as the Tri-State District because of extensive activity nearby in Kansas and later in Oklahoma. As the lead and zinc mines declined in Missouri and Kansas, many miners moved to Oklahoma where the rich field around Picher was thriving. This scene from Second Street looking south shows Picher's busy Main Street in the 1920s, as newcomers rushed in to seek their share of the wealth from the mines. Part of Picher later collapsed from undermining and was fenced off from use. (Photo Courtesy John Burrington.)

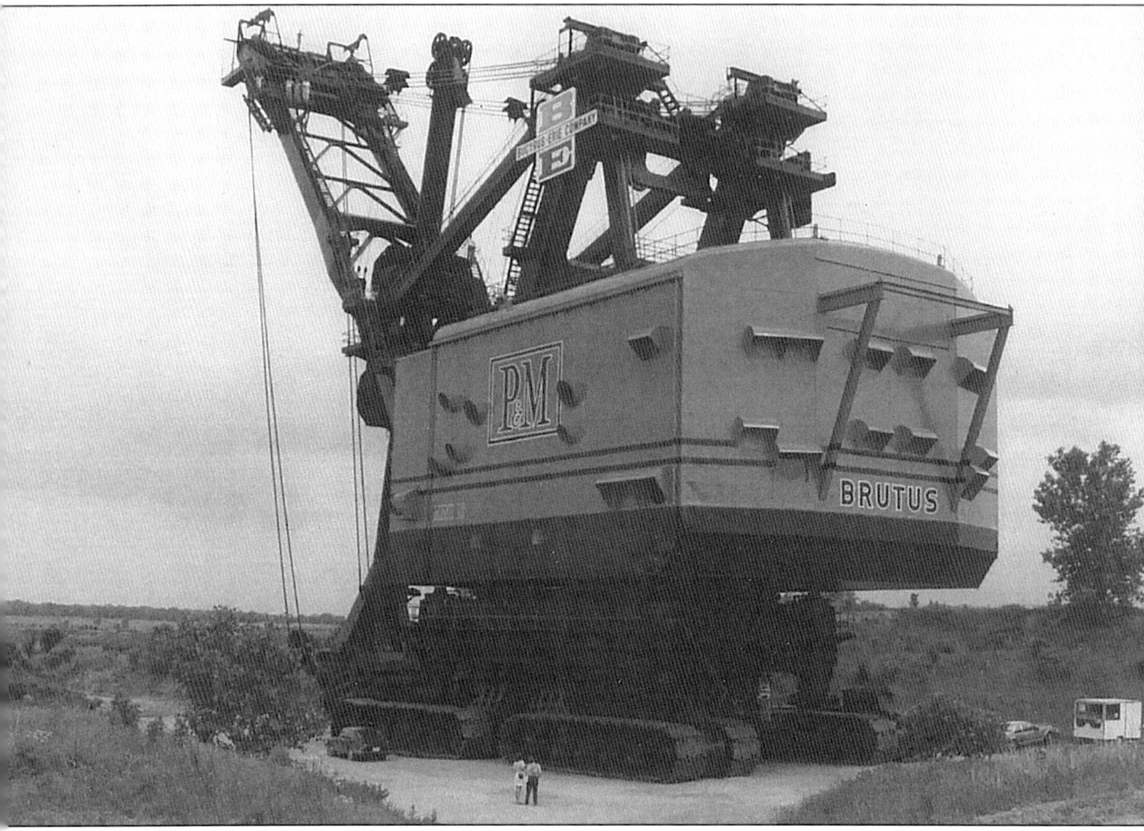

Galena, Kansas, is located just across the state line west of Joplin. The mines surrounding Galena were among the richest in the region, and the name itself is derived from the lead ore. Part of the town was undermined, and large pillars had been left to prevent cave-ins. The tale was told that men would slip down at night into the flooded drifts and "high-grade" ore from the pillars. Greed got the better of one pair. They extracted a chunk so large it crashed through the bottom of their boat, thereby abruptly ending the illicit venture. But coal mining was also profitable in southeast Kansas. At one time there was demand for coal for use in the smelters. Probably the most impressive remnant of the coal mining is Big Brutus, which operated from 1963 to 1974, removing overburden from the coal seams. It remains as a museum at West Mineral, Kansas. Towering 160 feet high, the giant machine looms over the flat countryside. It had a working weight of 11 million pounds and a dipper capacity of 92 cubic yards. An idea of the scale may be had by viewing the man standing atop the boom or the woman elsewhere on the superstructure. Do you see her? (Photo by author.)

Six

The Southwest Missouri Electric Railway and Lakeside Park

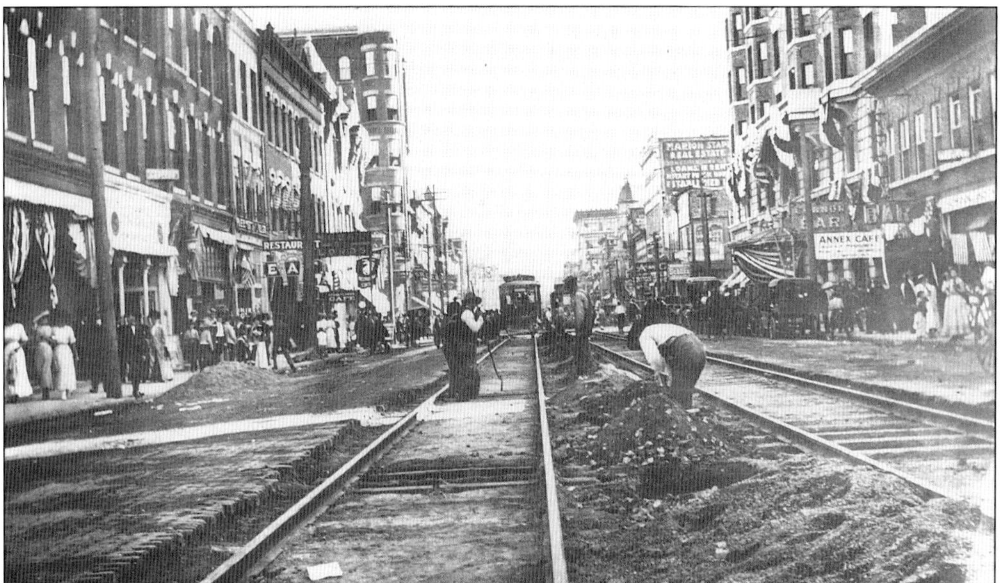

Some time after 1908, laborers make repairs between the Southwest Missouri Electric Railway tracks on Joplin's Main Street. The street was surfaced with bricks in 1900, and the edge of the surfacing can be seen where some have been removed adjacent to the ties. The House of Lords is visible on the left, and, at the end of the block, the Keystone Hotel to the left and the Connor Hotel on the right. The railway eventually extended through Kansas, ending at Picher, Oklahoma. (Unless otherwise indicated, all photographs in this section are Courtesy of Ron Mosbaugh.)

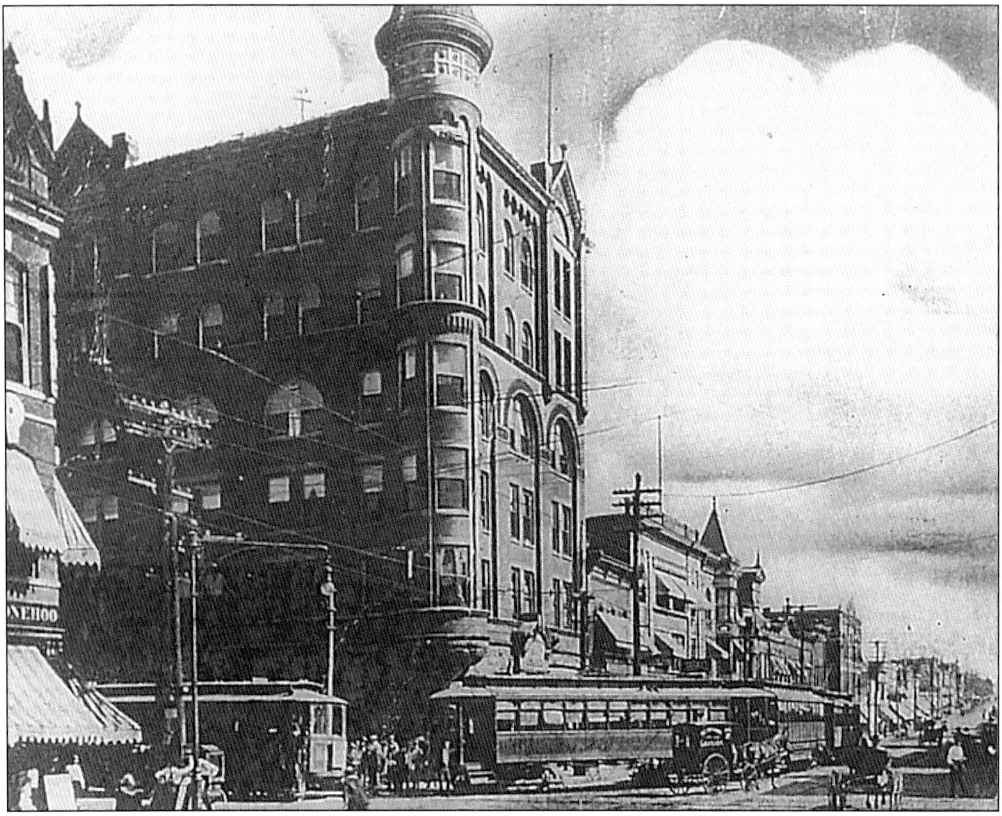

Cars turn east at Fourth Street around the Keystone Hotel Corner in Joplin.

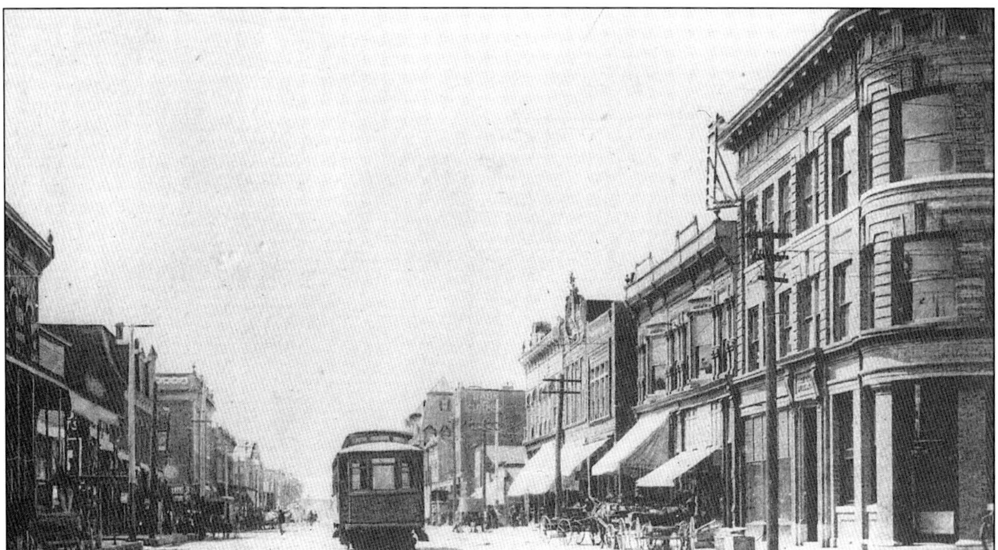
A car moves south on Webb City's Allen (Main) Street. The Merchants and Miners Bank building is on the right with the rounded corner. Webb City had an ordinance requiring the cars to slow to 3 miles-per-hour while turning corners.

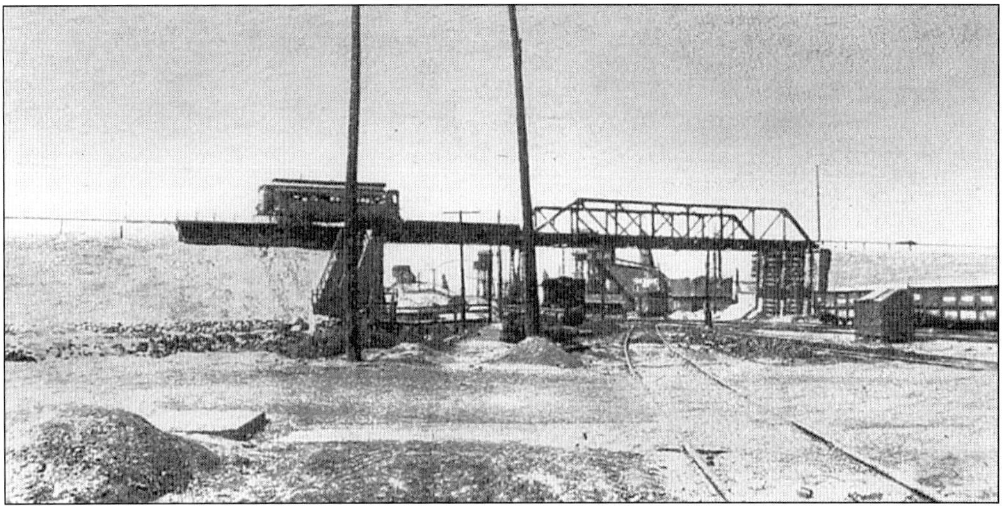

Between Webb City and Carterville, where the railway began as a mule-drawn line, the streetcars crossed a viaduct over the Missouri Pacific tracks. A platform was provided where passengers could off-load. They then made their way down the stairs, which are visible behind one of the utility poles, and along a walkway to the depot. John Webb made his original discovery of lead in the vicinity, and some of the richest mining land in the region can be seen beneath the viaduct in the distance. Webb City and Carterville were on the belt line that made a loop south to Duenweg, then back to Joplin.

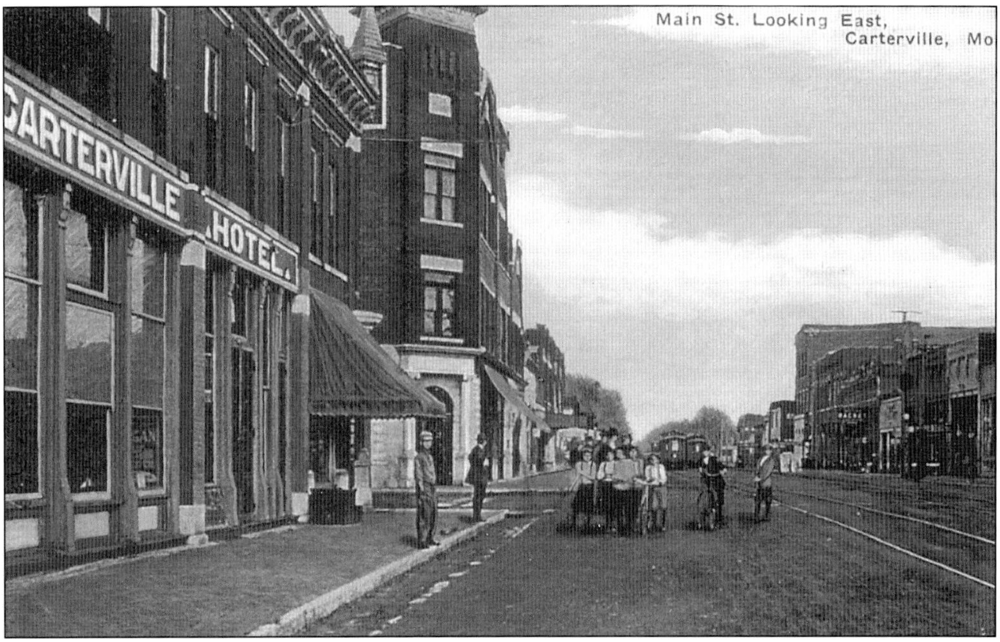

Three cars are visible on Carterville's Main Street as passengers wait to board and urchins stare in fascination at the photographer. A 30-minute schedule was maintained, except on weekends or holidays when traffic to Lakeside Park might require more frequent cars. Some of the line had two tracks, as pictured here, but some did not, and switches were required for the cars to pass. Carterville's ordinance mandated a maximum speed of 12 miles-per-hour. Hitting something as small as a dog could cause a derailment. (Photo Courtesy Clyde and Helen Clark.)

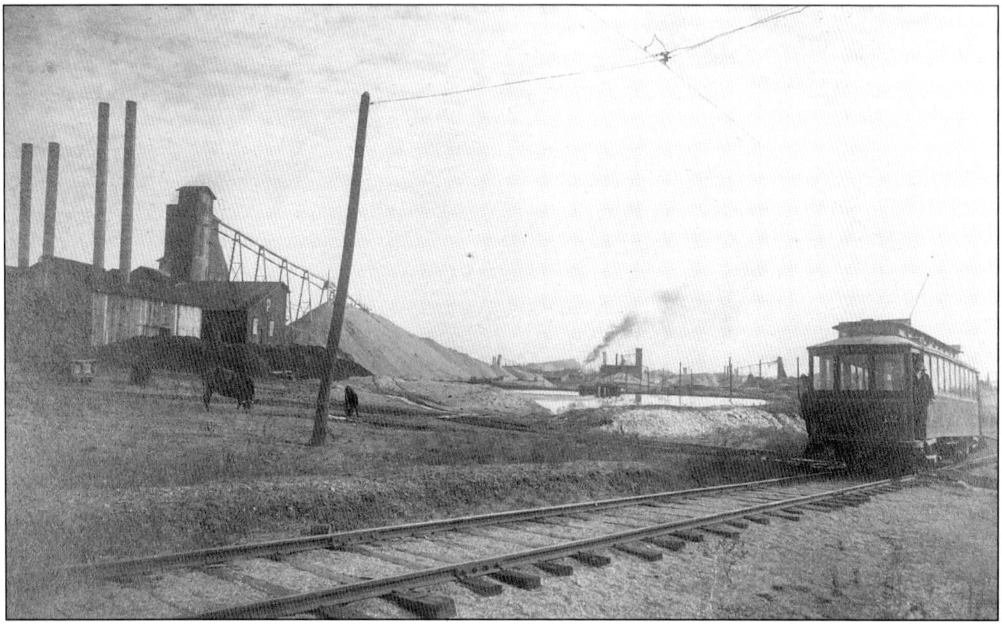

A streetcar passes one of the mines, which were the driving force behind the region's economy. Notice the cattle grazing nearby. In at least one town the people hired a drover who collected the critters in the morning, took them out to graze in the mining area, and returned them to their owners in the evening. Others were left untended and presented a hazard to the cars on the tracks.

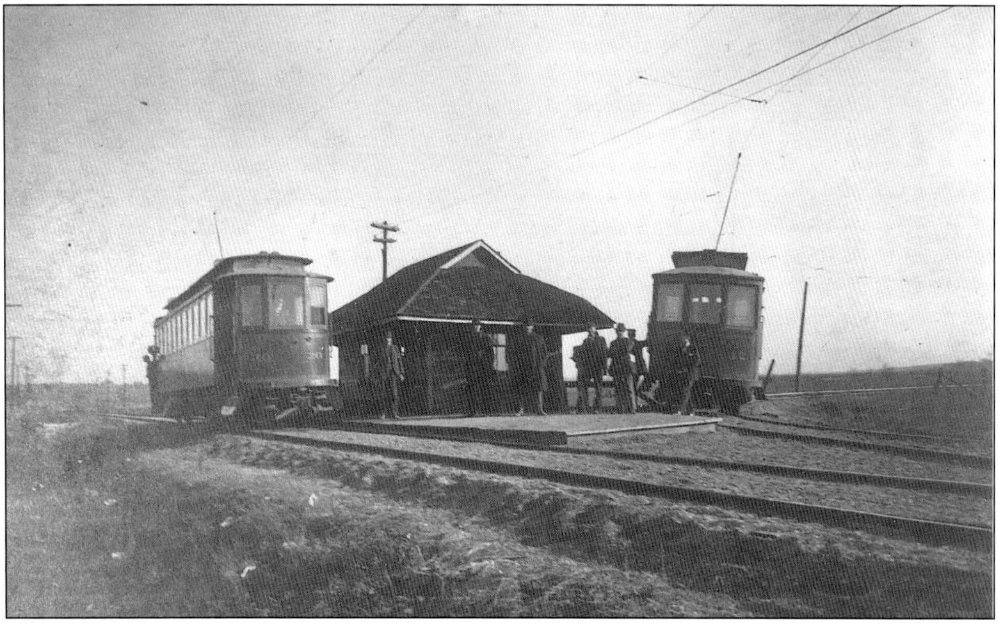

Prosperity Junction stood at the east edge of Carterville, and it was there that the belt line turned south through Prosperity to Duenweg, then back west to Joplin. The main line continued east to Lakeside Park and Carthage. Prosperity was thriving at the dawn of the twentieth century with numerous homes, stores, and churches. It was unincorporated, but the scattered population was estimated in the thousands.

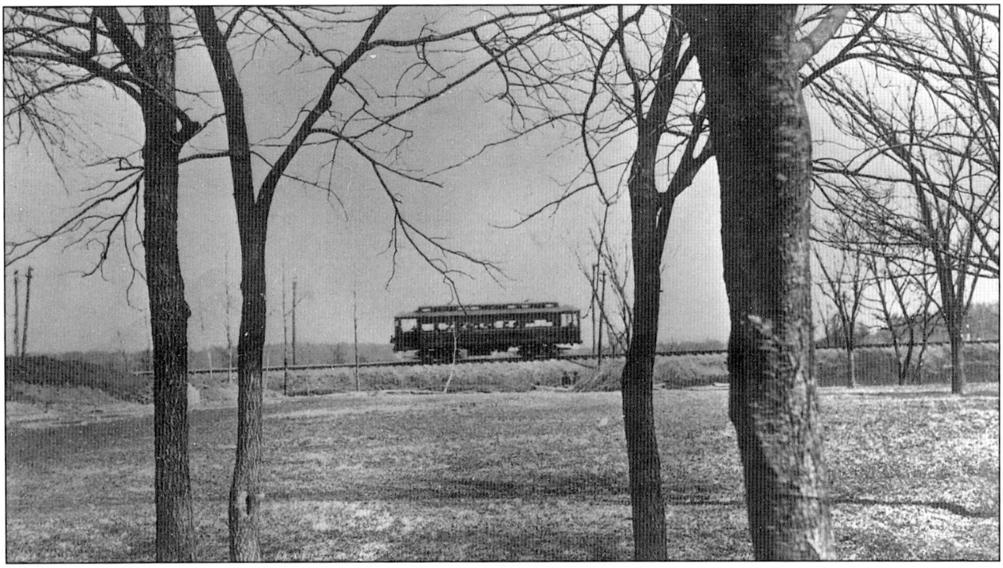

Framed by trees, a car travels through the countryside in 1908. The cars later had electric heat, but at first they used wood stoves, which meant that any accident—such as collision with a wandering cow—could result in fire.

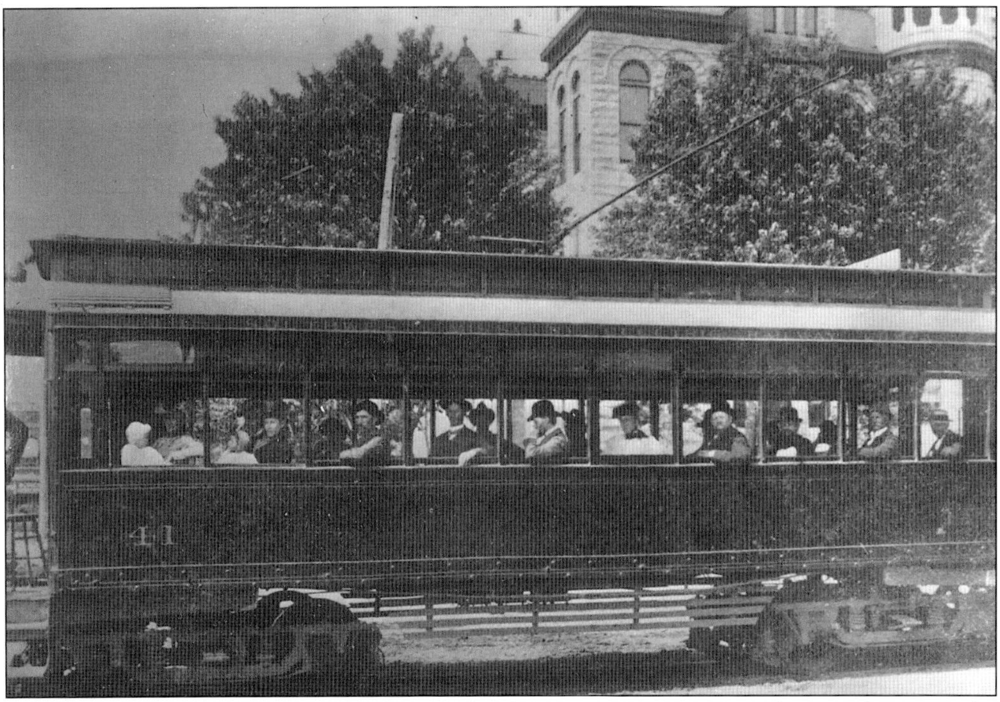

Two infants at the front stare intently at each other as Car No. 41 passes the courthouse on the Carthage square. One-way fares in the early 1900s were 15¢ from Carthage to Webb City and 25¢ to Joplin. The daily wage of a miner at that time ranged between $1 and $2, depending on the mine and the job.

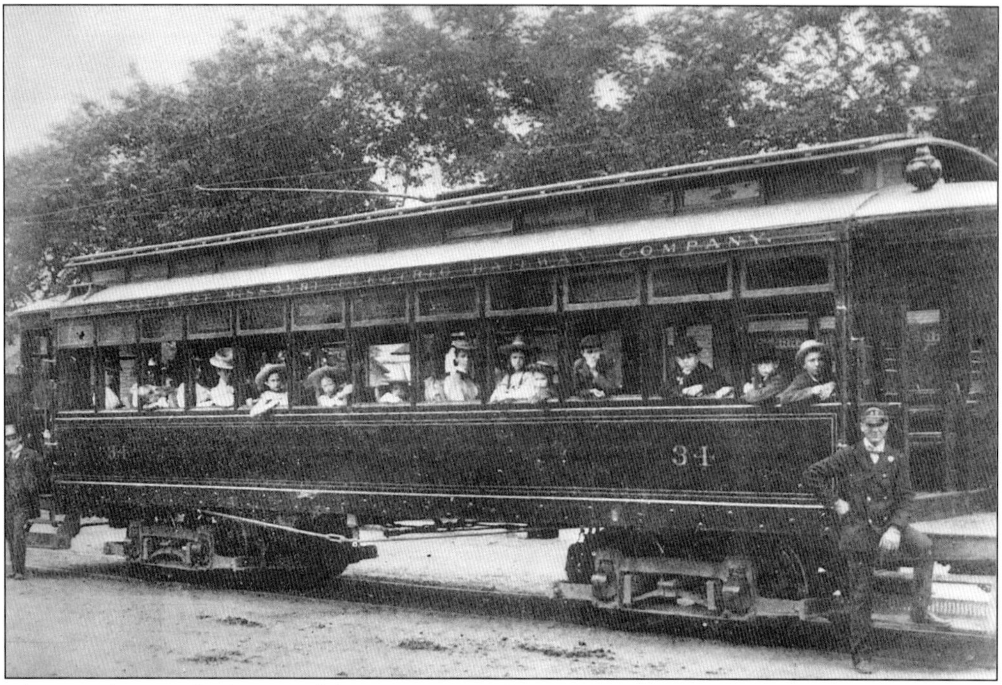

This was the standard car used for inter-city traffic and required a conductor and a switchman. Early cars were purchased elsewhere, but later were built in the company's own shops in Webb City—many of those were one-man cars.

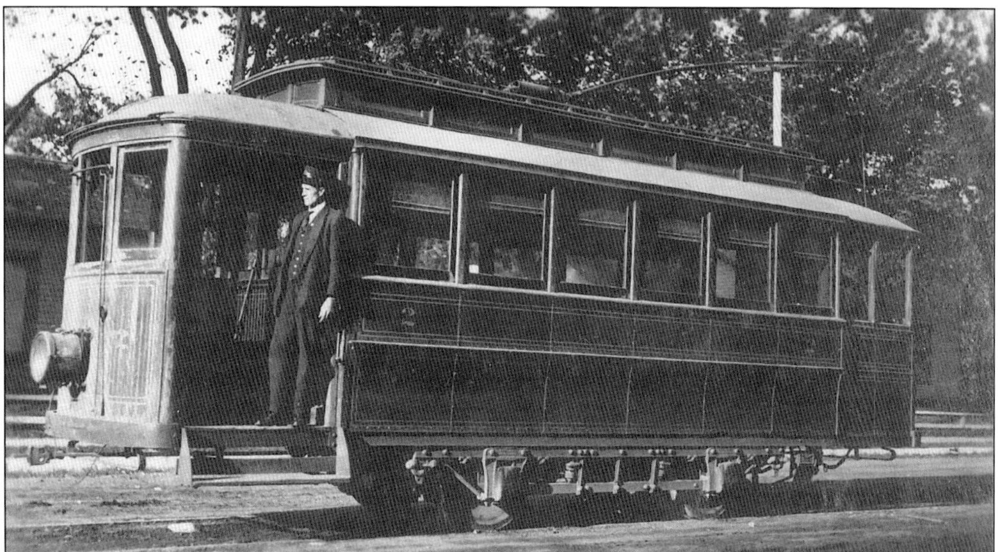

Smaller cars were sometimes used for short runs in town or for low-demand periods. Screened summer cars were also available for hauling passengers to Lakeside Park. As the company prospered and expanded, it became the largest inter-urban line west of the Mississippi.

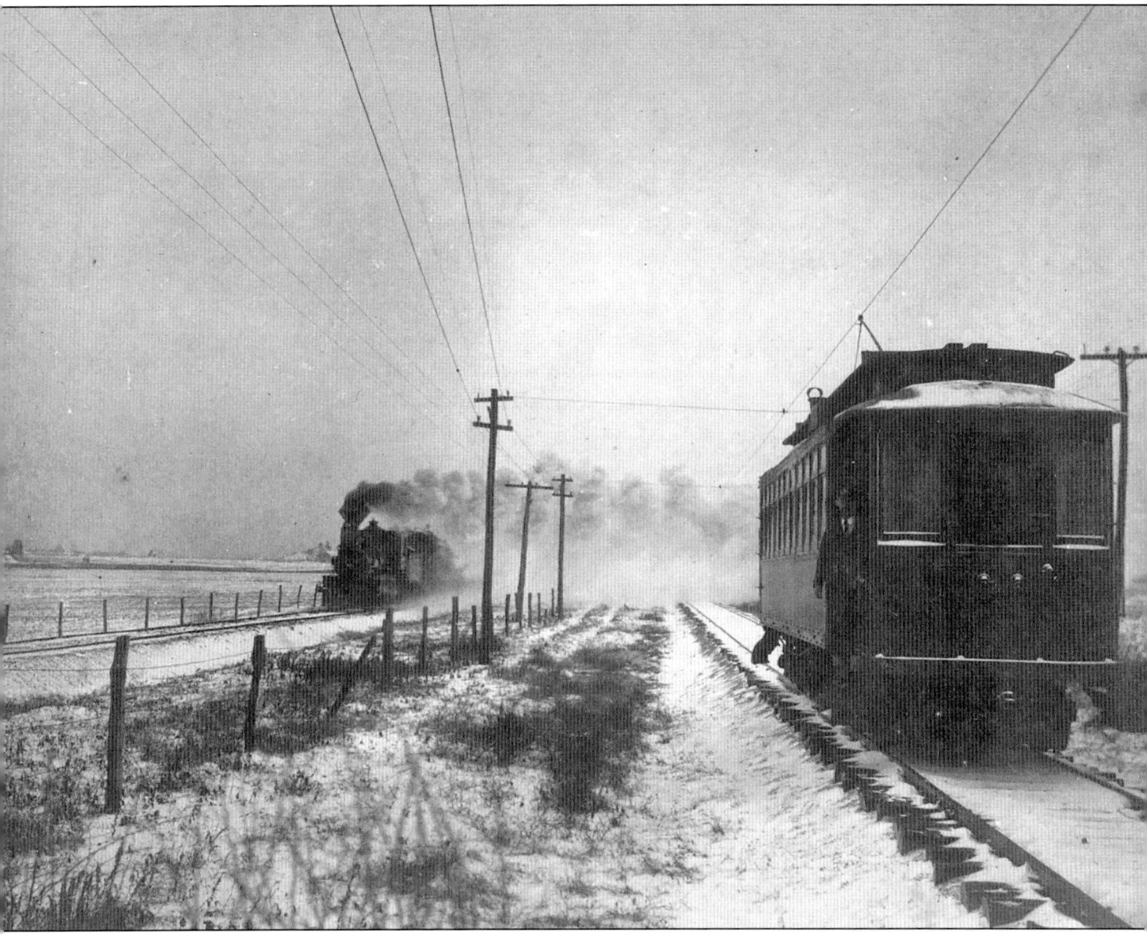

An early steam train with a flared stack and a streetcar run on parallel tracks like wraiths from the past in this haunting wintry scene frozen for posterity on the photographer's plate. It is difficult to overstate the impressive attraction the powerful locomotives exerted on people early in the century. It was written on the back of an early postcard that, in the summer of 1902 at 3:20 in the afternoon, a Missouri Pacific freight train crossed the Lakeside railroad bridge "running at 40 mi. per hr." Such speed was worthy of note in 1902. Shortly after World War II, people driving past depots and switching yards noticed with surprise that the last of the old steam engines had vanished, irretrievably vanquished by the much less romantic diesels. By that time, the Southwest Missouri Electric Railway was also only a memory, victim of changing times and modes of transportation. The automobile carried people to fun spots far more distant and exotic than Lakeside Park. The railway had survived until 1939, and was a freight-only line at the end. All are fading memories as surely as the faraway mournful whistle of a steam train in the darkness of a cold winter night.

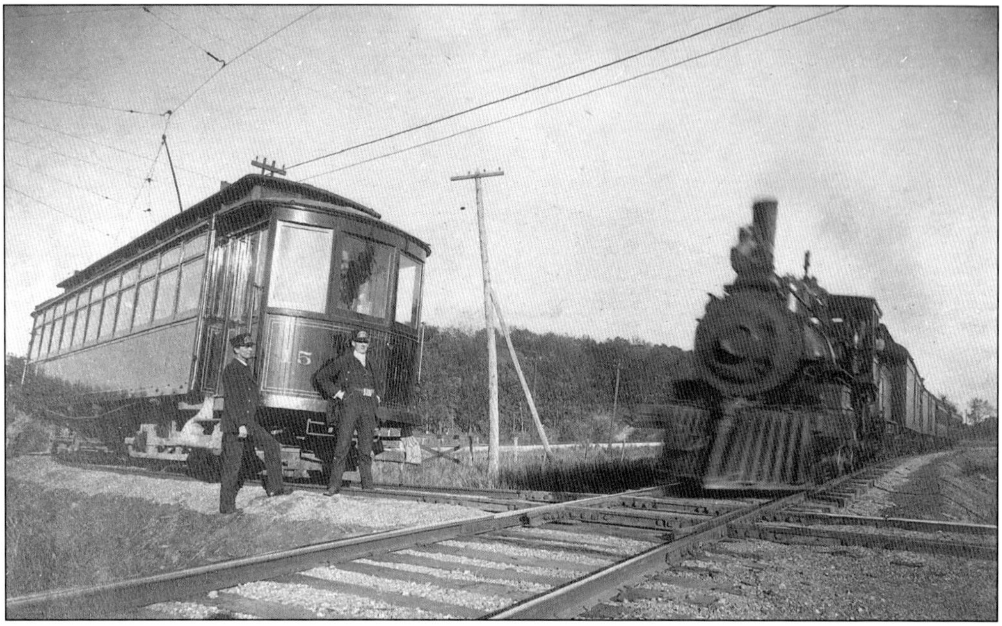

A streetcar stops to give the steam train right-of-way. Like the miners, the conductors and switchmen weren't above mixing a little fun with their work. One crew was Ellis Hummel, conductor, and Clem Ault, switchman, both from Carterville. Clem had an arrangement with the druggist at Prosperity where the track ran adjacent to the drug store. He would leap from the car, run in the back door, swig down a waiting drink, plunk down the payment and dash out the front door to rejoin the car which was at its Main Street stop. (Courtesy Leeman Dorsey.)

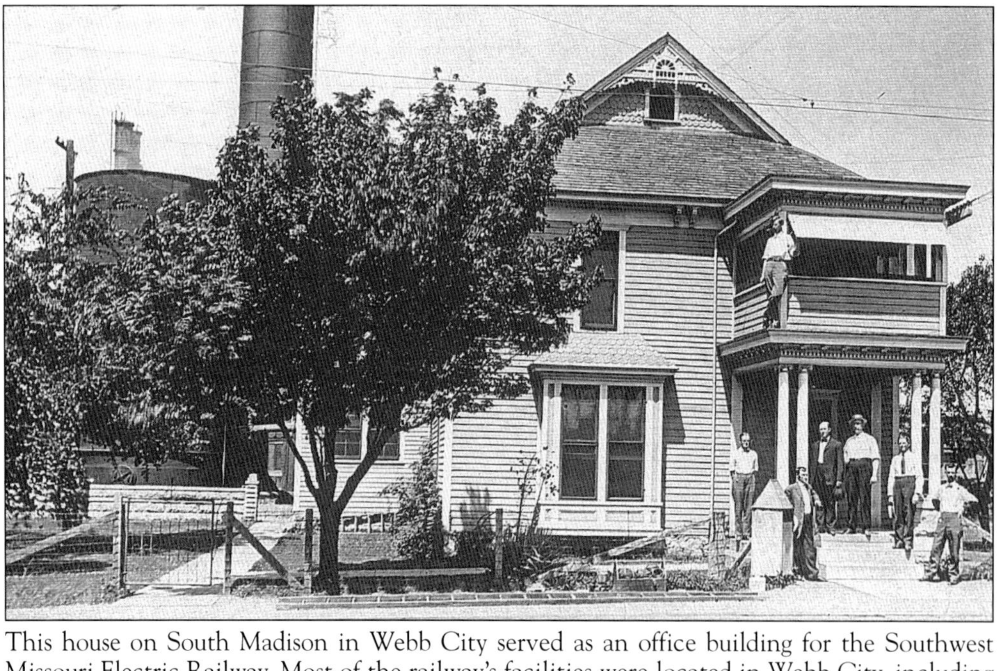

This house on South Madison in Webb City served as an office building for the Southwest Missouri Electric Railway. Most of the railway's facilities were located in Webb City, including the power-generating plant seen behind. (Photo Courtesy of Henry Hulett.)

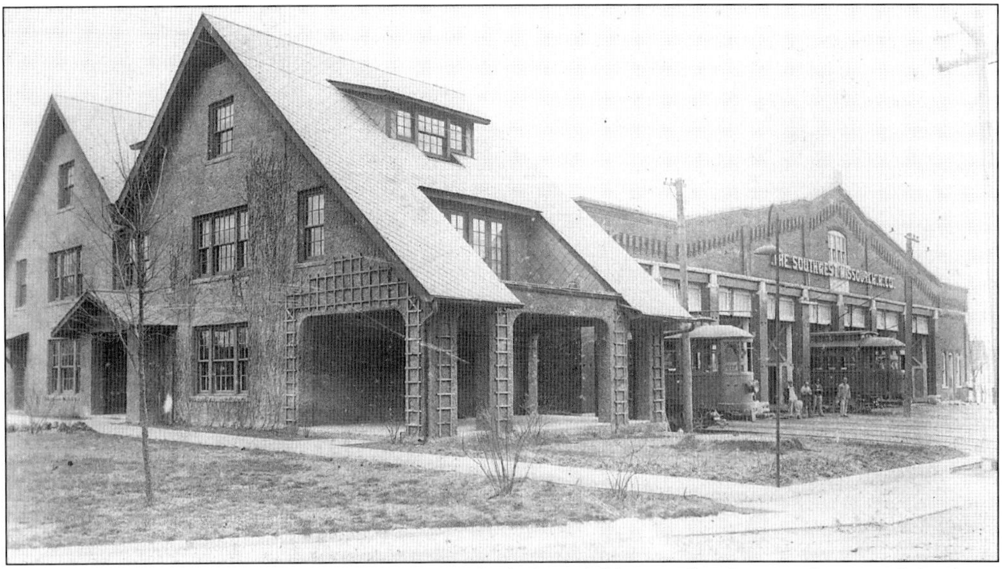

A.H. Rogers, who owned the railway, furnished this clubhouse for the employees of the line, whose numbers varied between two and three hundred. It is now the home of the Webb City Historical Society. The car barn is visible behind. Life was not always easy for the operators of the cars. Inebriated miners sometimes tried to liberate them for their own use. Clem Ault, mentioned before, had a hedge wood staff with a metal hook on one end for the switches and the other end weighted as a "drunken miner persuader." (Courtesy Leeman Dorsey.)

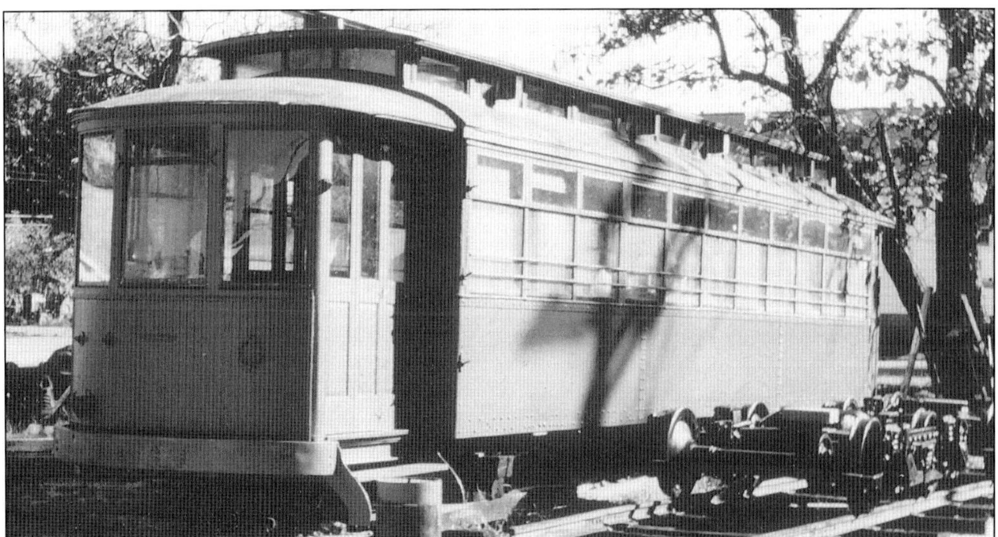

When the electric railway ceased operation before World War II, much of the equipment was sold in South America. In 1975, Fred Rogers, assisted by Jim Coffie, Ross Bivens, Whitson Kirk, and Cecil Veatch, began the arduous task of finding enough parts to restore one of the old cars. They found six remaining in the area—three in Oswego, Kansas, two on a farm south of Joplin being used for storage, and one which had been a diner in the west edge of Carterville. (Photo by author.)

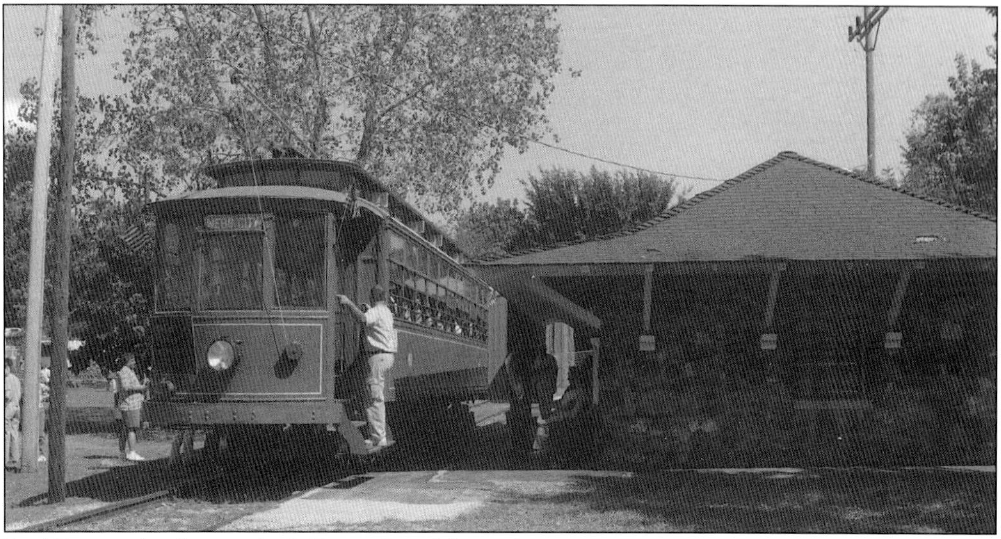

Fred Rogers (no kin to A.H. Rogers) and his volunteer assistants scoured the nation for parts and fabricated what they couldn't find. Twelve years later, in 1987, the restored car, No. 60, was complete, and a barn to house it had been built. On September 18, 1999, during Webb City's Mining Days celebration, the fully loaded car left the old Prosperity Junction building—relocated and given new life—and took the passengers around a track that loops through King Jack Park. (Photo by author.)

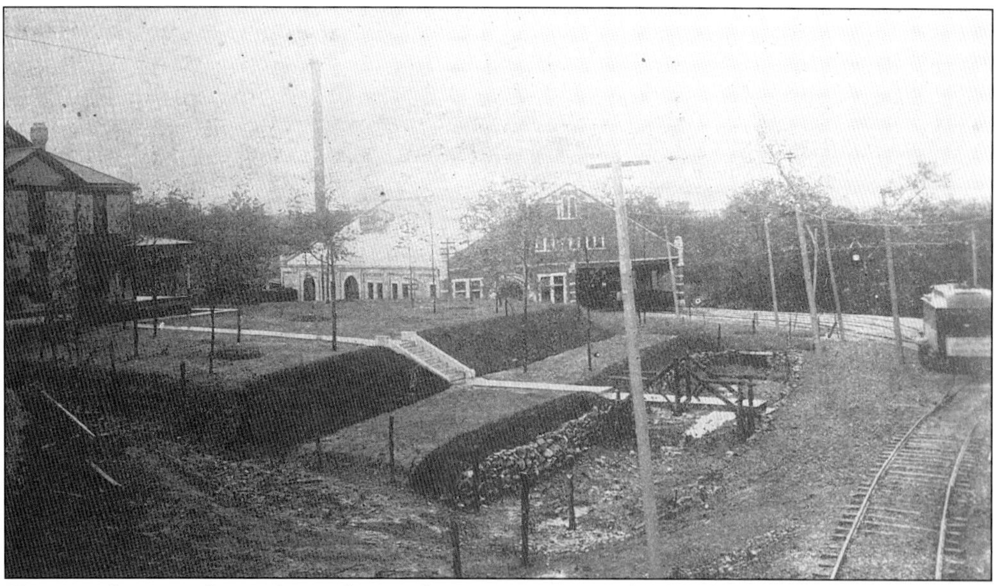

This was the powerhouse located near the west entrance of Lakeside Park. The hotel, or inn, is located at the extreme left. Many employees of the mines and the railway were fresh off the farm. One man recalled being laughed at on his first trip to town as a youngster when he told the counterman in a cafe he didn't want a stack, he wanted pancakes. The only stacks he knew about were haystacks, and he certainly did not want any of that to eat.

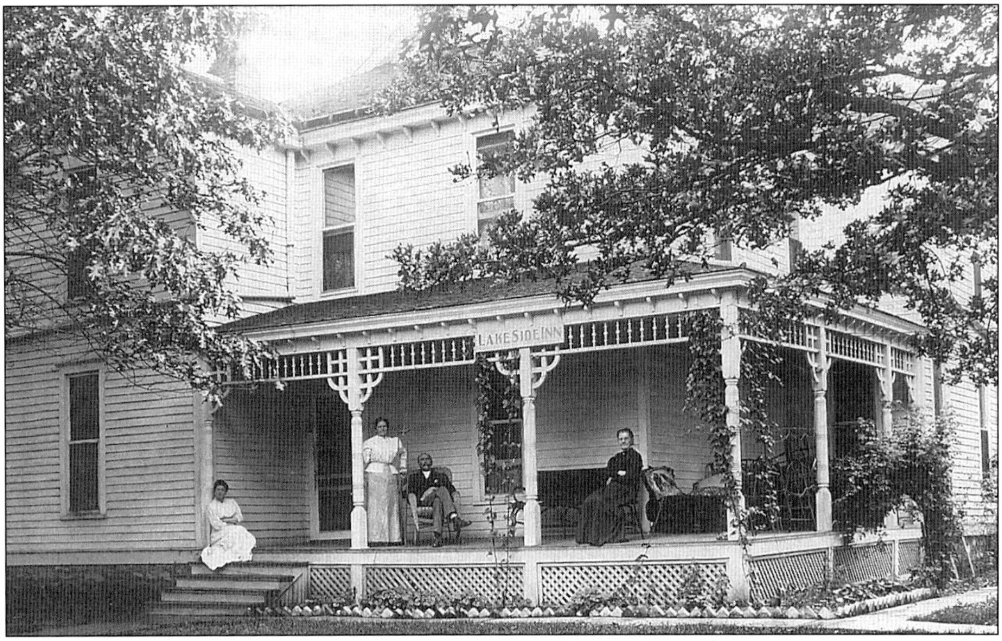

Guests relax on the porch of the inn after a pleasant day at the park. The west entrance to the park was just to the east. John Seth Carpenter and his wife Mahala Melinda kept the inn at the dawn of the century, and this photograph, also used on postcards, was in their collection. (Photo Courtesy Jim and Jean Horn.)

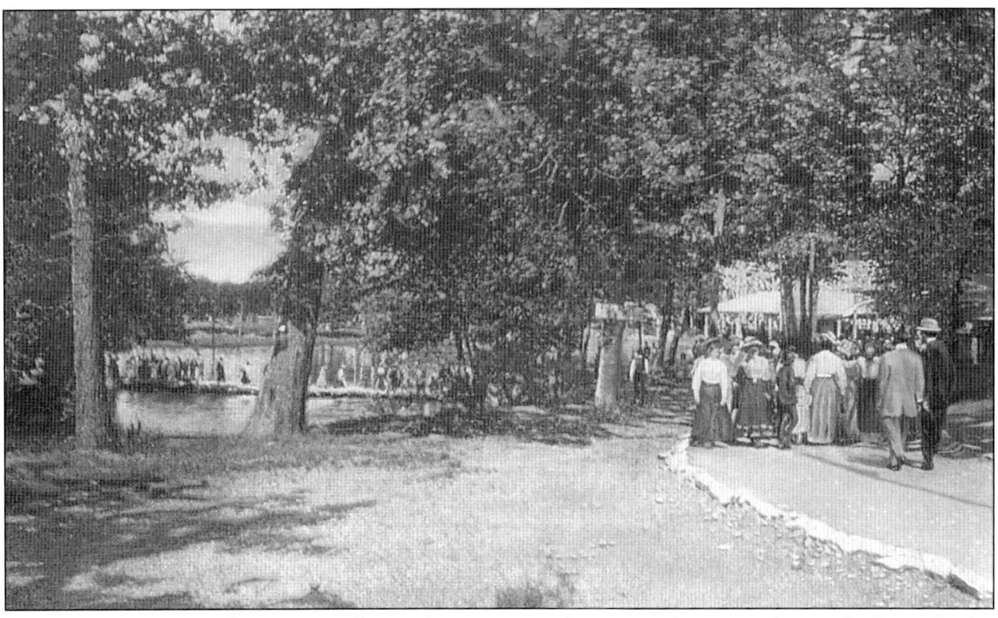

On a late summer afternoon early in the century, these people enter the park through the west entrance, while others in the distance cross Center Creek on the causeway or dam. The bathhouse, water slides, tennis courts, ball field, children's playground, and picnic area were on the left or north side of the river in this view. (Photo Courtesy Clyde and Helen Clark.)

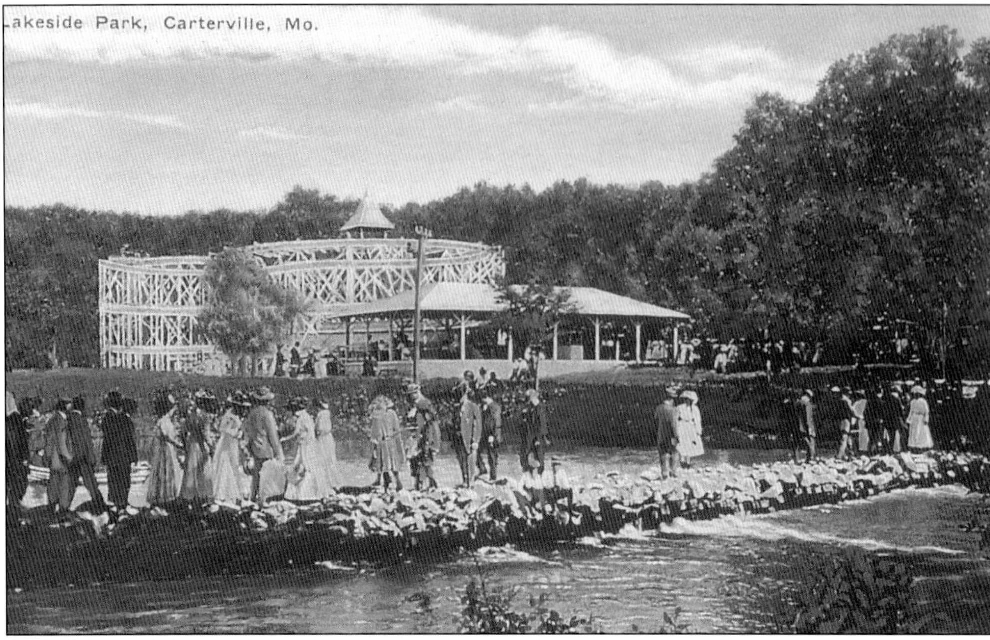

It is likely a pleasant mid-summer Sunday afternoon as gentlemen and their ladies stroll on the dam. Across the river is the "figure eight." Tame by today's standards, it was typical of the roller-coaster-like thrill rides of the era. Almost hidden in the crowd, two young men sitting on the dam watch the photographer capture this view of the park, as well as themselves. (Photo Courtesy Clyde and Helen Clark.)

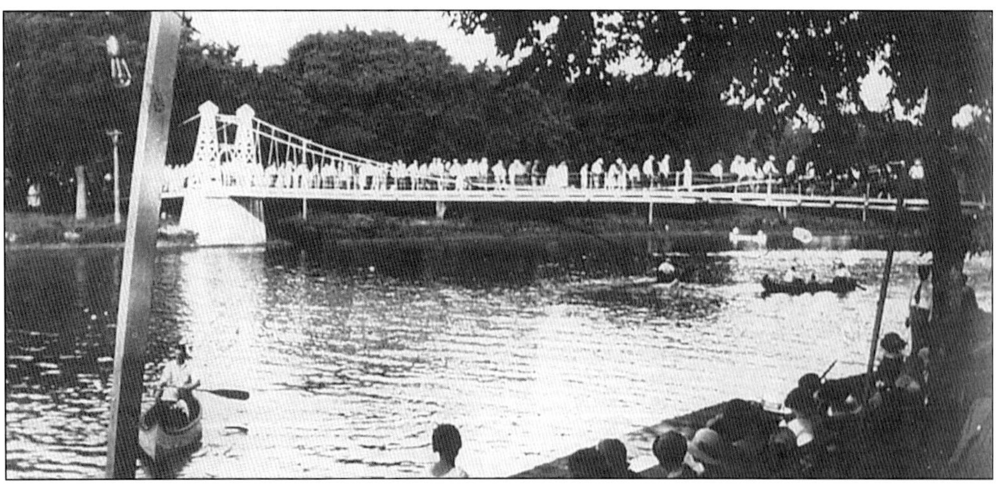

The swinging bridge is packed with people probably leaving a balloon ascension in the 1920s. Earlier in the century, thousands of people packed the grandstand and stood and sat on the grass to watch Trolley League games. Wilson Smith Becker operated the park in the '20s. (Photo Courtesy of the H.W. Becker Collection.)

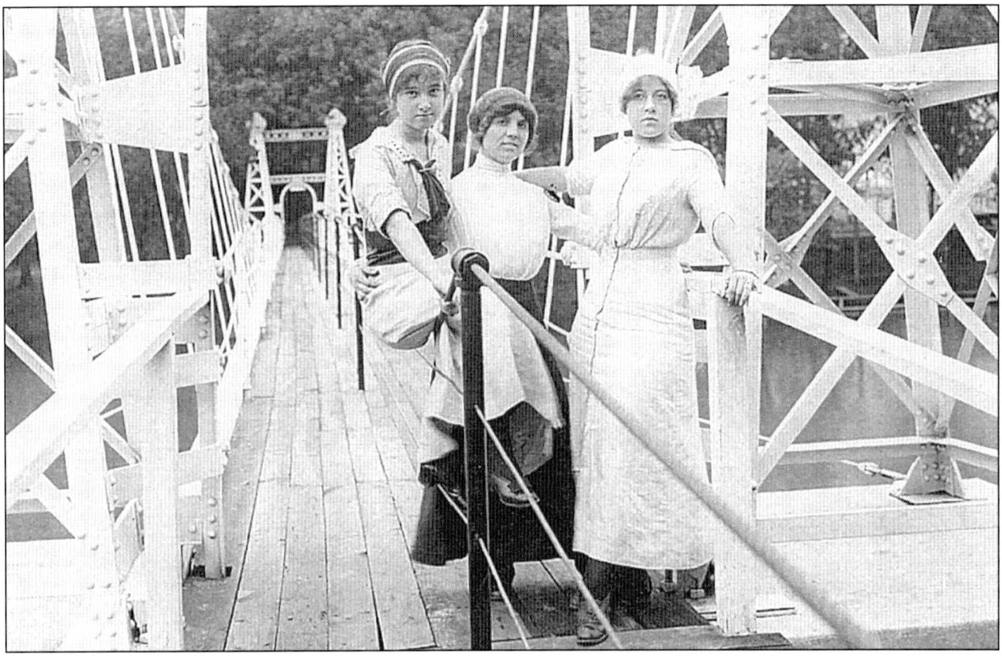

Three early-century belles pose on the swinging bridge. The bridge provided "two-way traffic," with a dividing cable down the middle. One man recalled the day a crowded wooden approach ramp at one end collapsed in a great cloud of dust. No one, he said, was seriously injured. (Photo Courtesy Henry Hulett.)

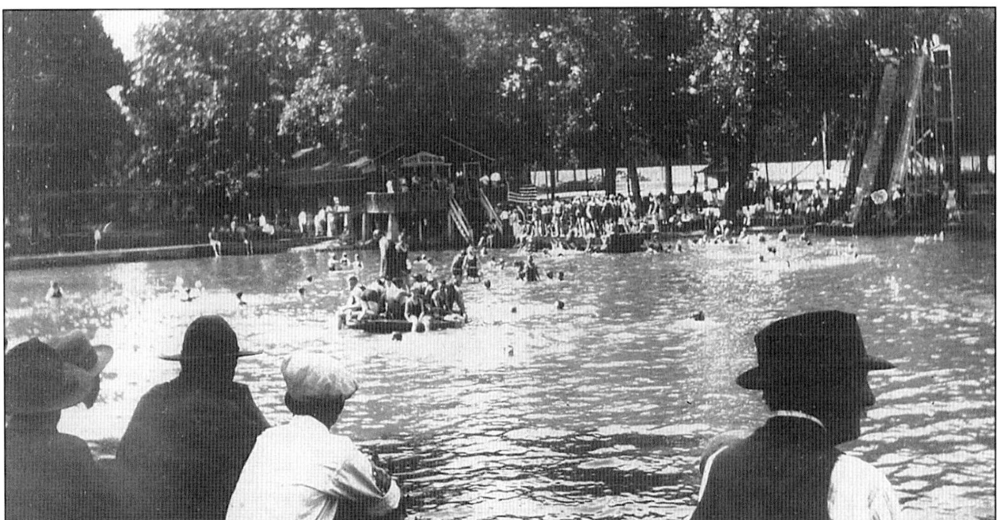

Probably around the time of World War I, two soldiers in their distinctive campaign hats join the onlookers across from the bathhouse. Behind the crowded dock is the open field that would fill with people gathered to watch balloon ascensions. Boat and canoe rentals and swimming were park attractions. Drifting on the ebony water at night was particularly pleasant, with the reflections of the myriad lights ashore twinkling through the trees and the muted sound of mandolins or music from the pavilion wafting on errant breezes through the darkness. (Courtesy H.W. Becker.)

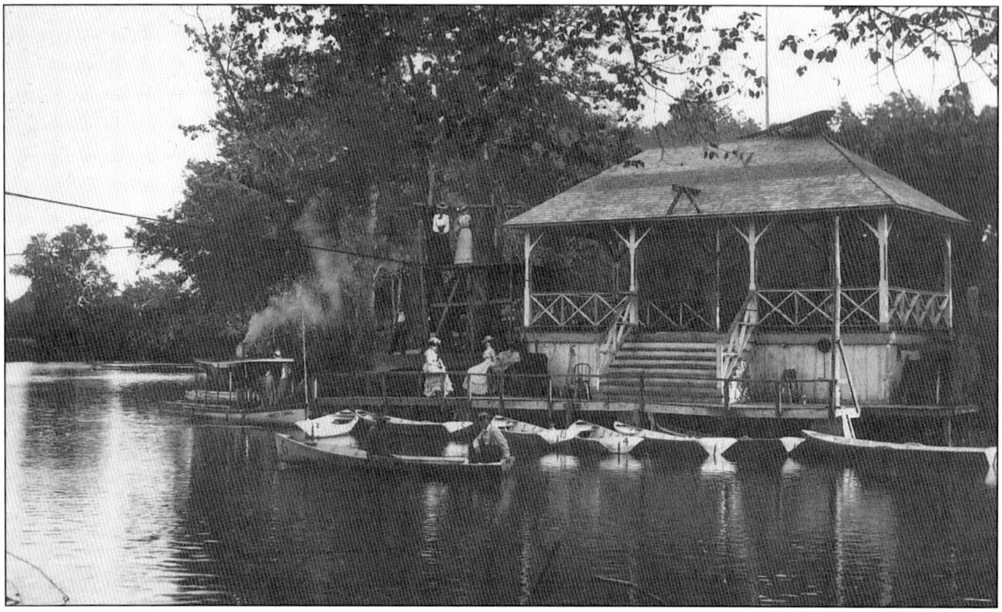

With the steam launch puffing smoke in the background, early arrivals at the boathouse and landing await the host of people who will arrive later. Crowds at the park were legendary. One person said 33,000 people were estimated in attendance at a John Philip Sousa concert. Other enthusiastic estimates speak of throngs approaching 50,000 on Fourth of July holidays. One visitor recalled a sudden deluge soaking many people who could not find sufficient shelter in the park.

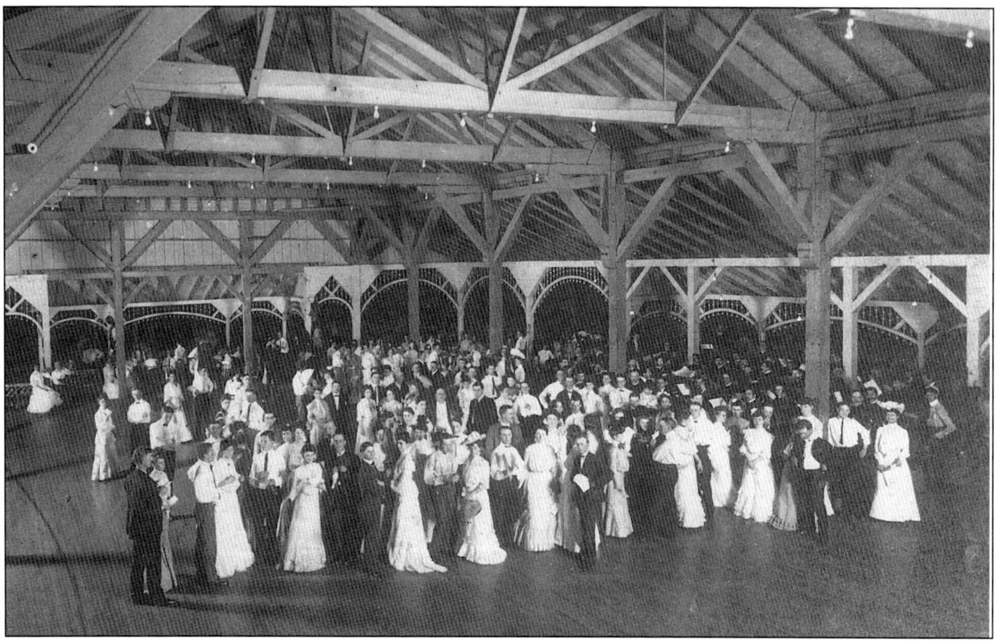

Dancers in the pavilion pause for the photographer on the warm summer evening of July 16, 1903. Music was provided on that occasion by Ellery's Royal Italian Band. Outdoor movies, perhaps among the first in the nation, later also became popular at night.

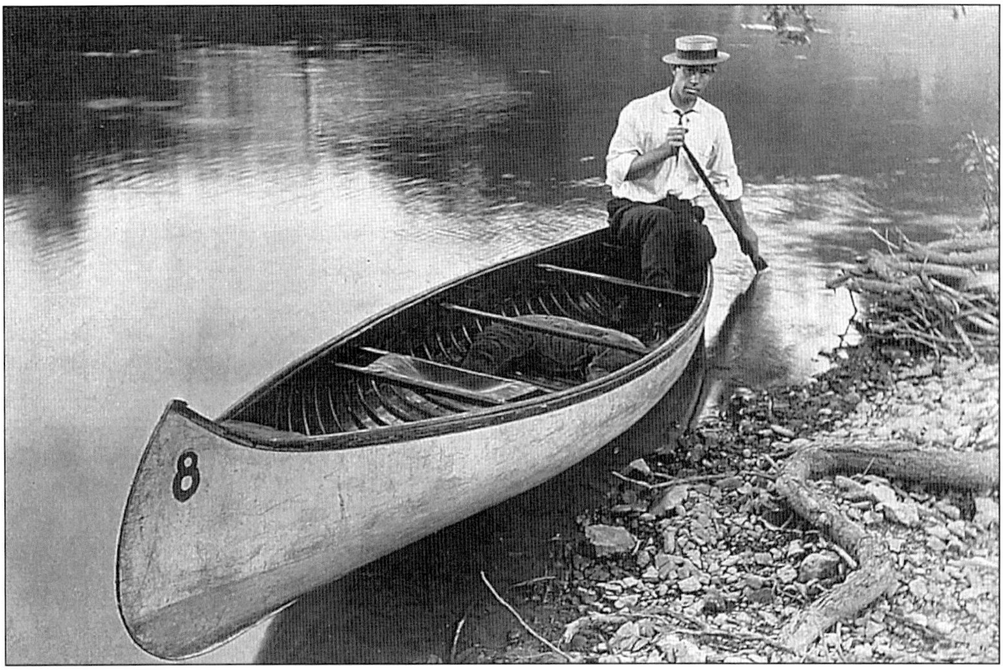

Steven Hurlbert Carpenter canoes on the lake about 1903. "Bert," as he was known, was the son of the innkeepers, Seth and Mahala Carpenter. He ran the boathouse and rentals early in the century. (Photo Courtesy Jim and Jean Horn.)

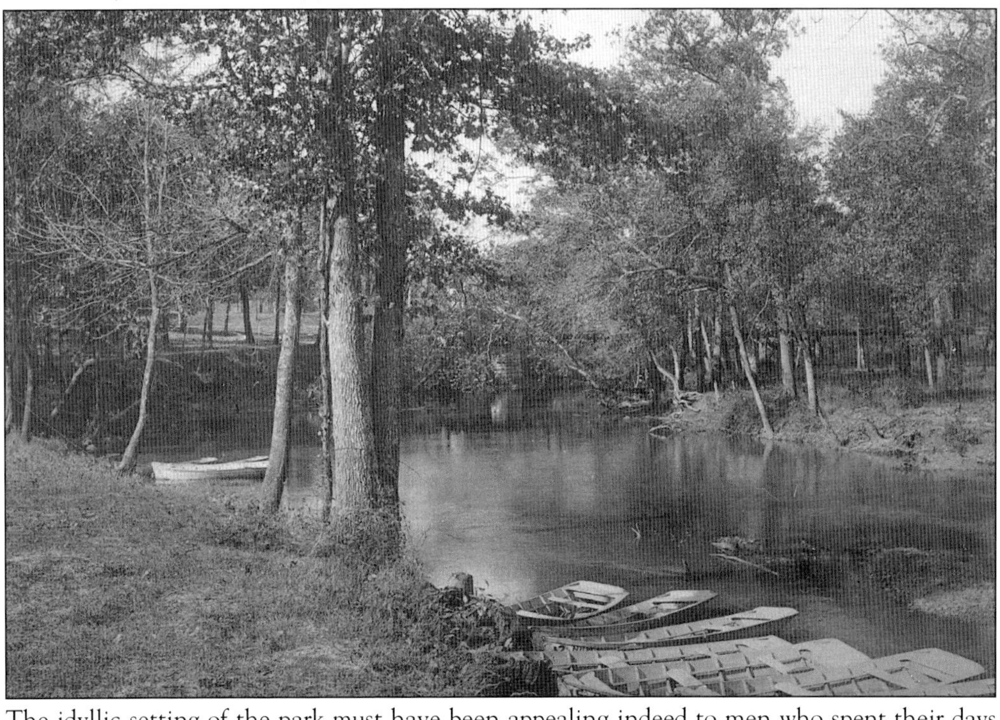

The idyllic setting of the park must have been appealing indeed to men who spent their days laboring in near-darkness far below ground. This is either upstream Center Creek far enough above the dam to remain within its normal banks or a view below the dam.

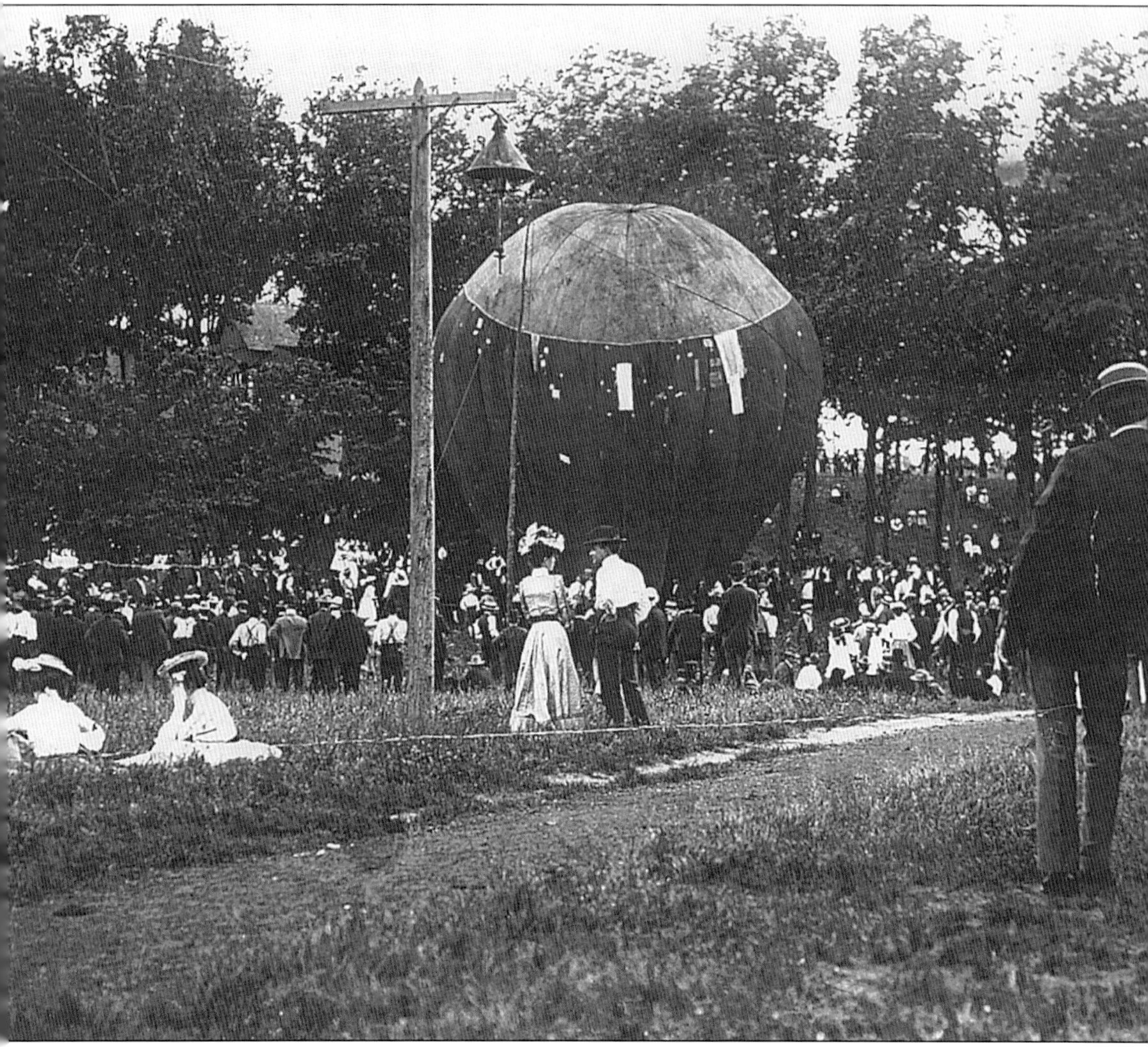

On November 21, 1783, the first manned free flight of a balloon took place in Paris. It attained a height of 3,000 feet, and remained aloft for 25 minutes. Andre Jacques Garnerin, who was Napoleon's balloonist, tested the first parachute in 1797. Over one hundred years later, the fascination for balloons and parachutes remained strong around the world. People are gathering in Lakeside Park about 1910 for an ascension and parachute jump. The oft-patched balloon is clearly a veteran of such events. One balloonist, Fred Ambrose of Carthage, was noted for diving from a tall pole into the river, swimming ashore to strap on the parachute and then being carried aloft. Occasionally on breezy days, the balloon with its dangling passenger would be swept out of view, and men on horseback would set off to track the wayward craft. Sometimes the unfortunate parachutist would be found dangling from a tree. (Photo Courtesy Henry Hulett.)

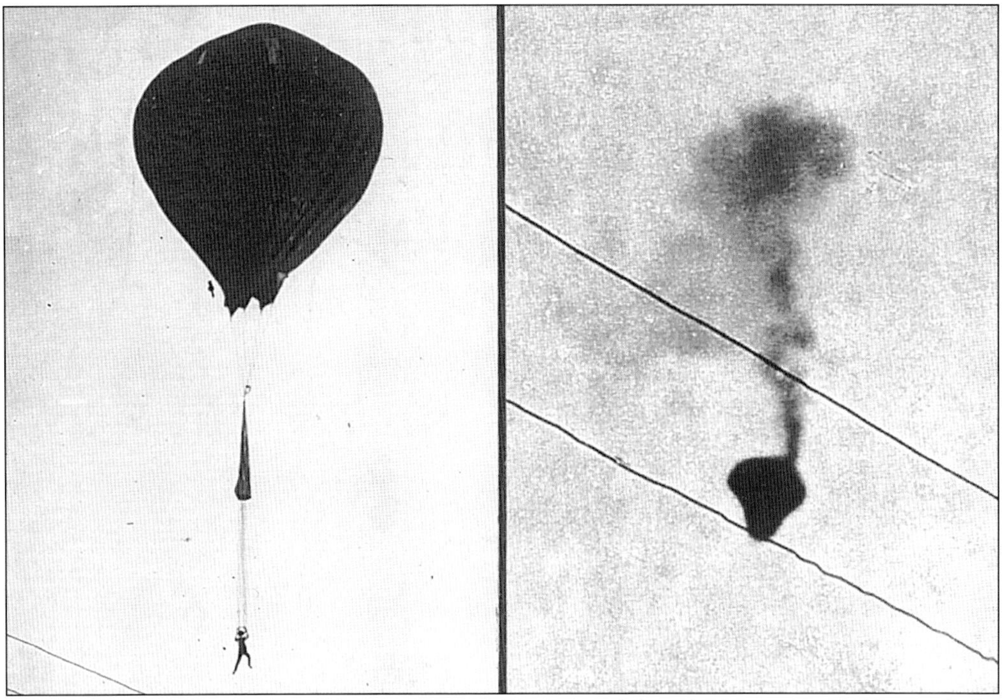

One is likely to think of a parachutist jumping from a gondola or basket suspended beneath the balloon. That was clearly not the case here. After the dangling jumper cut loose, the balloon spilled its smoky heated air and fell back to earth. The balloon was initially inflated with hot air from a large fire being forced through an underground tunnel and into the opening at the bottom of the balloon. (Photo Courtesy Henry Hulett.)

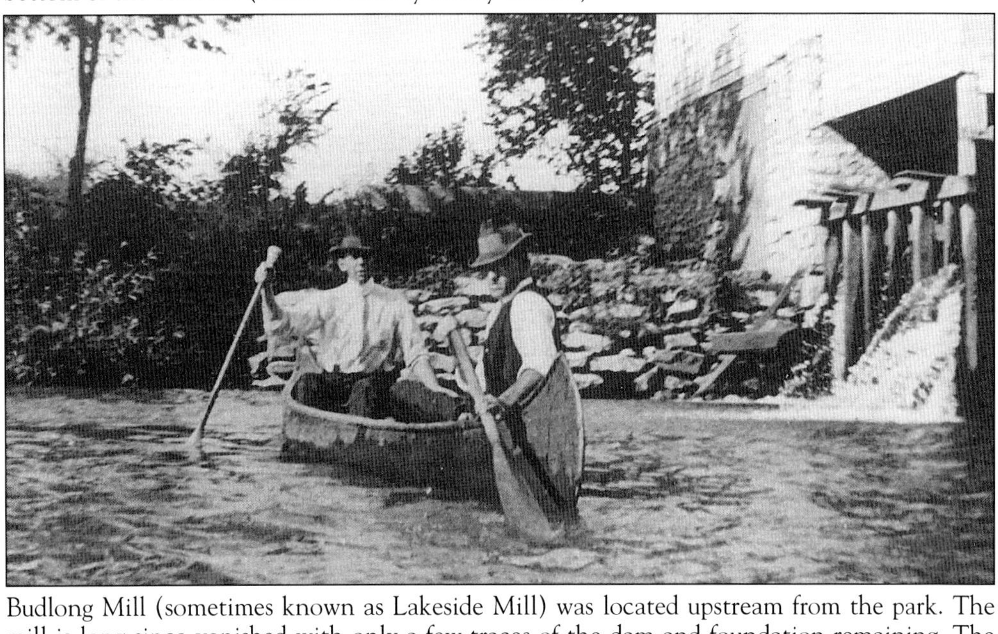

Budlong Mill (sometimes known as Lakeside Mill) was located upstream from the park. The mill is long-since vanished with only a few traces of the dam and foundation remaining. The nearer man in the canoe is Jess Carpenter, another of the Carpenter sons and also a park concessionaire. (Photo Courtesy Jim and Jean Horn.)

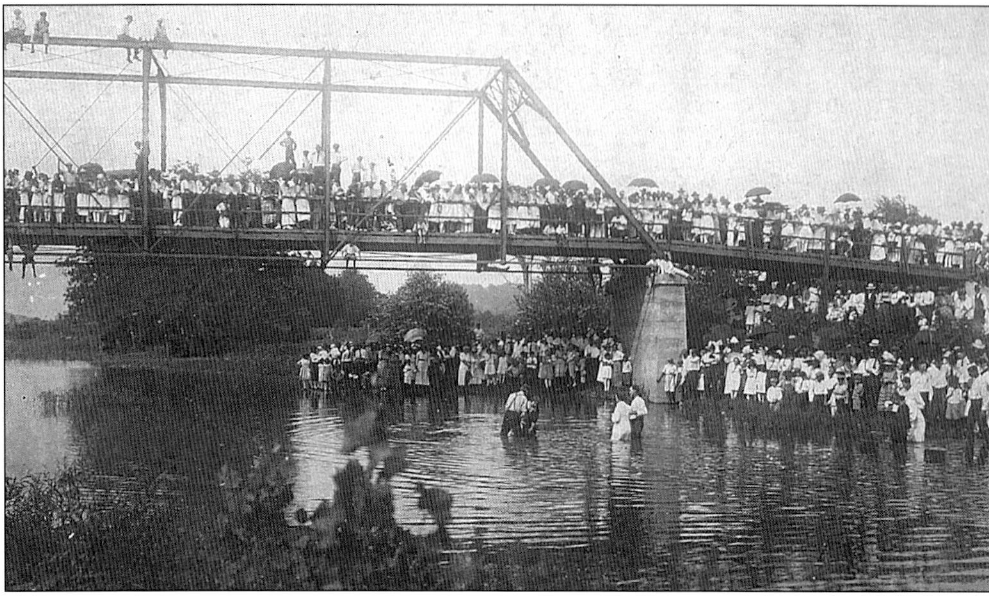

At the downstream edge of the park, c. 1920, a large crowd views a baptism. Two candidates in the stream await their turn while others are gathered at the river's edge on the right. Such professions of faith, usually to the a cappella strains of "Shall We Gather at the River," were common occurrences in the first half of the century. (Photo Courtesy Webb City Historical Society.)

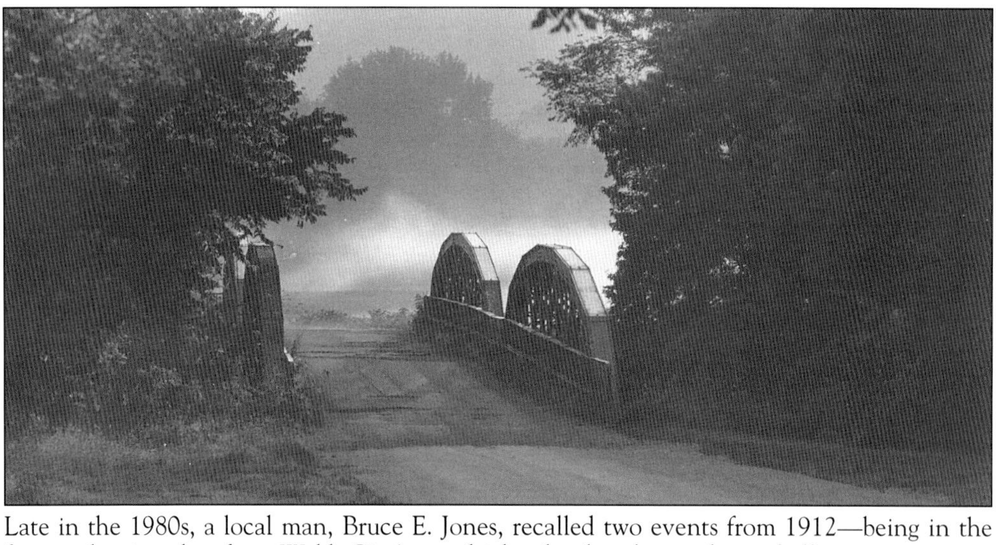

Late in the 1980s, a local man, Bruce E. Jones, recalled two events from 1912—being in the first graduating class from Webb City's new high school and attending a balloon ascension at the dedication of the new bridge over Center Creek just north of the park at Lakeside. It served only local traffic then, but in 1926 it became part of historic Route 66, and for 25 years the narrow steel truss bridge carried every vehicle between "Saint Looiee and Joplin Missouree" on the Mother Road. In 1928, the runners in the great Transcontinental Foot Race crossed its deck on their way to New York, and generations of kids swam in its shadow. Near the end of its existence, early morning mist rising from Center Creek provides a spectral backdrop as the old bridge awaits dismantling. It was replaced in 1995 with a much-less-picturesque county bridge. (Photo Courtesy Bob Foos/*Webb City Sentinel*.)

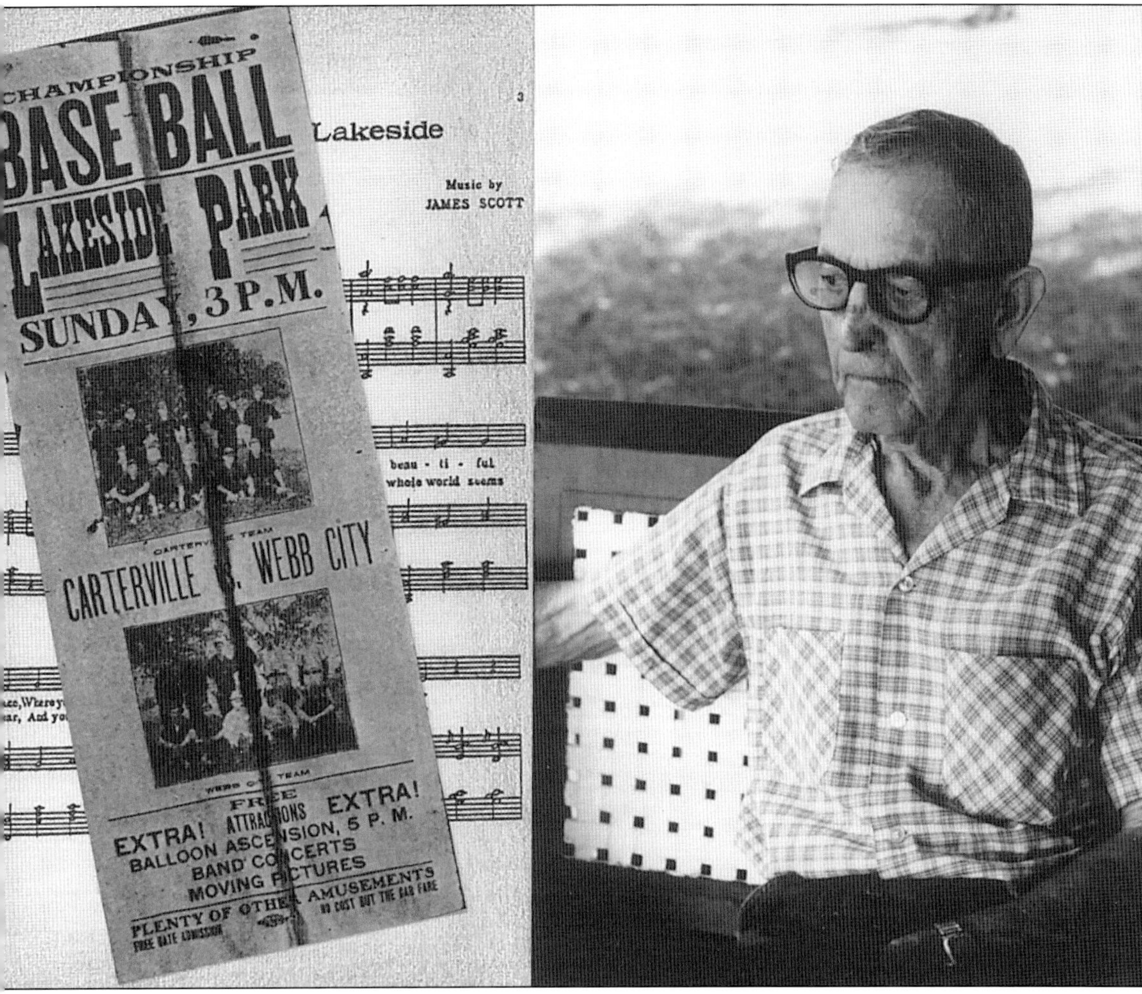

A handbill for a ball game and the sheet music for "Take Me Out to Lakeside" flank the late C.E. "Charlie" Burke, who paused in the mid-1970s to reminisce about Lakeside Park and his life. Born in 1894, he relished his experiences as a young man. He and his wife ran the Blue Ribbon Bakery in Carterville. He shook his head in wonder as he remembered the Fourth of July when they hand-formed and baked 40,000 buns to be used at the park. He recalled that, in peak holiday periods, the screened summer cars would run every five minutes, as many as 150 per day, with the interiors loaded and passengers hanging on the outside. A looped spur track was built to the south entrance, and it would sometimes have as many as eight cars loading and unloading at once. Admission to the park was free. The only cost was the carfare and 25¢ for entrance to the 160-foot-long grandstand if so desired.

His fondest memories concerned playing baseball in the Trolley League when he was a boy. The league had serious talent. Some of the men were paid to play, and a number moved on to the major leagues. Charlie was good enough in 1910 at age 16 to be the catcher for the Carterville team that had topped the league the year before. In autumn of that year, crouching in the dust behind home plate, he caught the pitches of Walter Johnson of the Washington Senators, who along with Zack Wheat was barnstorming after the major league season had ended. Sixty-five years later, the old catcher wondered if he were the last man alive who had caught the pitches of the immortalized "Big Train."

ACKNOWLEDGMENTS

This project would not have been possible without the generosity of Ron Mosbaugh, John Burrington, and Henry Hulett who made their collections of postcards, photographs, and memorabilia available. My special appreciation to them.

Many other people were unfailingly obliging and helpful, and I must thank as many of them as possible: Bob Foos, editor of the *Webb City Sentinel*; Eileen Nichols of the Webb City Historical Society; Sue Oliveira, Lucille Kent and the other fine people at the Webb City Genealogical Society and at the library. Also, the Genealogical Society Board deserves special thanks for permission to excerpt their publication on mining accidents; Dr. Ron Lankford, Webb City schools; Ray Savage, Granby Miner's Museum; Brad Belk, Joplin Museum Complex; Jim and Jean Horn, Joan and Don Johnson, Helen Chickering, Oletha Crocker, Clyde Clark, Eddie and Julia Snyder, John Furry, Leeman Dorsey, Fred Rogers, and Bill Duensing for memories, anecdotes, or photos; H.W. Becker for permission to use his old photographs of Lakeside Park; Michele Hansford of the Powers Museum at Carthage, park information; Roy Campbell, ex-miner who answered questions about mining practices; Keith Ulrich, Arcadia's Missouri editor; and the good people at Quik Print and Photo Place, always patient and obliging. Others shared memories or insights, including: Bob Basnett, Wesley Bivens, John and Jane Campbell, Raymond Ervin, Diana Fleischaker, Hugh Hatcher, Bob Heiniger, Doug Jackson, Jo Kilbane, Stanley and Pat McDonald, June Madden, Carl and Virginia Miller, Charles and Melvin Misner, Jeanne Newby, Kenny Oliver, Tim Pearson, Russell Pryor, Charles Rand, Gary Roney, Randy Roy, Louis and Les Sage, Karl Shaner, Joy Shirk, Ernest Stratton, Jan Stanley, C.A. Wade, Doyle White, Bertie Wahlstrom, Pat Lawrence, and Kenny Molloy.

Facts came from old newspapers, church, city, state, and county histories (published and unpublished), city directories, old advertising circulars, school annuals and publications, as well as people's memories.

I am grateful to Diane and Jack Barnes who have walked this path before, and especially to Kelli Pryor, Andrew Rosenstein, and Robb Pryor, professional writers all, who must have been distressed at my clumsy efforts but let me steer my own course anyway. To all who assisted—your help is invaluable. Any mistakes are my own.

Finally, my warmest regards go to the old timers, long vanished, whose stories, tall tales, and, perhaps lies are fading echoes. How I wish I could hear them again.

GIDEON MENDEL UN PAISAJE ROTO

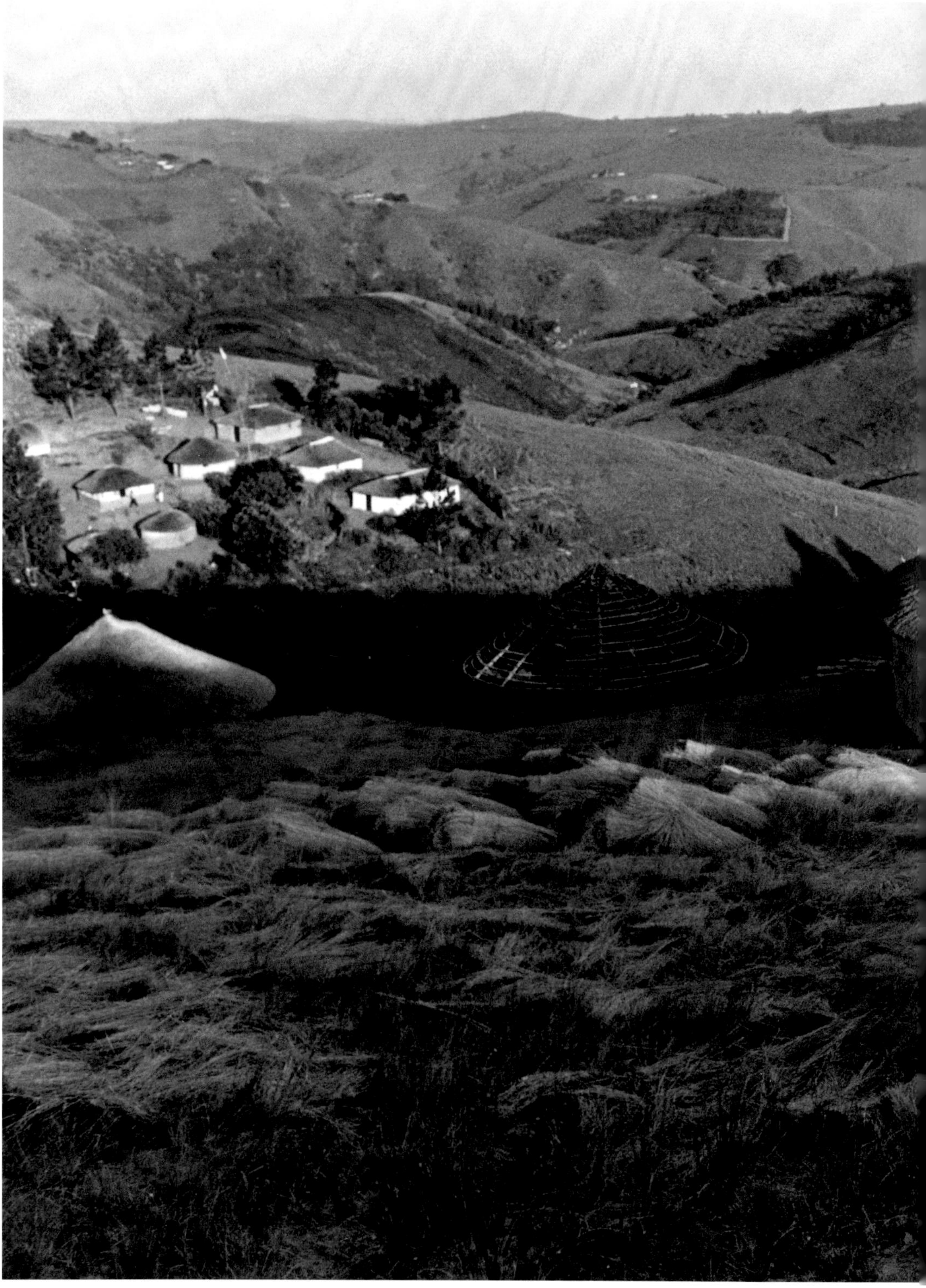

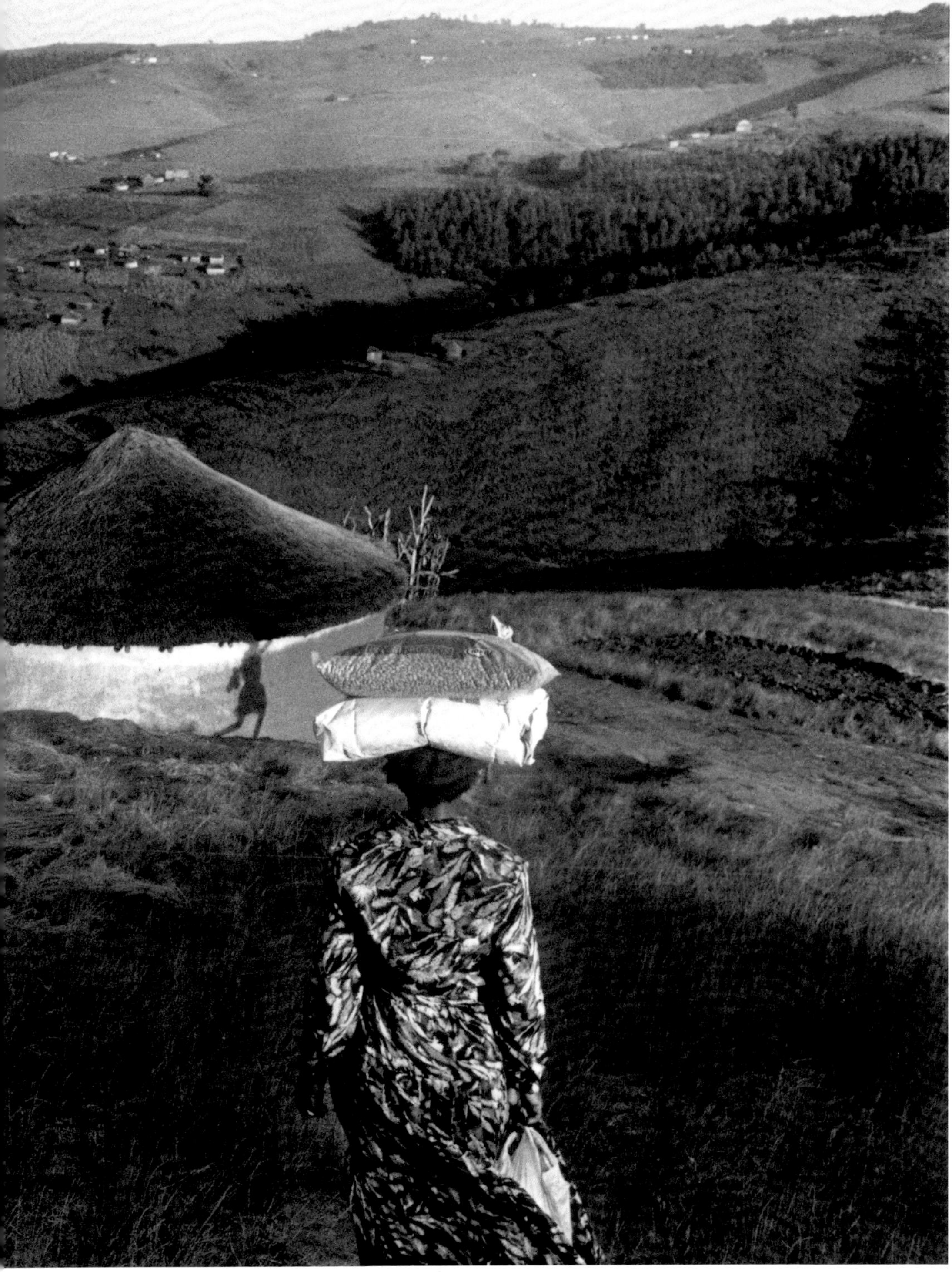

Noerine Kaleeba. Introducción

Comencé a involucrarme en temas relacionados con el VIH un día de junio de 1986, cuando el virus entró por mi puerta. Un año más tarde, la enfermedad se había llevado a mi marido, dejándonos a mí y a mis cuatro hijas luchando contra las repercusiones sociales de esta epidemia. A partir de la fundación de la organización de ayuda a los afectados de sida (AIDS Support Organisation) en 1987, me he enfrentado a la infinidad de preguntas que nuestro mundo se ve obligado a hacerse a causa del VIH y el sida, sobre la franqueza, la compasión o sobre cómo resolver un problema que afecta a todo el planeta.

El sida es la epidemia más devastadora de la historia. Desde 1982, más de 22 millones de personas han perdido la vida a causa de esta enfermedad y más de 36 millones viven en el mundo con VIH y sida. Por lo tanto, a nivel global, aproximadamente una persona de cada 100, con una media de edad entre los 15 y los 49 años, está infectada por el VIH. Muchos adultos y padres de hijos pequeños fallecerán antes de llegar a los 35 años. Se estima que más del 90 % de las personas morirán sin saber si han contraído el VIH (o si lo han transmitido). El impacto de esta epidemia sobre la humanidad será devastador si el sida no se controla.

El 95 % de las personas que han muerto de sida vivían en países en vías de desarrollo, y de ellas, casi 15 millones procedían del África subsahariana. La cruda realidad es que cada 15 segundos muere una persona en África a causa de una enfermedad relacionada con el sida. Las implicaciones para el continente son inmensas, pues el número creciente de enfermos y muertos de sida va en detrimento de décadas de desarrollo a todos los niveles.

En África viven hoy más de 12 millones de niños en hogares afectados por el sida o en lucha por sobrevivir después de la muerte de uno o ambos progenitores. La población adulta joven disminuye rápidamente, y la demanda creciente de servicios sanitarios desborda los sistemas públicos de asistencia. En los últimos años, los pacientes infectados con el VIH han ocupado la mitad de las camas disponibles en los grandes hospitales de las ciudades de Burundi y Kenia. En cuanto a la educación, el sida está mermando al profesorado. Por ejemplo, en la República Centroafricana, el 85 % de los maestros que murieron entre 1996 y 1998 eran seropositivos.

El efecto sobre la agricultura y la economía puede apreciarse en una plantación de caña de azúcar que sufrió una reducción del 50 % en su producción, lo que causó mayores costes por horas extraordinarias, la quintuplicación de funerales y la multiplicación por diez de los costes sanitarios. En el otro extremo de la escala, cuando un agricultor pobre enferma, los efectos son devastadores. La familia pierde los ingresos por la venta de la cosecha pero debe comprar alimentos, por lo que venden el equipo de trabajo y sus enseres domésticos para sobrevivir. Los niños no pueden ir al colegio porque deben trabajar y colaborar con estas necesidades adicionales.

He llegado a conocer muy bien al sida. Para mí, las estadísticas implican amistades, familia, vecinos y colegas. Recuerdo sus historias, su lucha y, lo más importante de todo, su coraje. Esto es lo que impulsa mi trabajo. Esto es lo que me hace darme cuenta de que todos debemos escuchar las historias de todos aquellos que mueren y viven con el VIH y el sida.

Este libro nos explica la historia del sida en África. Nos muestra lo que esta tragedia humana está provocando, y la fuerza y la capacidad de mujeres, hombres y niños, los cuales se enfrentan a problemas inimaginables.

África se enfrenta a las enormes y costosas consecuencias del VIH y la pobreza, unidas en una espiral descendente de privaciones y desesperación. Las personas pobres y desnutridas, cuyos sistemas inmunitarios ya estan debilitados, sucumben con mayor facilidad. No pueden ganar dinero y sus necesidades nutricionales y sanitarias se incrementan. En la mayoría de los casos no pueden pagar ni una buena alimentación ni los tratamientos médicos necesarios.

Aunque el número de personas infectadas por el virus se acerca a los 25 millones en África, afecta a cerca de 150 millones de mujeres, hombres y niños africanos. Las mujeres se llevan la mayor parte de la responsabilidad de cuidar a aquellos que están enfermos o quedan desamparados. Las abuelas ven morir a sus hijos e hijas y deben hacerse cargo de sus nietos económica y emocionalmente.

Se ha despojado a las mujeres y niños de sus derechos de sucesión y herencia. Su pobreza económica y la falta de posibilidad de elección las expone a más peligros, y esta desigualdad es uno de los factores que más influyen en la extensión de esta epidemia. Las mujeres son también más propensas a la infección que los hombres. Se estima que hay unos 13,3 millones de mujeres con sida en África, frente a los 10,9 millones de hombres. Me preocupan mis hijas y otras mujeres jóvenes que son cada vez más vulnerables a las violaciones y los abusos sexuales, ya que se las supone libres del VIH.

El sida se alimenta de la desigualdad, la ignorancia, el miedo y la culpabilidad. Los estigmas y las actitudes negativas en las comunidades y centros de trabajo que experimentan las personas infectadas por VIH y sida aumentan la discriminación.

Mi primer encuentro con una persona afectada de sida fue muy breve, pero permanece vivo en mi memoria. Me encontraba en el hospital de Mulago explicando cómo trasladar a un parapléjico de una cama a una silla de ruedas. Un hombre joven —de no más de 30 años— dio permiso para que mis alumnos aprendieran de él. Su historial médico indicaba que sufría paraplejia debido al síndrome de inmunodeficiencia, pero yo no estaba familiarizada con el diagnóstico.

Más tarde pregunté a la enfermera de guardia sobre el horario previsto para la demostración. Ella me advirtió: «Yo no lo tocaría si fuera usted. Tiene sida. Nosotros no lo tocamos, únicamente enseñamos a su madre lo que debe hacer».

Cancelé la clase y busqué otro paciente voluntario sin dar a aquel joven ningún tipo de explicación. No volví a pensar en él hasta que diagnosticaron sida a mi marido. Aún me pregunto lo que le pasó y si él y su madre tenían a alguien que les apoyara. Supongo que nunca lo sabré.

Ahora sé que la discriminación conduce a la gente al secretismo y a la negación. El miedo al ostracismo por parte de la familia y los vecinos, unido al riesgo de perder el puesto de trabajo, continúa confundiendo y complicando los esfuerzos para tratar esta epidemia. Todos desempeñamos un papel importante a la hora de favorecer el respeto y una mentalidad abierta, dos aspectos que resultan cruciales en esta lucha.

Por eso este libro es tan importante. Gracias a él valorará los riesgos que las personas afectadas con el VIH han corrido al revelar su situación. Su fenomenal coraje al compartir con los demás el modo en que este virus afecta sus vidas es una aportación que otorga significado a su sufrimiento y valentía. Quisiera rendir homenaje a su sinceridad y fortaleza al aceptar el compartir imágenes muy personales, así como sus sentimientos acerca de su situación.

Haber trabajado en el ámbito del VIH y el sida durante 15 años me ha permitido ver el éxito de Uganda. He aprendido muchas lecciones sobre lo que funciona y lo que no, y sé que ahora nos enfrentamos a uno de los mayores retos de la historia: proporcionar vida y esperanza a las generaciones futuras con respecto a la epidemia.

El VIH puede prevenirse, y ésta debe ser nuestra prioridad. La prevención no sólo consiste en informar y repartir preservativos, sino también en reforzar la comunicación y la sinceridad en las relaciones personales y profesionales. Debemos encontrar una manera para que los países en vías de desarrollo accedan a los medicamentos que han convertido el sida en una enfermedad manejable y crónica y no en una sentencia de muerte. Y debemos asegurarnos de que las personas tengan acceso a los centros de detección.

Pero ¿cómo lo lograremos? Para tener éxito debemos alejarnos de las acusaciones —ya sean a nivel individual o nacional— con el fin de llegar a una situación en la que se asuma una responsabilidad global compartida. A nivel individual debemos cambiar nuestro comportamiento y actitudes, sobre todo en lo que respecta a los estigmas y la discriminación. A nivel nacional debemos asegurarnos de que los gobiernos tomen medidas efectivas y cumplan los compromisos adquiridos. A nivel internacional debemos corregir los desequilibrios de poder y recursos entre el mundo desarrollado y los países en vías de desarrollo.

He sentido gran orgullo al observar a la gente de Uganda, y a los africanos en general, al sobreponerse al miedo paralizador y adoptar una actitud organizada y positiva ante la enfermedad. Veo las formas efectivas en que los ciudadanos, los voluntarios y los profesionales contribuyen a paliar los efectos de la epidemia. Pero la mayoría son insuficientes y se necesita urgentemente una ayuda a mayor escala.

La lección más importante que he aprendido es que resulta vital mostrar una mente abierta hacia las personas, compartir las habilidades, trabajar en equipo y asegurarse de que todos aquellos que viven con el VIH y están afectados por el mismo sean el centro de nuestro trabajo. Debemos crear un clima que anime a las personas que conviven con el sida a buscar y apoyar las acciones colectivas, así como a participar en ellas.

Este libro nos acerca a las vidas de personas reales. Estas historias nos pueden ayudar a entender el impacto diario de este virus. Porque en el centro de esta epidemia, junto a todo el sufrimiento, la pena y el dolor, hay esperanza, valentía y aliento. Y es esta fuerza y esta tenacidad lo que nos ayudará a superarla.

Joseph Gabriel. Mwanza, Tanzania

A Joseph le cuidan en casa su madre Dorika y su hermano Daniel, de 12 años.

Hace 10 años que estoy enfermo.

Me despierto por la mañana y tomo el té. Mi hermano me lava. Cuando hace calor y no llueve, como hoy, paso el día sentado fuera de casa, bajo el alero que construyó mi hermano. Mi madre o mi hermano me llevan en brazos hasta aquí por la mañana y me quedo hasta la tarde. Me gusta escuchar la radio; disfruto de la música, el fútbol y las noticias. A veces mi hermano me da dinero para comprar baterías. Lo gana limpiando zapatos en una caseta en el exterior de la casa. Mis amigos vienen y me cuentan cómo van sus vidas. Leen la Biblia conmigo.

Por el momento mi vida es rica. No siento dolor. El creer en Dios me da felicidad y tengo el amor y el cuidado de mi familia.

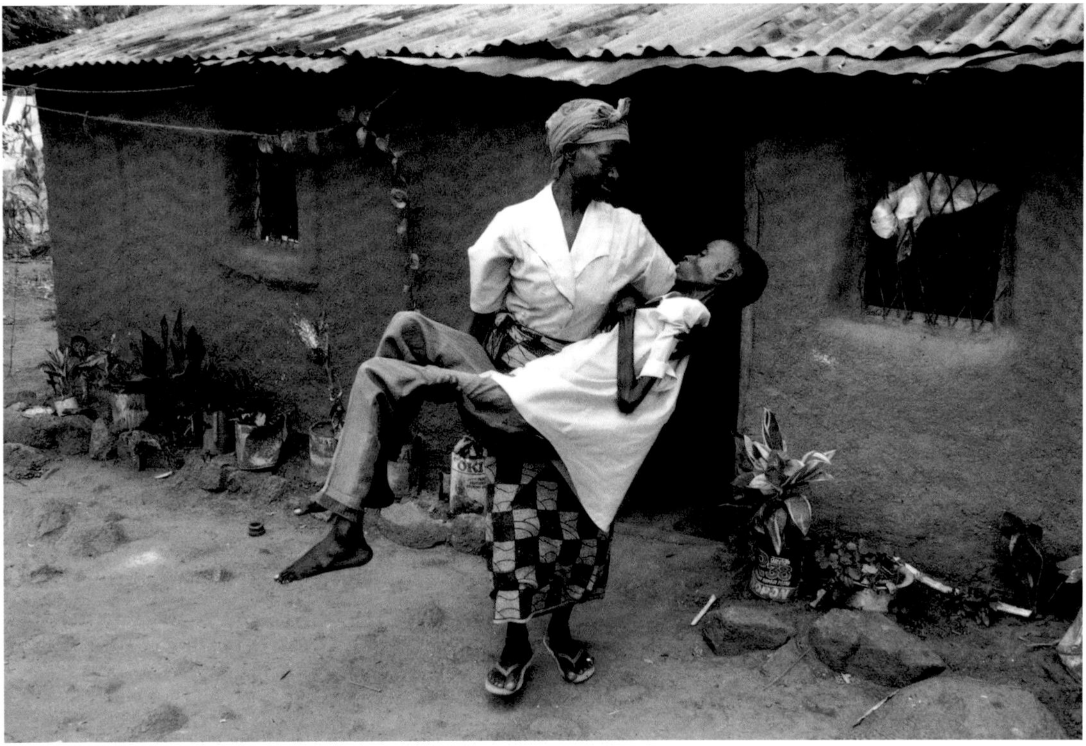

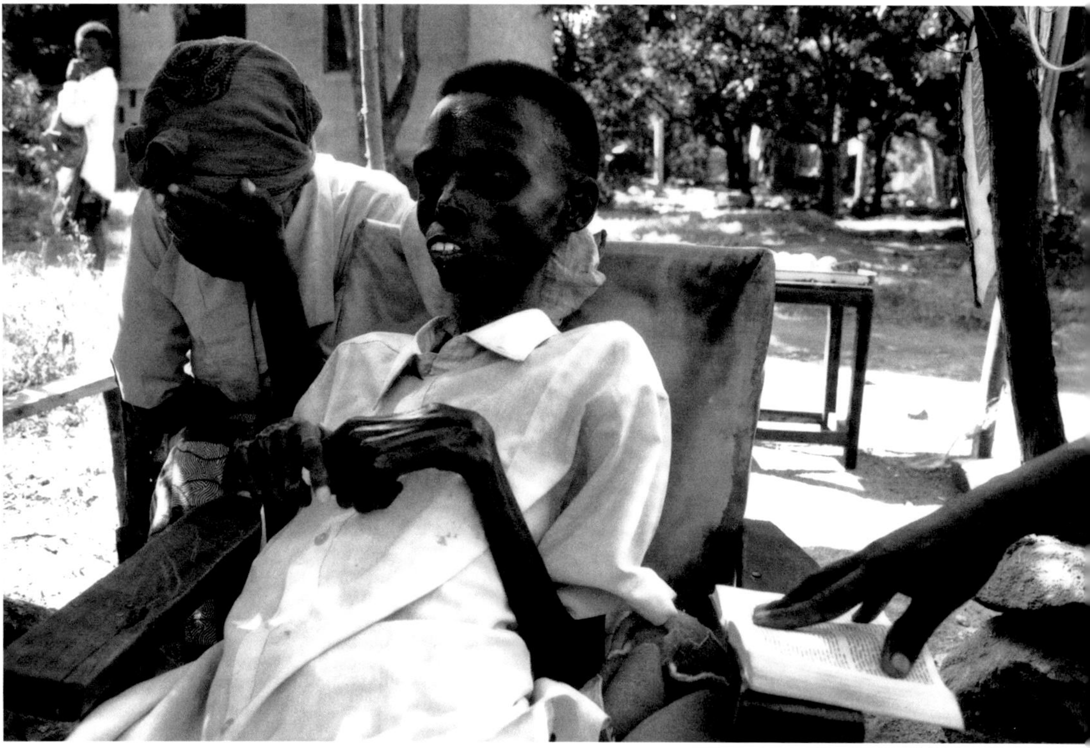

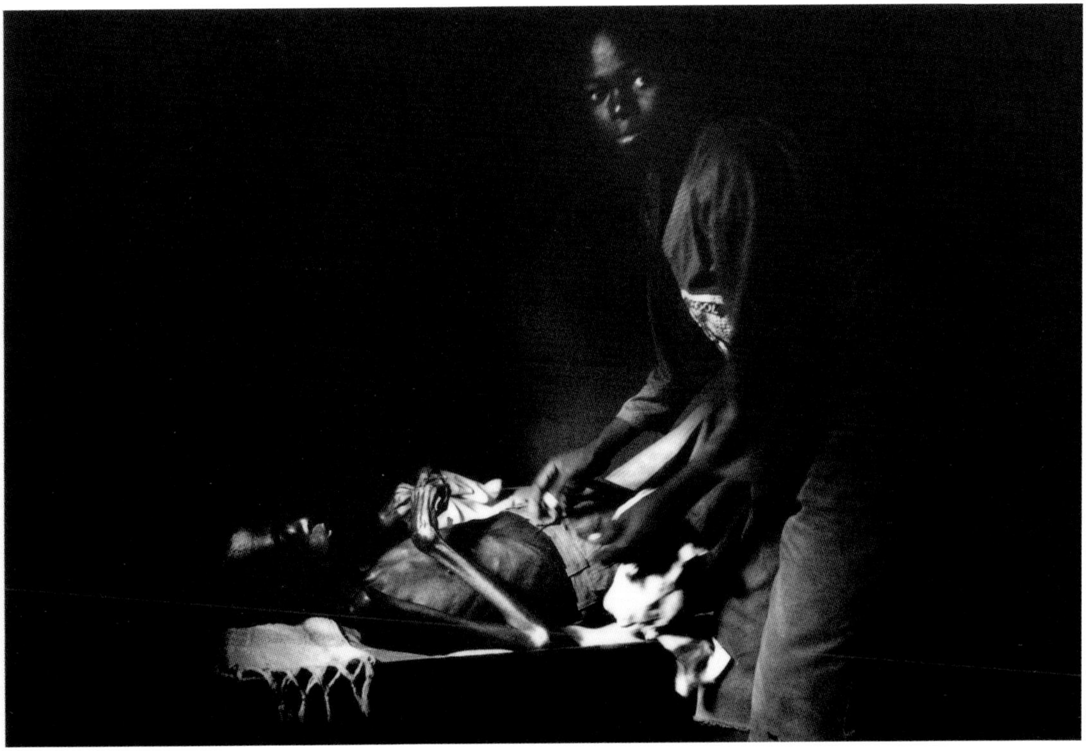

Martha Mukaratirwa. Headlands, Zimbabue

Martha cuida a su hija Eva en casa.

Mi día comienza a las 06.00 horas, momento en que me despierto y limpio el patio. Voy a buscar agua y luego preparo unas gachas de avena para el desayuno de Eva. Lavo los platos y la ropa y trabajo en el jardín. Doy la comida a Eva a las 13.00 horas. La lavo después de comer y hago una siesta. Cuando me despierto doy a Eva algo de beber y la pongo en la cama. Durante la noche debo vigilarla continuamente para cambiarle la ropa de cama.

Sentí una gran pena cuando mi yerno abandonó aquí a mi hija. En aquella época mi hija tenía un bebé y apenas podía caminar. Mi yerno no dejó ni un céntimo para que pudiera ver a un médico. Ahora tiene otra mujer y un hijo de dos años. Ocasionalmente viene a visitarnos, pero viene con las manos vacías, aunque es totalmente consciente de la escasez de comida en esta casa. A veces dos de los hermanos de Eva, que trabajan en Harare, me dan algo de dinero.

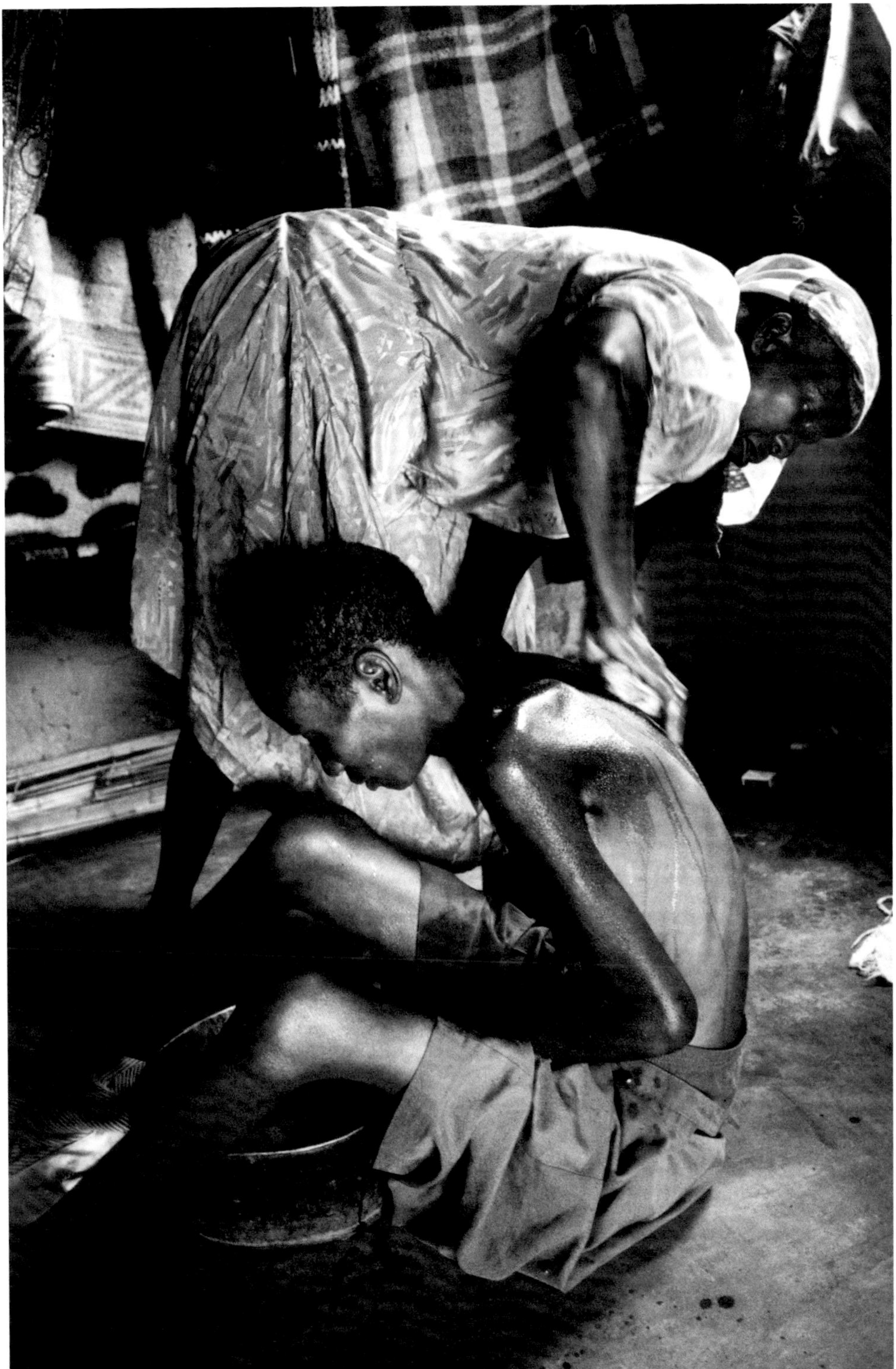

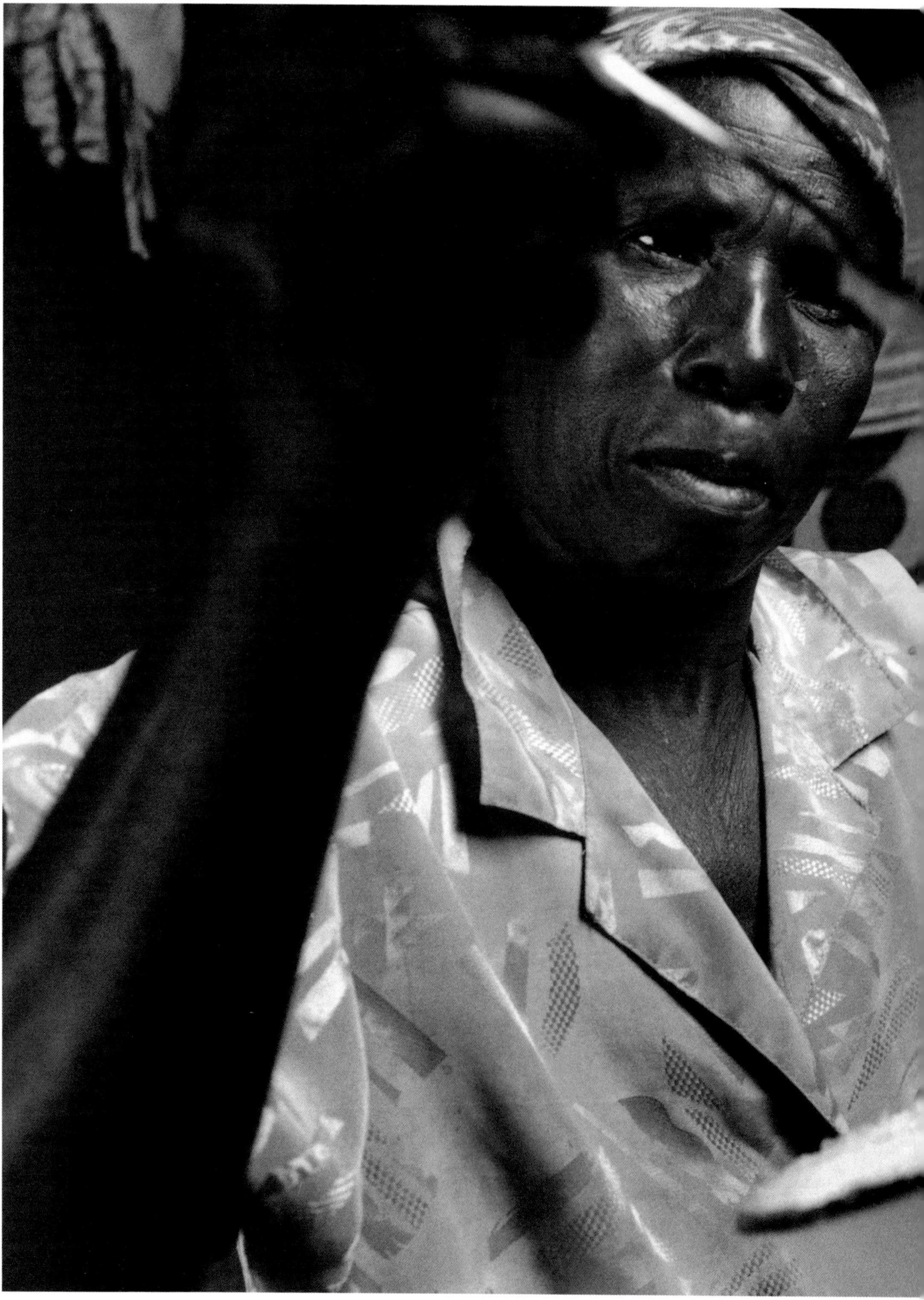

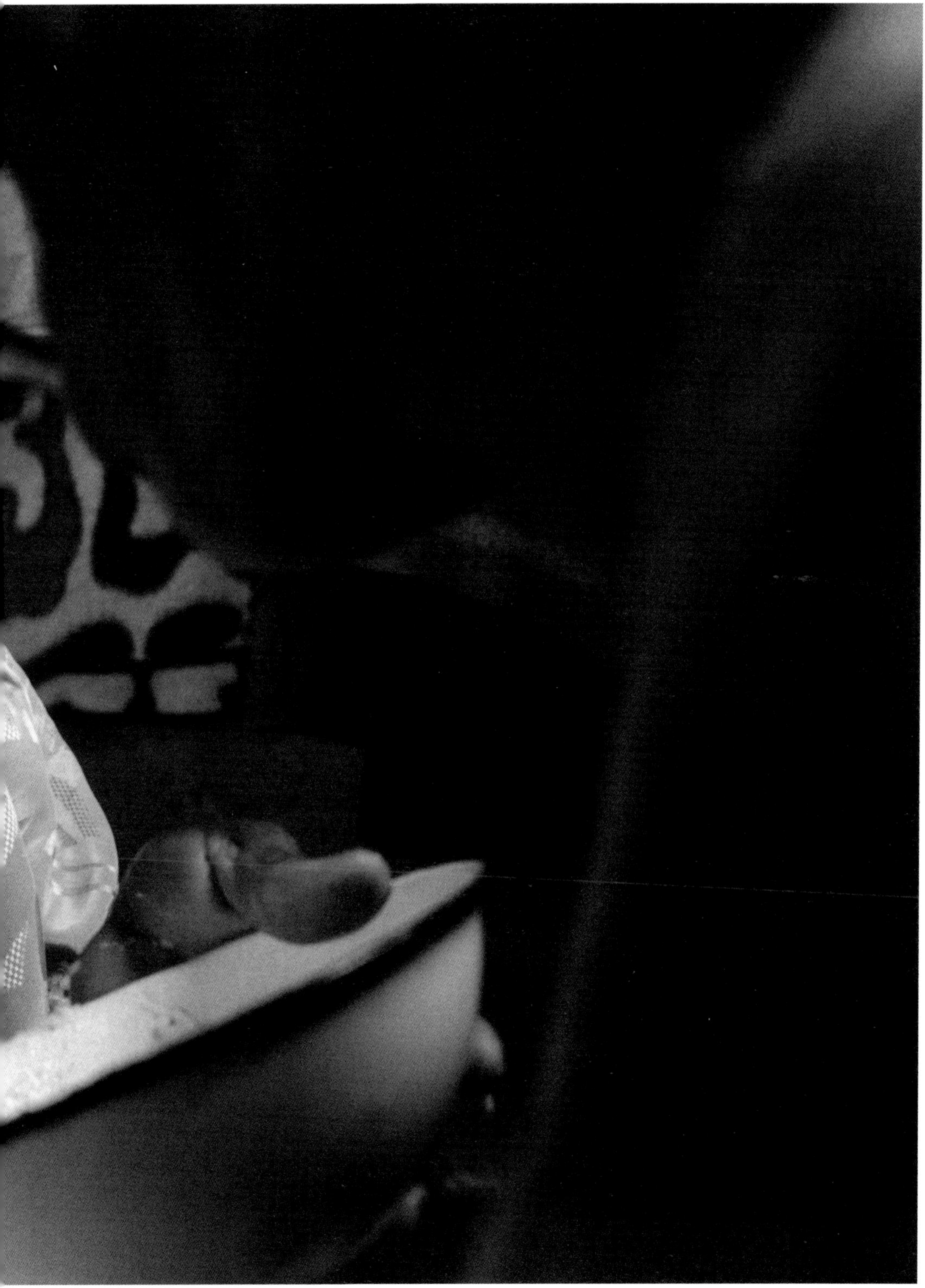

Wiseman Thabede. Municipio de Enseleni, Sudáfrica

La abuela de Wiseman, Elenena Myeni, que es portadora del VIH y participa en actividades locales relacionadas con el sida, lleva a Wiseman a la clínica de sida del hospital de Ngwelezane para recibir su tratamiento.

Ahora hace cerca de un año que estoy enfermo. Estudiaba ingeniería eléctrica en una universidad técnica de Newcastle, pero tuve que abandonar el curso debido a mi enfermedad. En marzo de este año me hicieron las pruebas y encontraron que tenía VIH.

Cuando puedo, voy a ver a un sanador tradicional. Reza por mí, y calienta y bendice agua que vierte sobre todo mi cuerpo. Sufrí una gran úlcera en la espalda y el sanador aplicó unas hierbas que eliminaron la llaga. Muchas personas van a ver a este sanador, especialmente aquellos que están enfermos como yo. La medicina tradicional es mejor para mí que el hospital. El sanador me permite tomar los medicamentos que me recetan en el hospital para la tuberculosis. Reza a Dios, no a los ancestros, para que me cure. Hará que me encuentre mejor, pero no será capaz de curarme del virus.

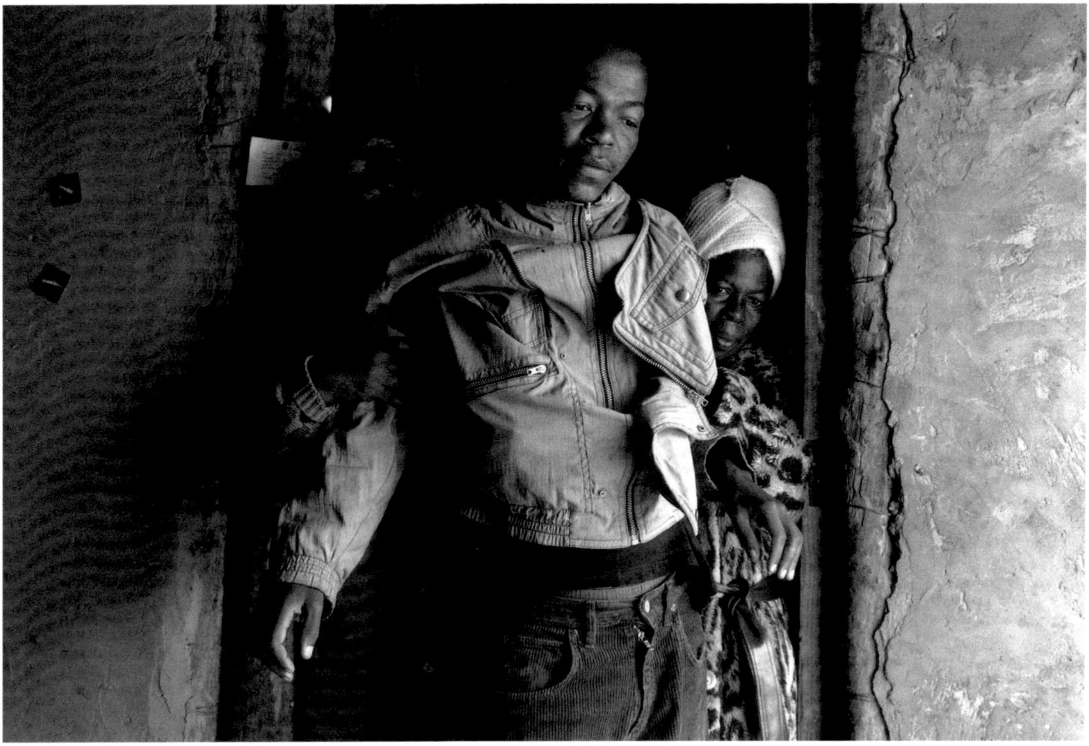

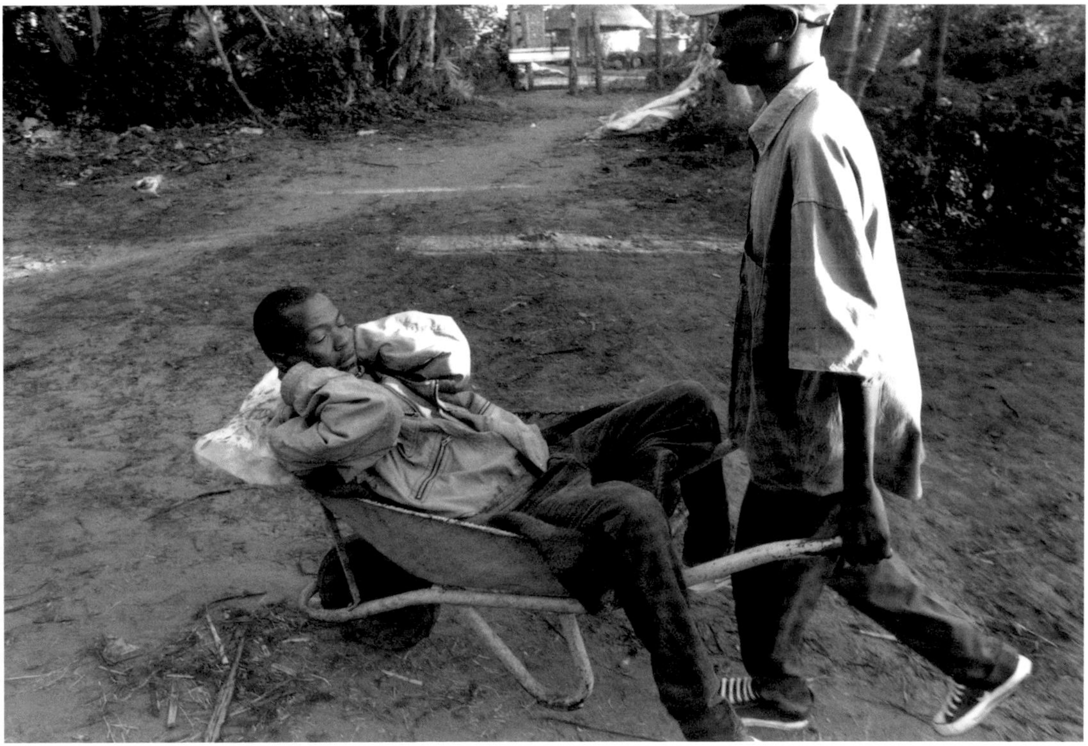

Samkelisiwe Mkhwanazi. Ngwelezane, Sudáfrica

A Samkelisiwe la cuida su madre, Nesta.

Mi hija Samkelisiwe sufre tuberculosis recurrente sin perspectivas de curación. Le han hecho pruebas y le han diagnosticado sida. Ha decidido hacer pública su situación a la comunidad para ayudar a educar a los demás. Yo la apoyo porque la quiero y la respeto.

La vida es muy difícil porque tengo una pensión de 500 rand mensuales. Con este dinero tengo que mantener a mis dos hijas, Nomhlahla y Samkelisiwe, que están enfermas, a mis otros dos hijos que no tienen empleo y a siete nietos de los que me ocupo. Nomhlahla tiene tres hijos y Samkelisiwe tiene uno. A mis 59 años es difícil ser una madre para todos estos niños, pero intento darles todo mi amor.

Si tengo un deseo en la vida es que mis hijos estén bien y no mueran jóvenes.

NESTA MKHWANAZI

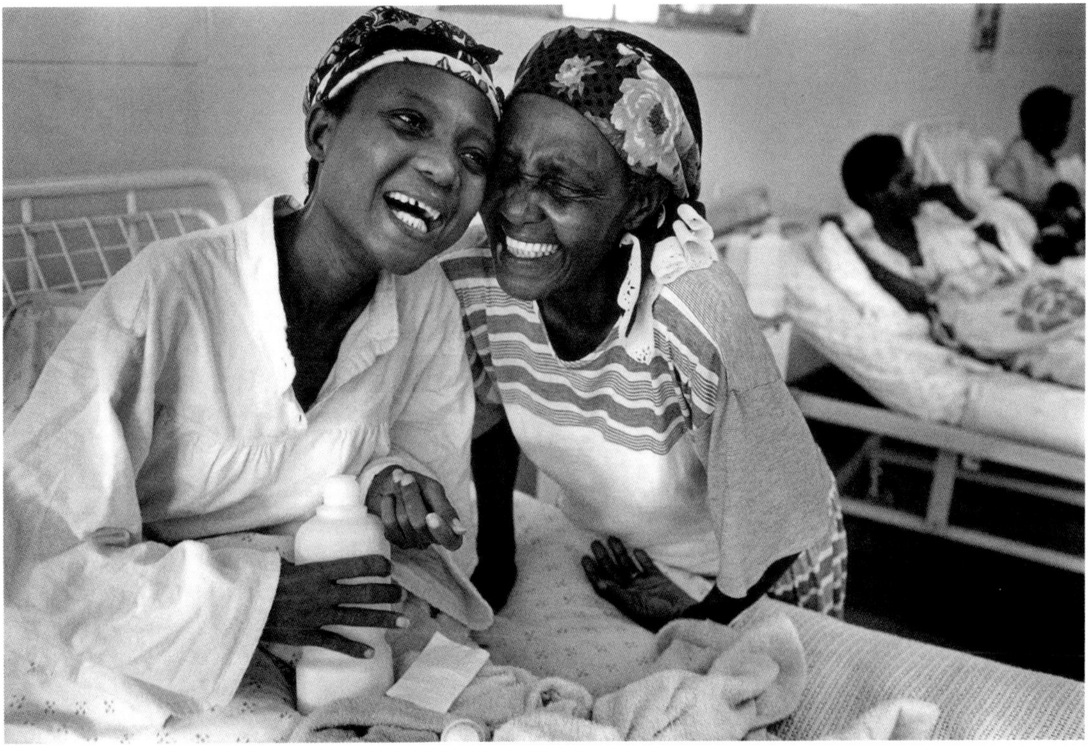

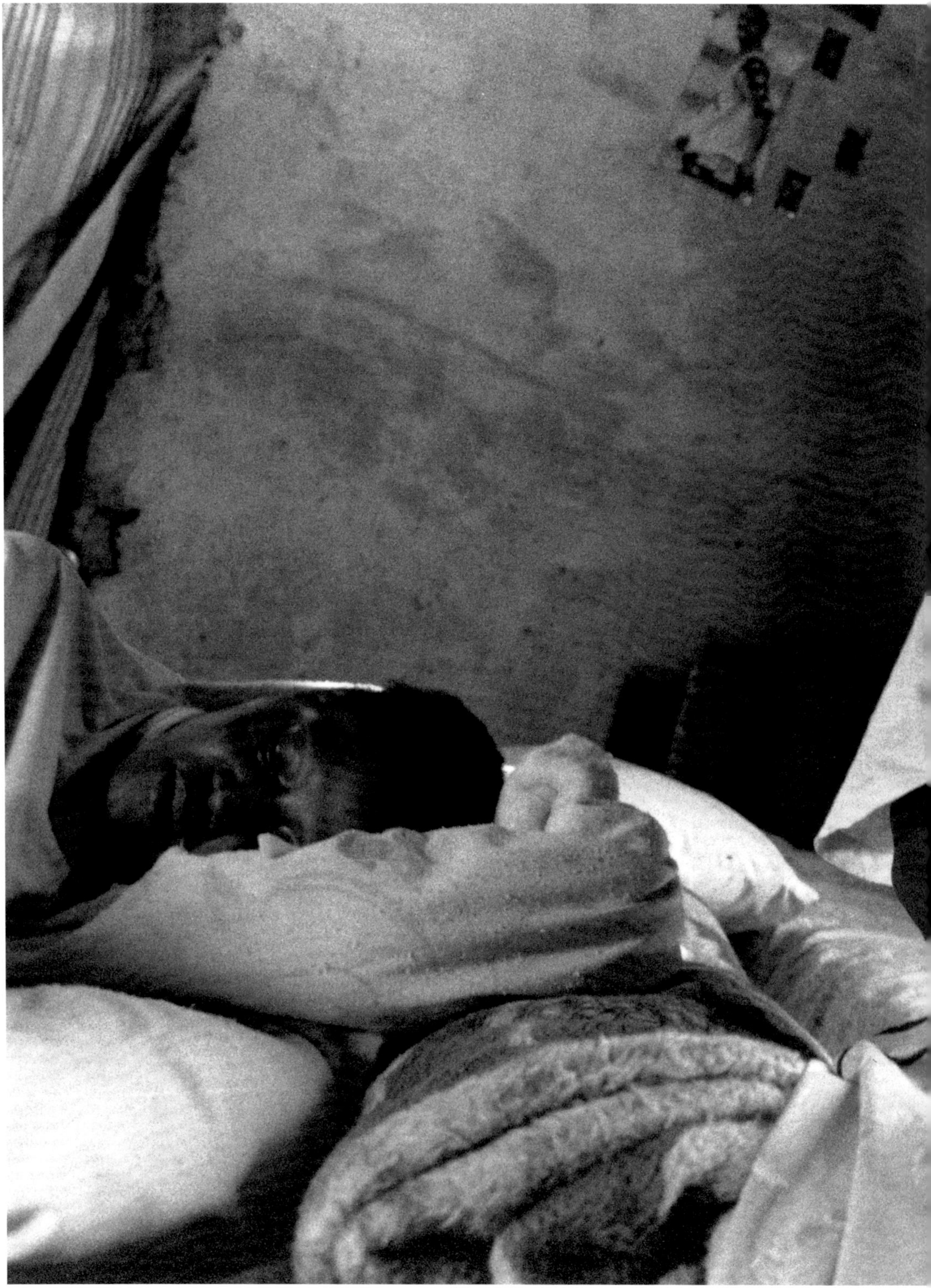

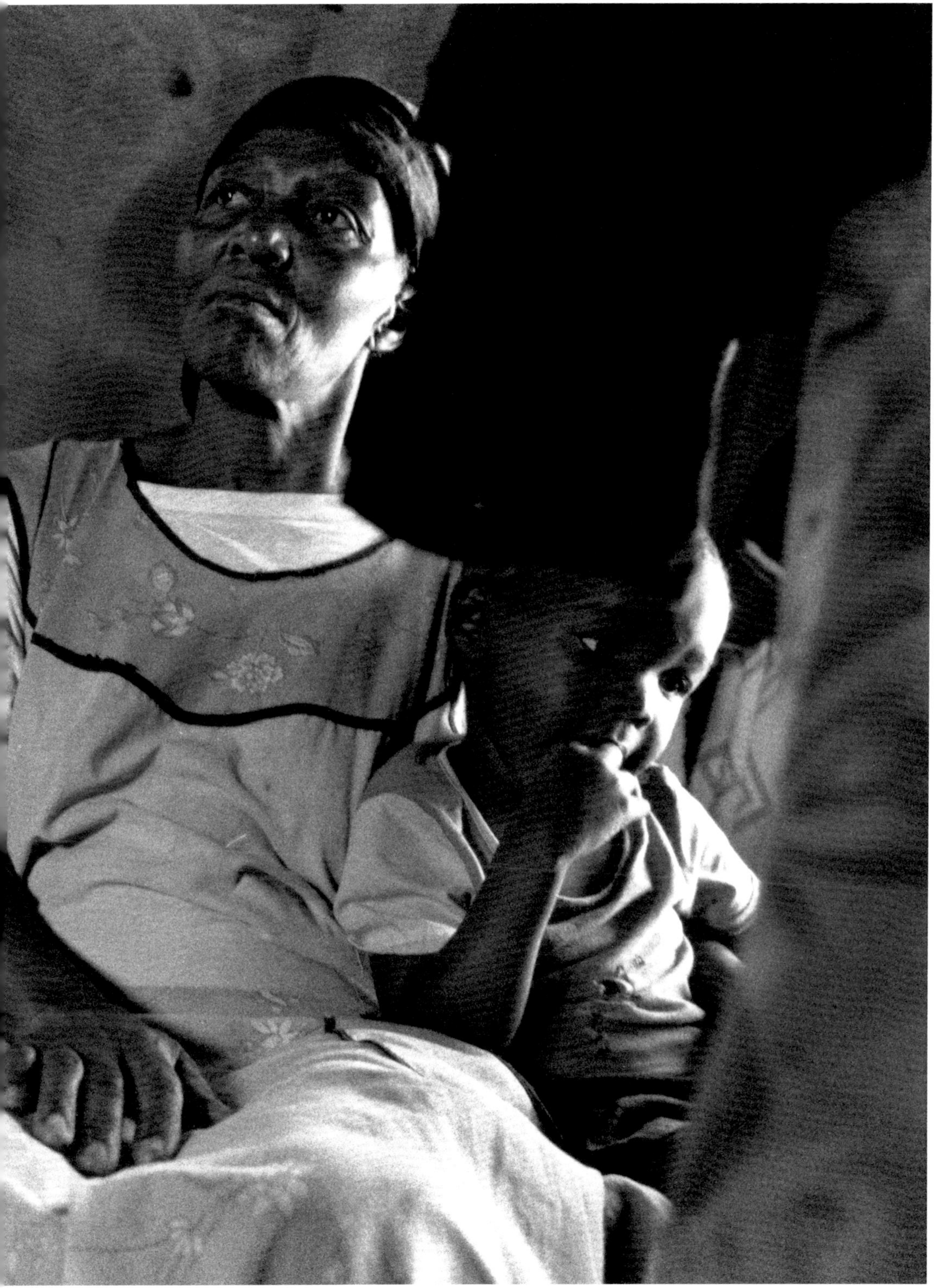

Estoy muy enferma desde diciembre del año pasado. Me diagnosticaron tuberculosis y me ingresaron en el hospital de Ngwelezane. Estaba muy enferma y estuve ingresada tres semanas. Mientras estuve allí me hicieron pruebas y me diagnosticaron VIH. Fui a un sanador tradicional. Me quedé con él durante tres meses. Me trataba con hierbas y agua y medicina natural, pero no tuvieron efecto alguno.

Amo a mi madre. Con todos los problemas que tengo, cuando me abraza y me cuida siento que ella es la solución. Yo nací en 1973 y debería ser yo quien cuidara de mi madre, pero ahora vuelvo a ser como un bebé. Ella cuida a toda la familia. Desde que mi hermana murió el 3 de junio, ella también cuida a sus hijos. Ella cuidará a mi bebé. Quiero estar con ella hasta mi muerte.

SAMKELISIWE MKHWANAZI

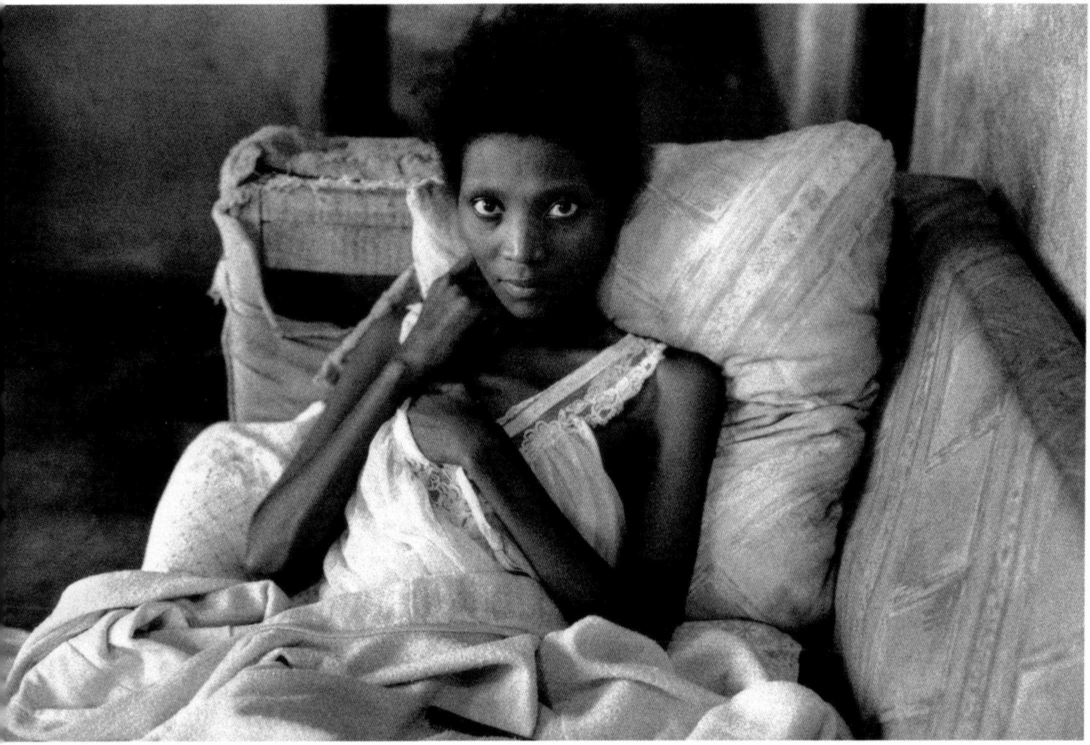

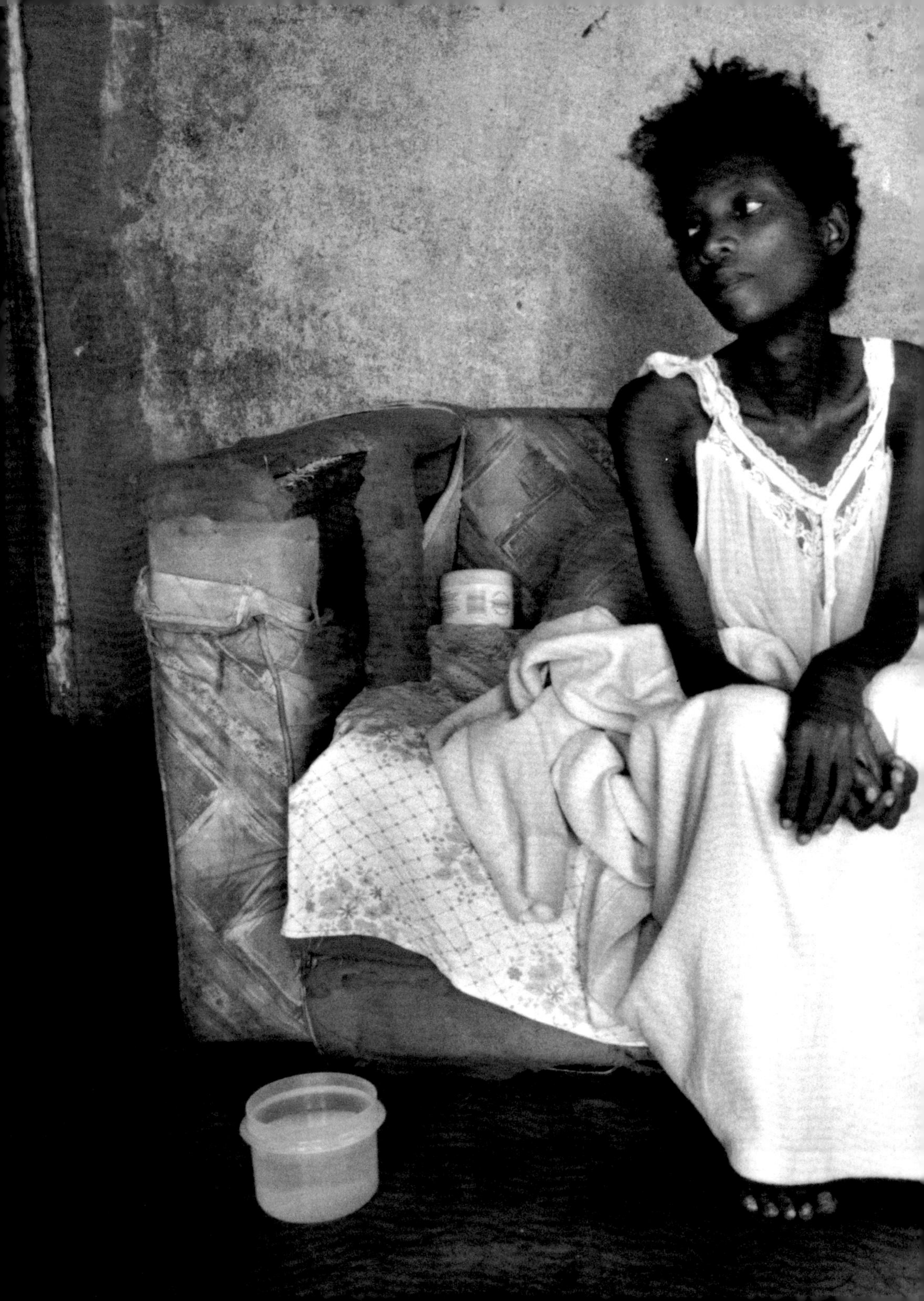

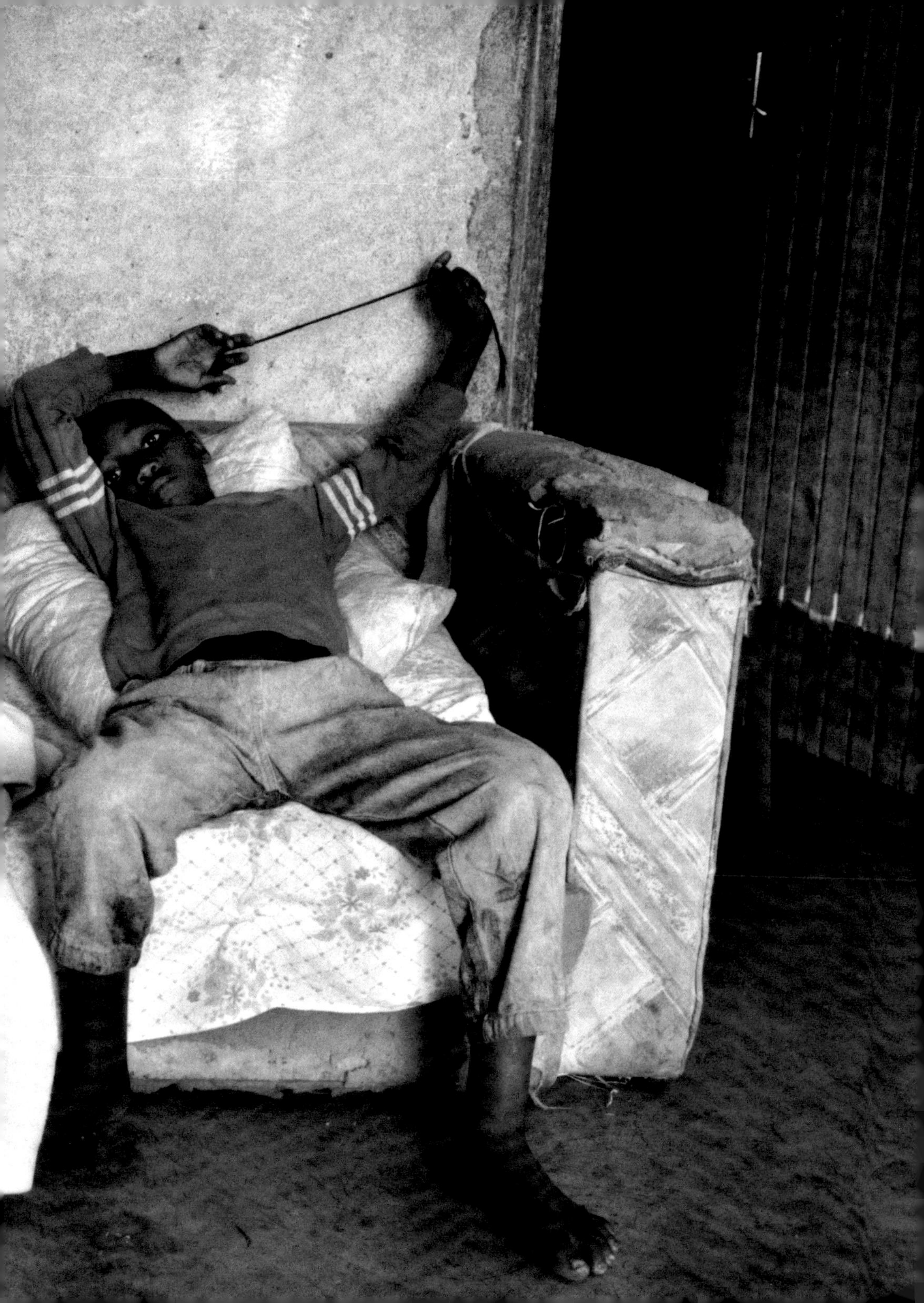

Hospital de la Misión. Matibi, Zimbabue

Un remoto hospital rural situado en una región en la que cerca del 30 % de las mujeres embarazadas dan positivo en la prueba de VIH.

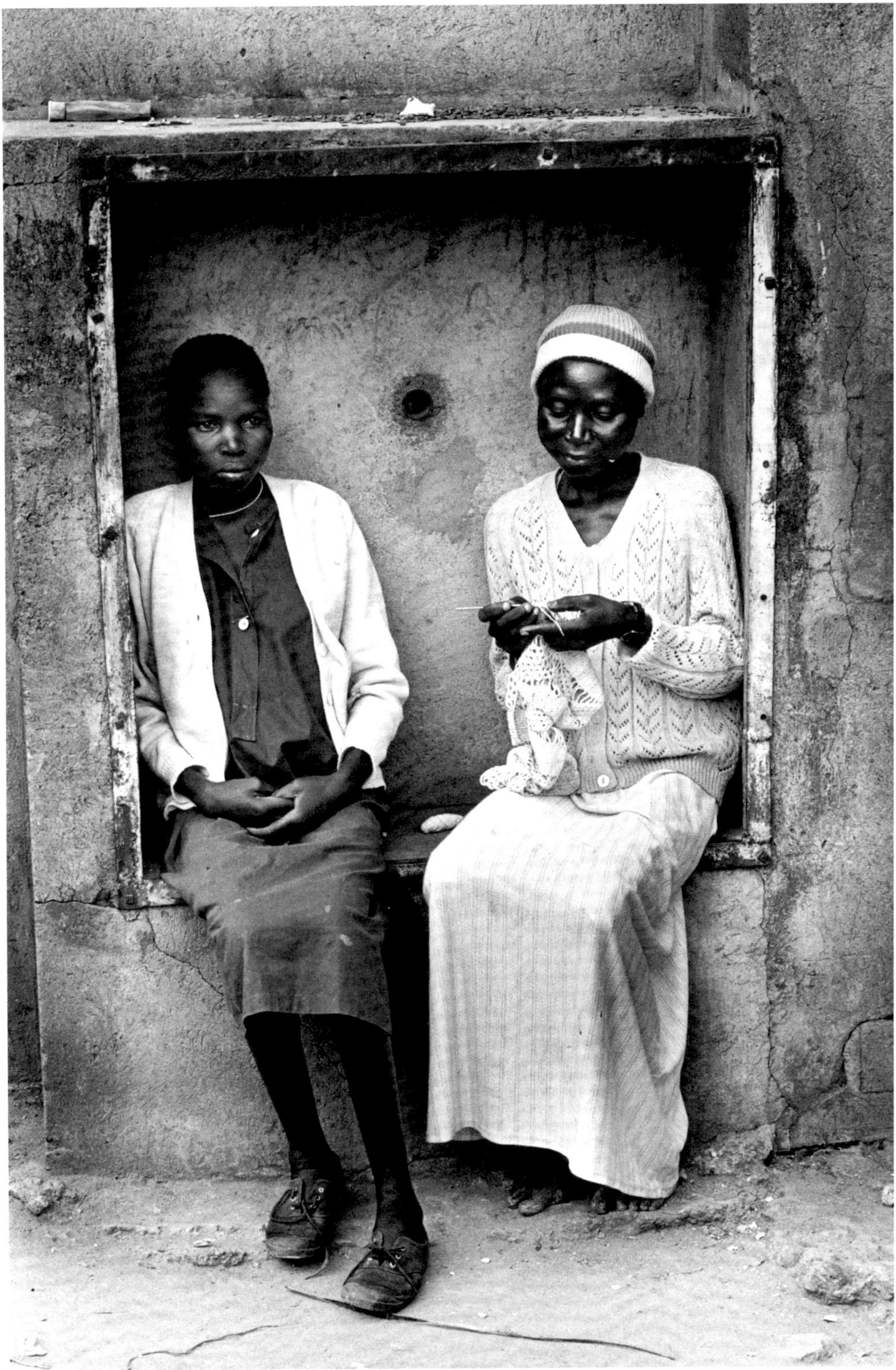

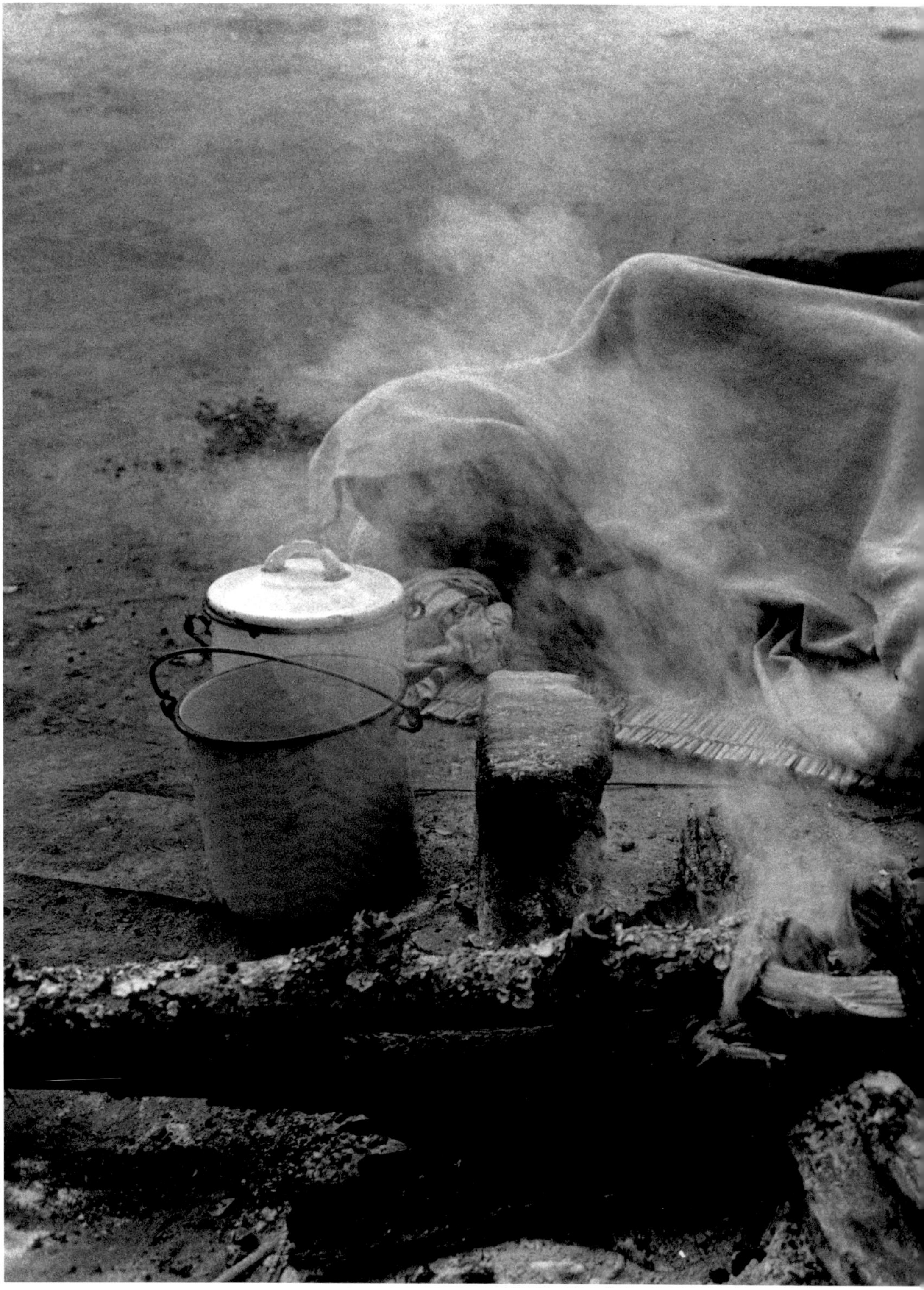

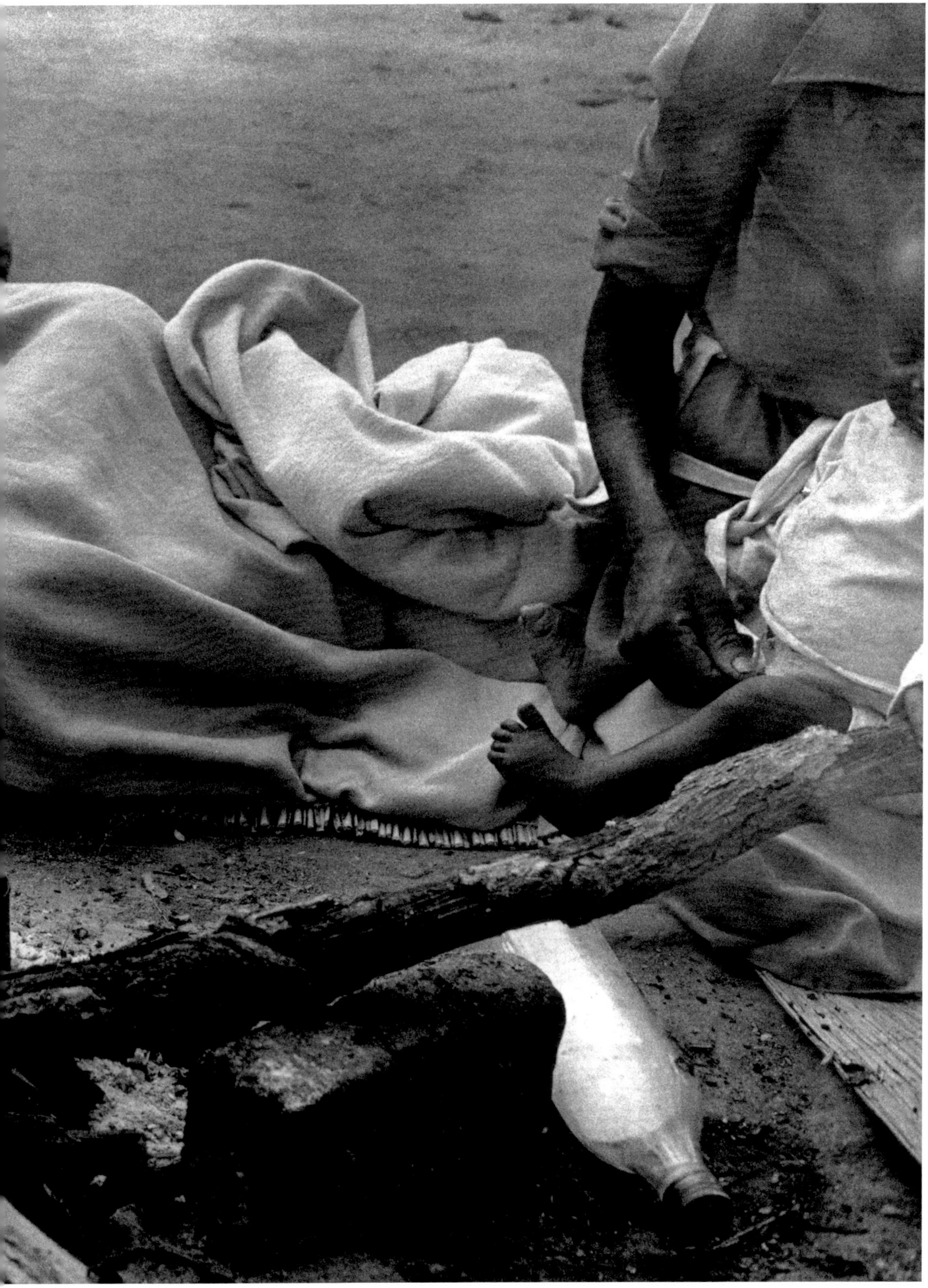

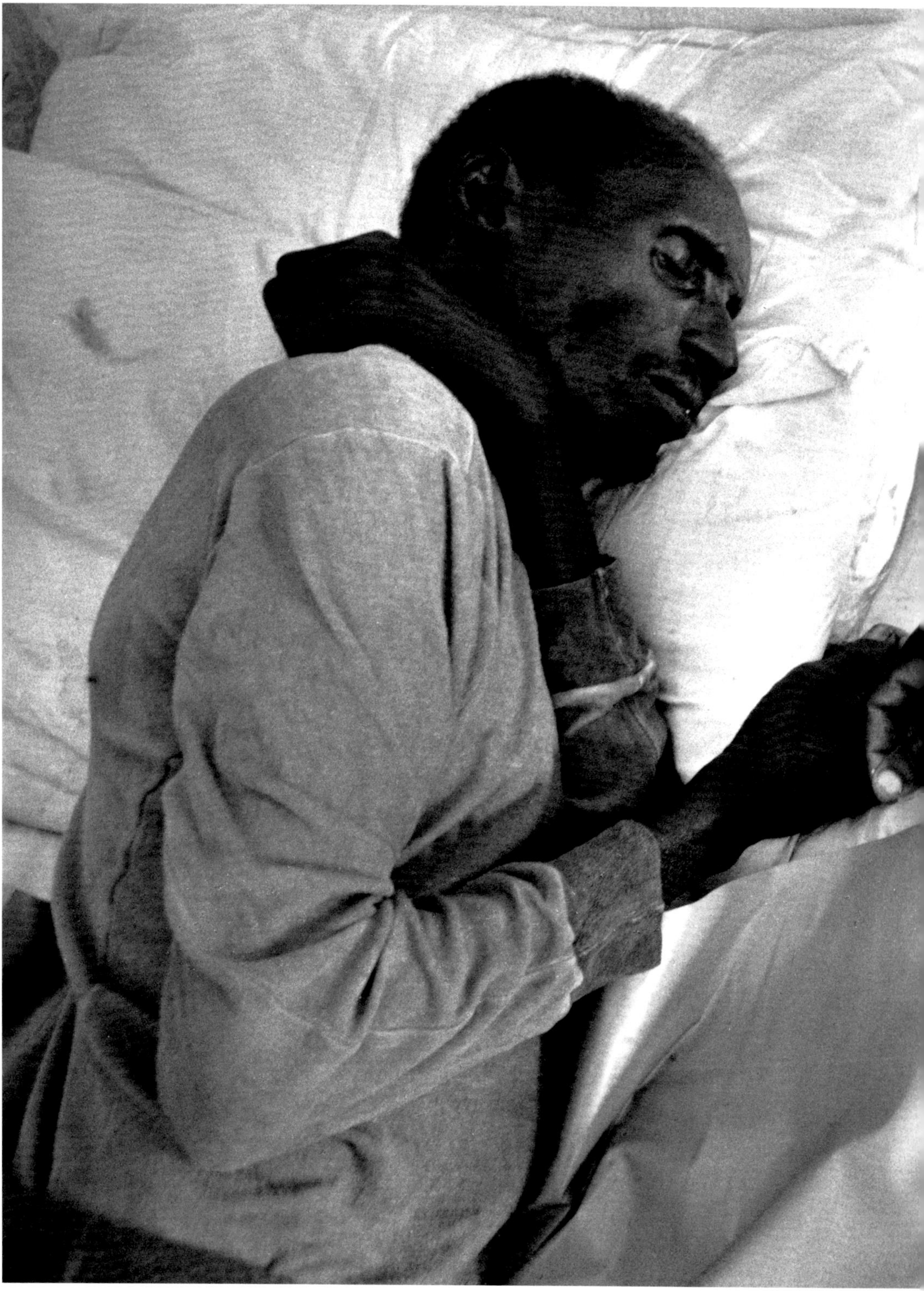

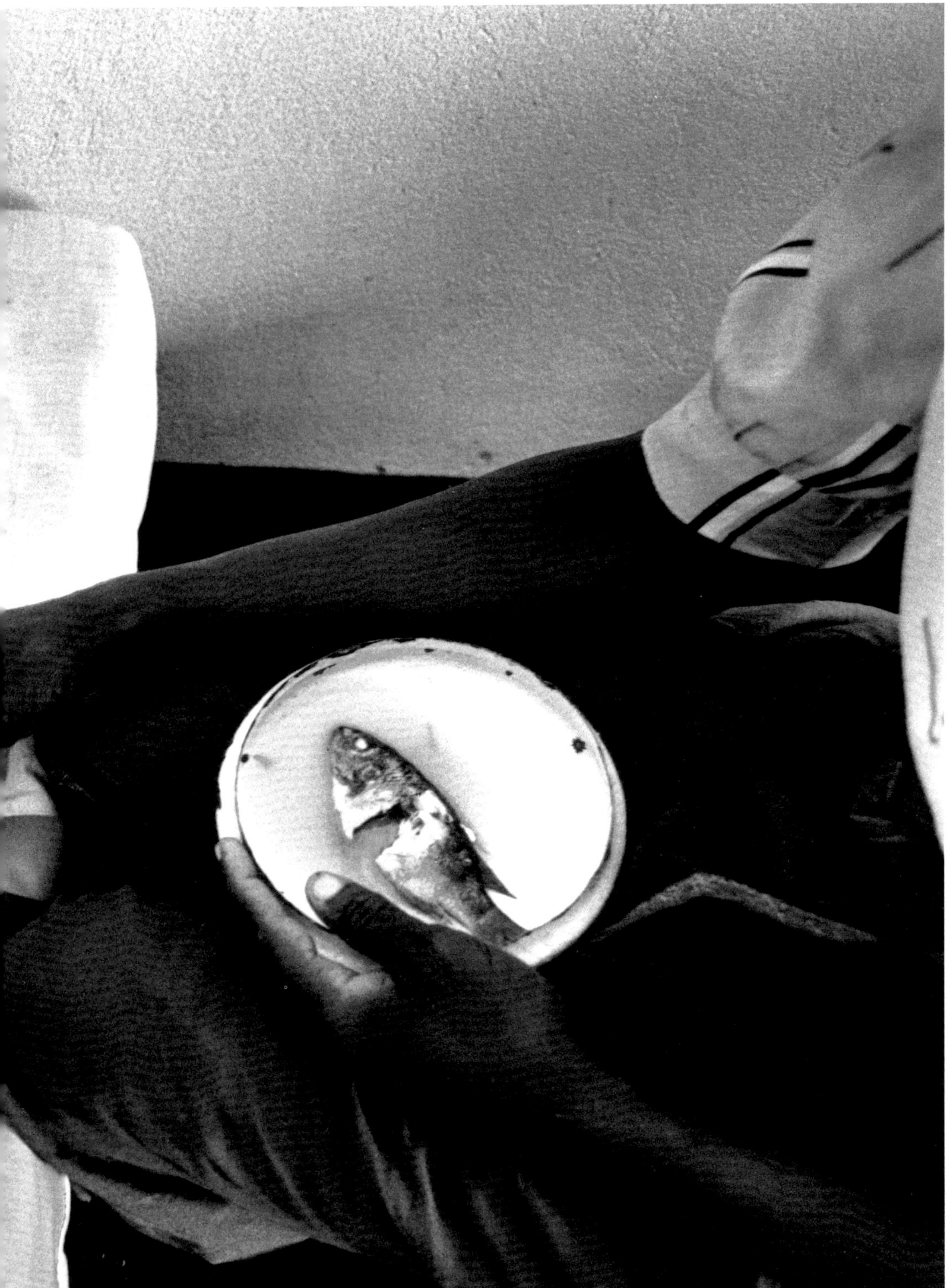

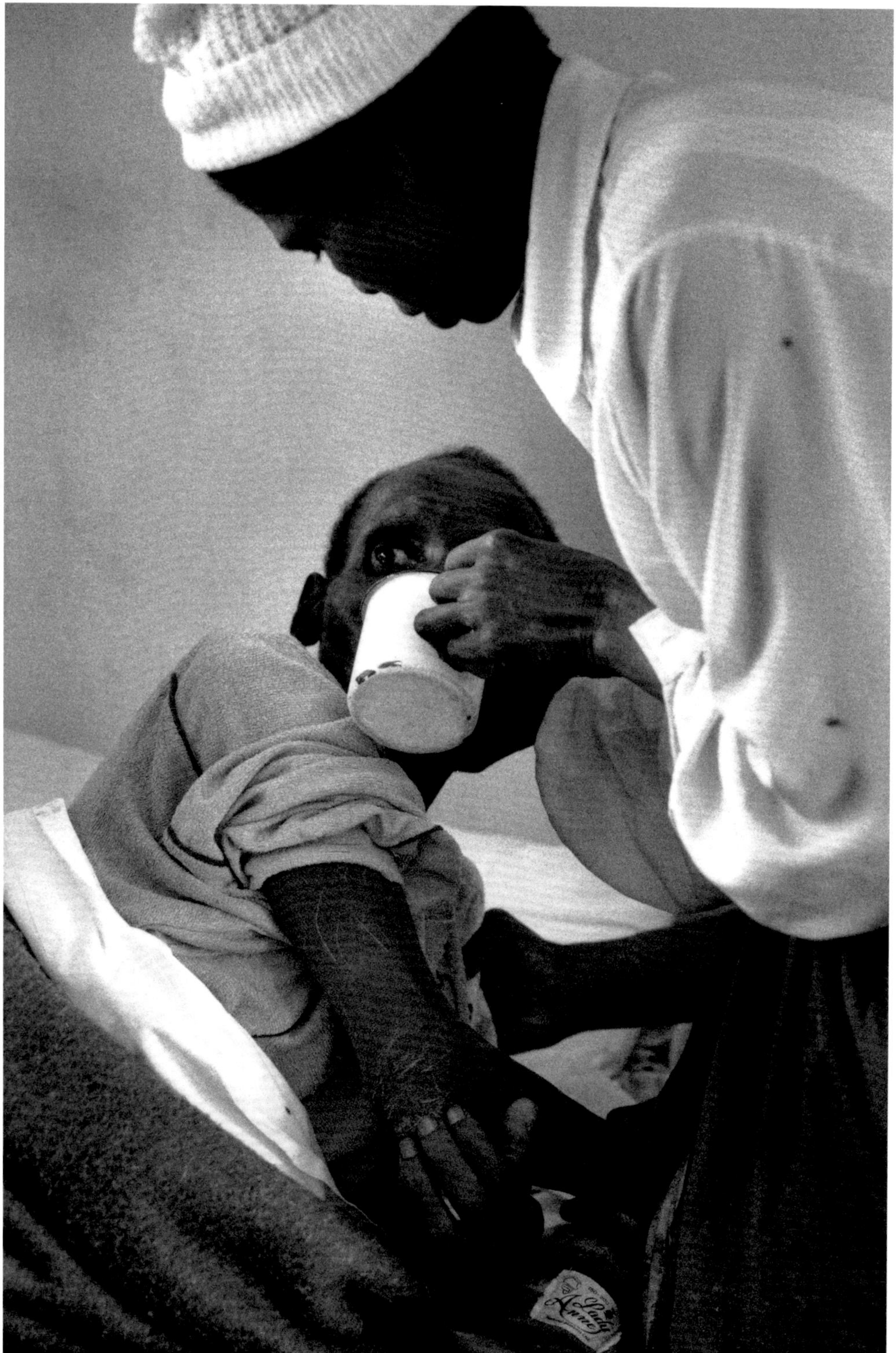

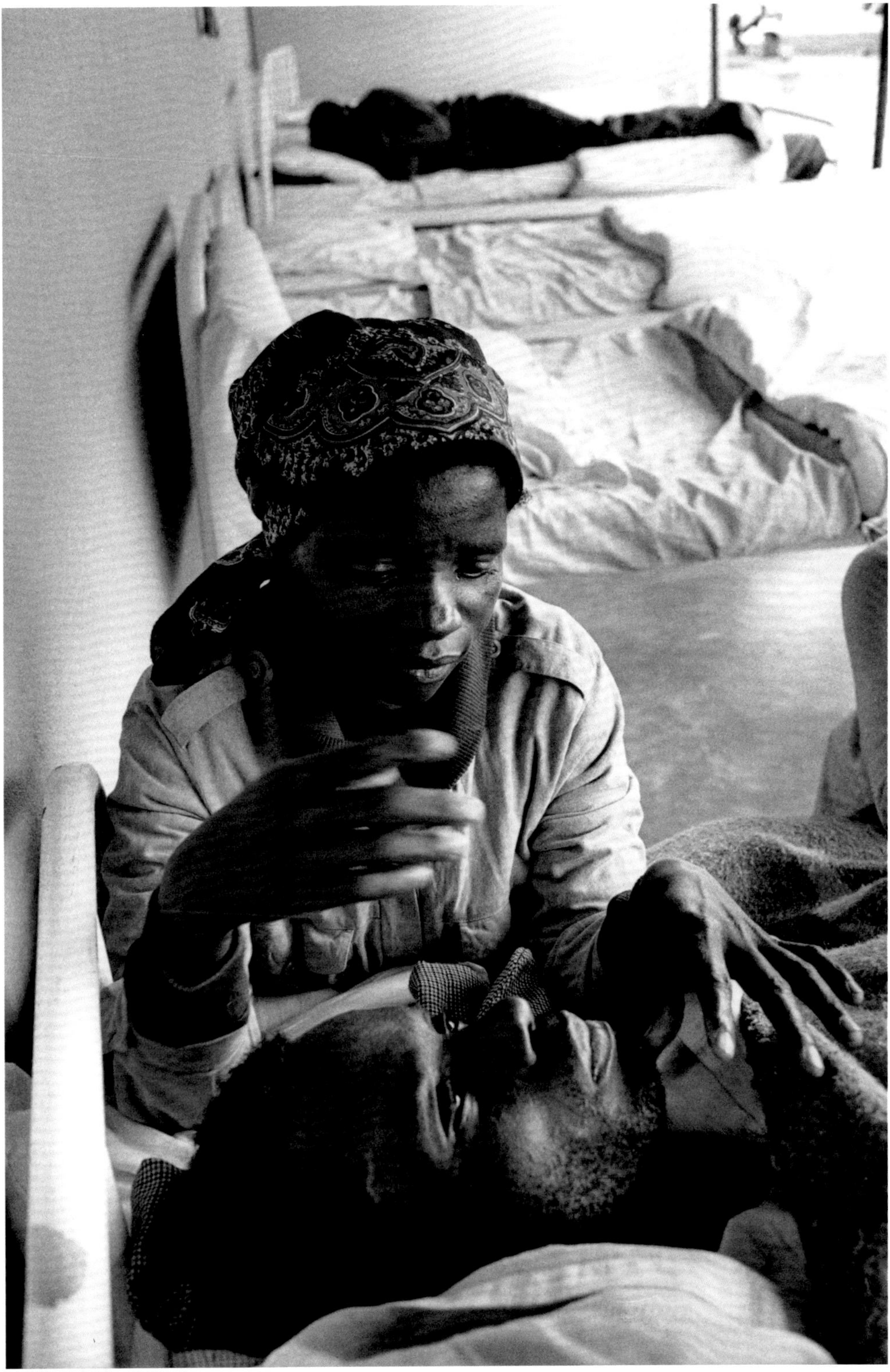

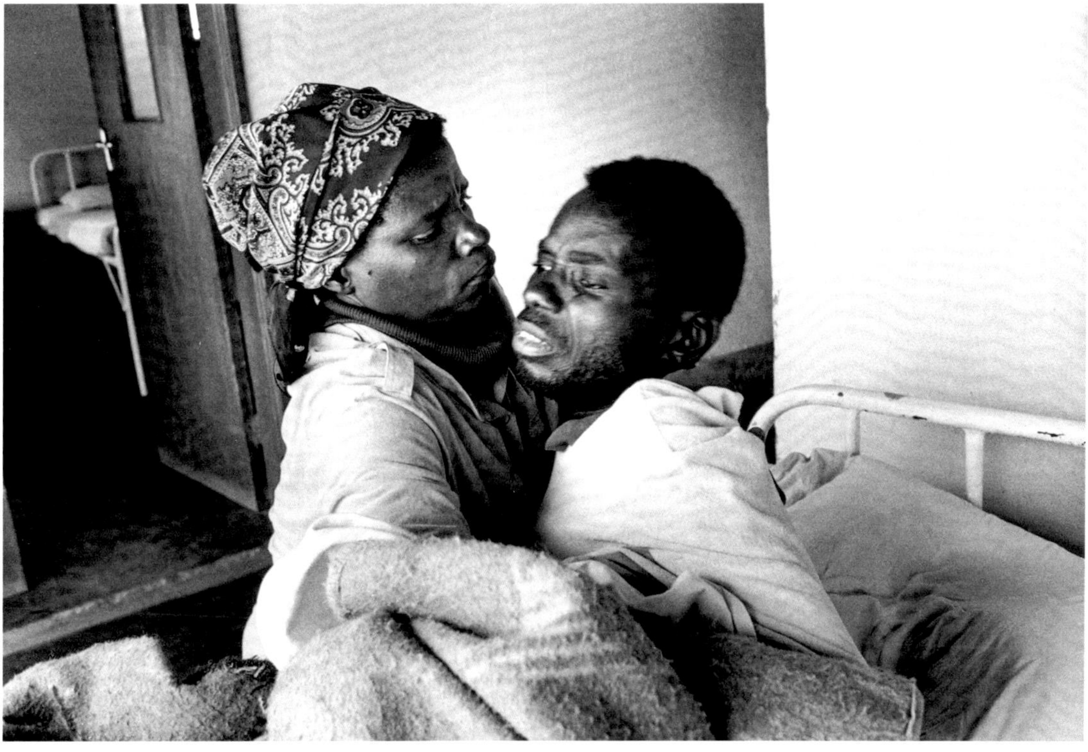

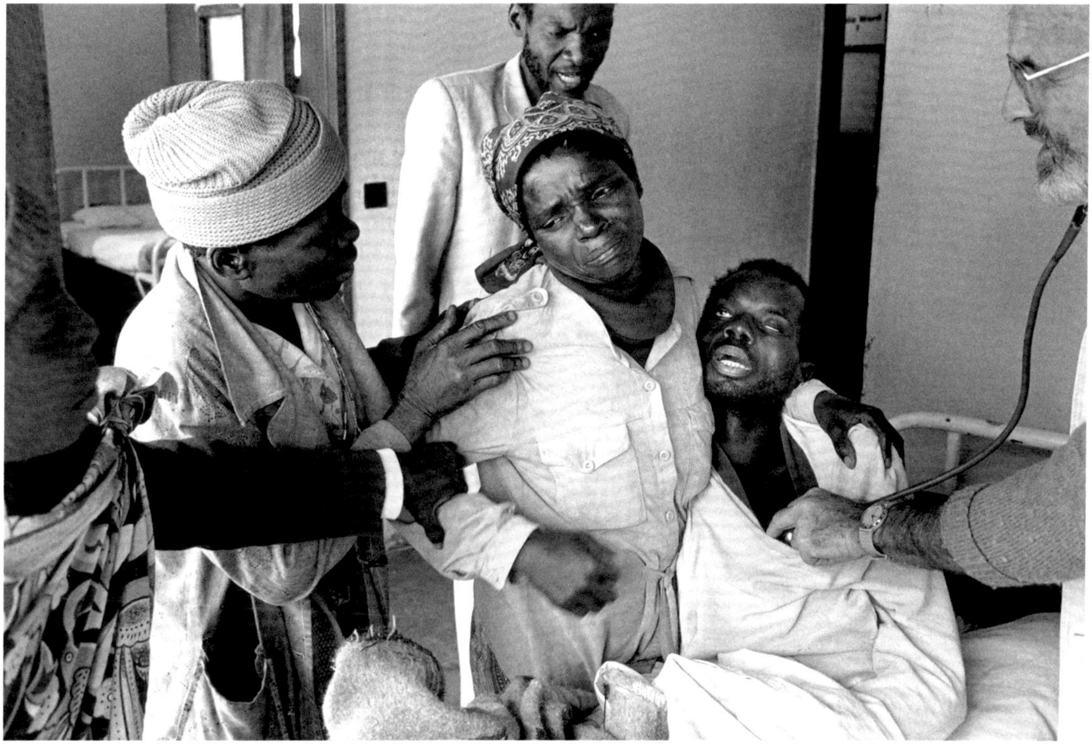

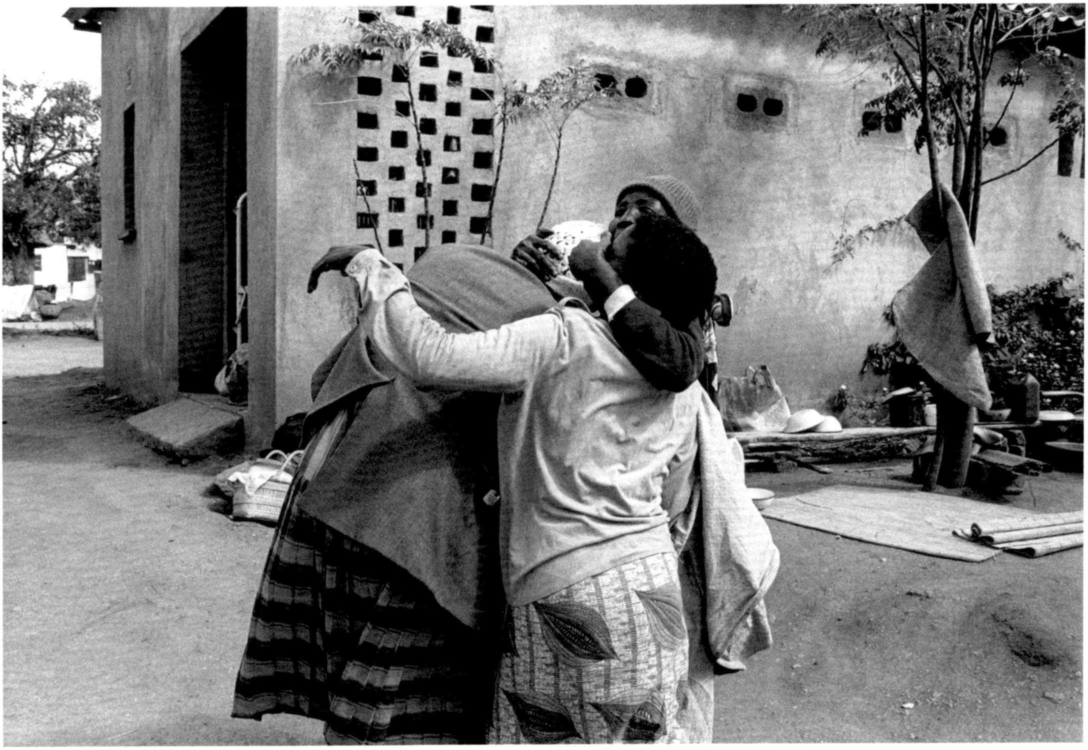

Hospitales. KwaZulu-Natal, Sudáfrica

Los hospitales de Ngwelezane, Hlabisa y Edendale se encuentran en la provincia de Sudáfrica que presenta la mayor incidencia de VIH y SIDA.

Este hospital cuenta con 604 camas y la mitad de nuestras 700 admisiones mensuales están relacionadas con el VIH. En marzo de 2000 se realizó la prueba del VIH a 733 personas del hospital, y 461 dieron positivo. La tuberculosis se ha duplicado en los últimos tres años. Las condiciones actuales no difieren mucho de las de hace unos años; es más, las enfermedades ya existentes afectan aún más a las personas. Antes dábamos el alta a un tuberculoso al cabo de entre cinco y diez días, pero ahora muchos están demasiado débiles incluso para alimentarse por sí mismos. Perdemos entre uno y tres miembros de personal cada mes debido al VIH.

La negación es un problema importante. A menudo los pacientes pasan de un hospital a otro sin darse cuenta de su situación de inmunodeficiencia. O esperan a estar gravemente enfermos antes de buscar un tratamiento. Mi equipo me infunde una gran fuerza, al igual que los pacientes que asisten a la clínica de seguimiento de VIH. Los que aceptan el diagnóstico adoptan una actitud positiva y luchan contra la enfermedad.

DOCTOR PETER HASELAU, DIRECTOR DEL HOSPITAL DE NGWELEZANE

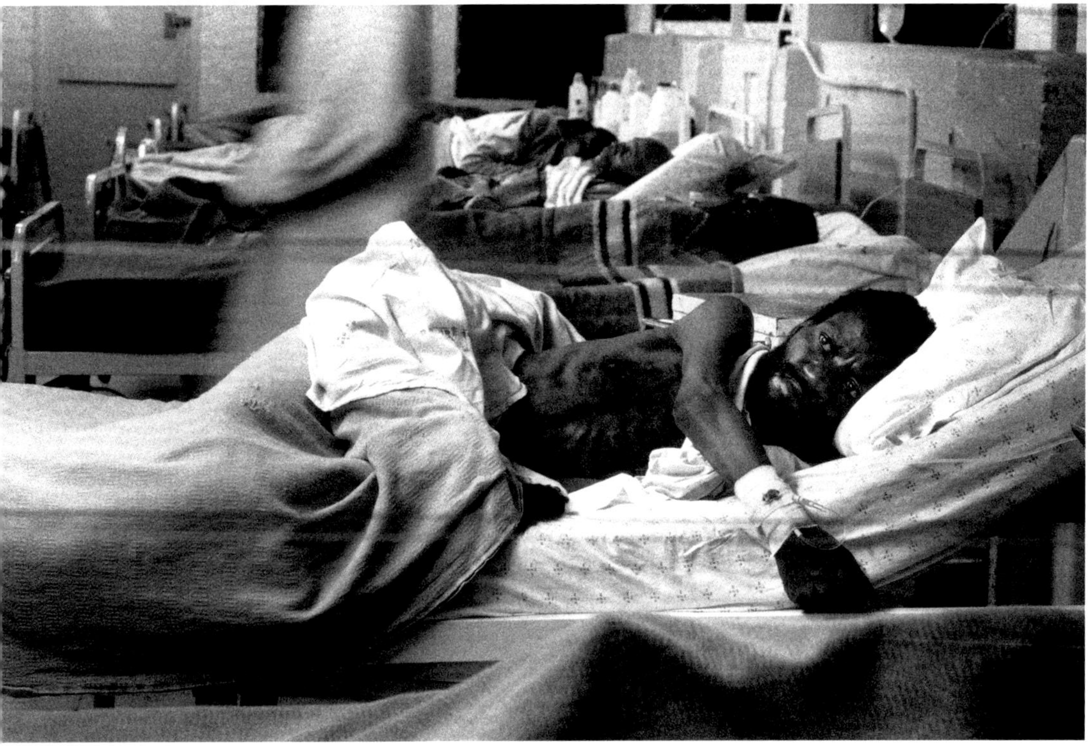

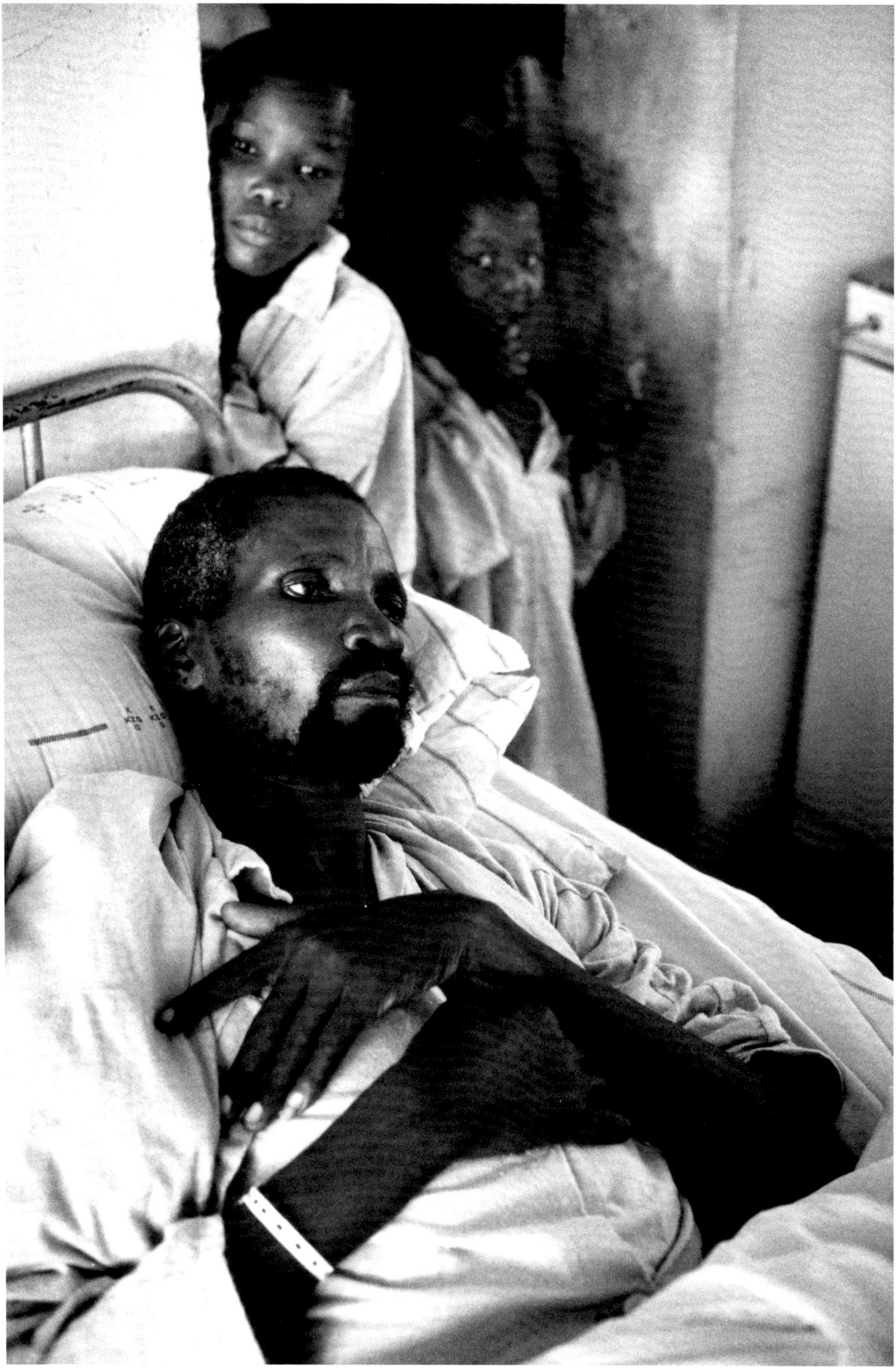

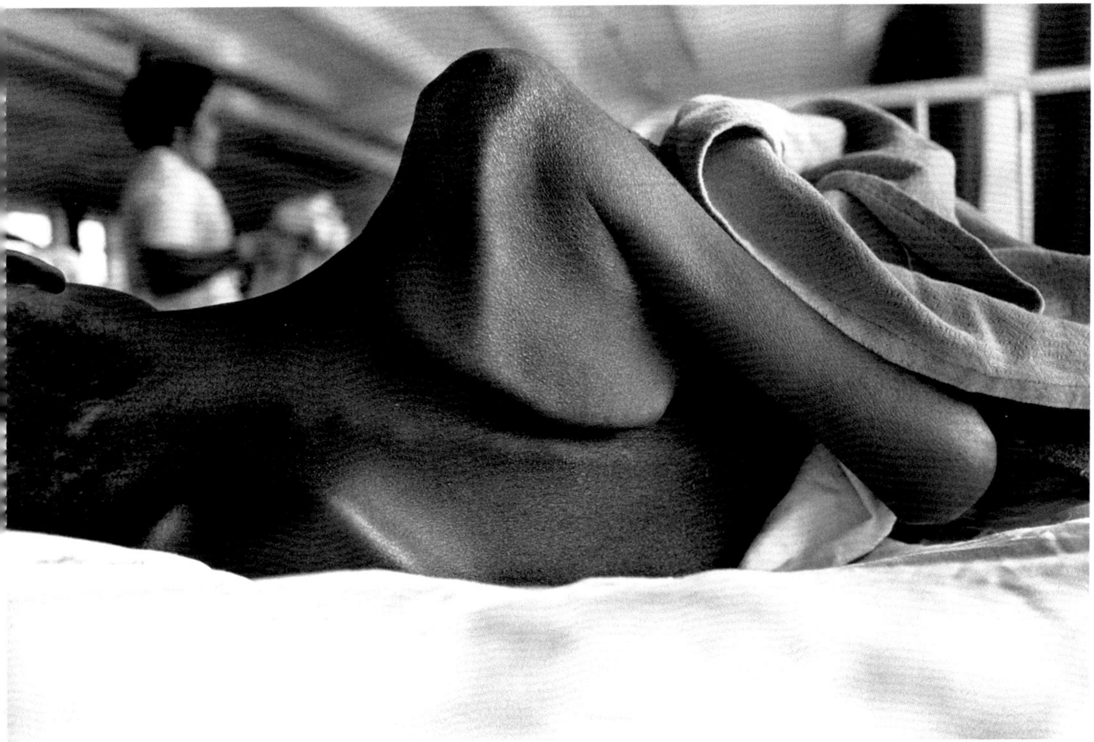

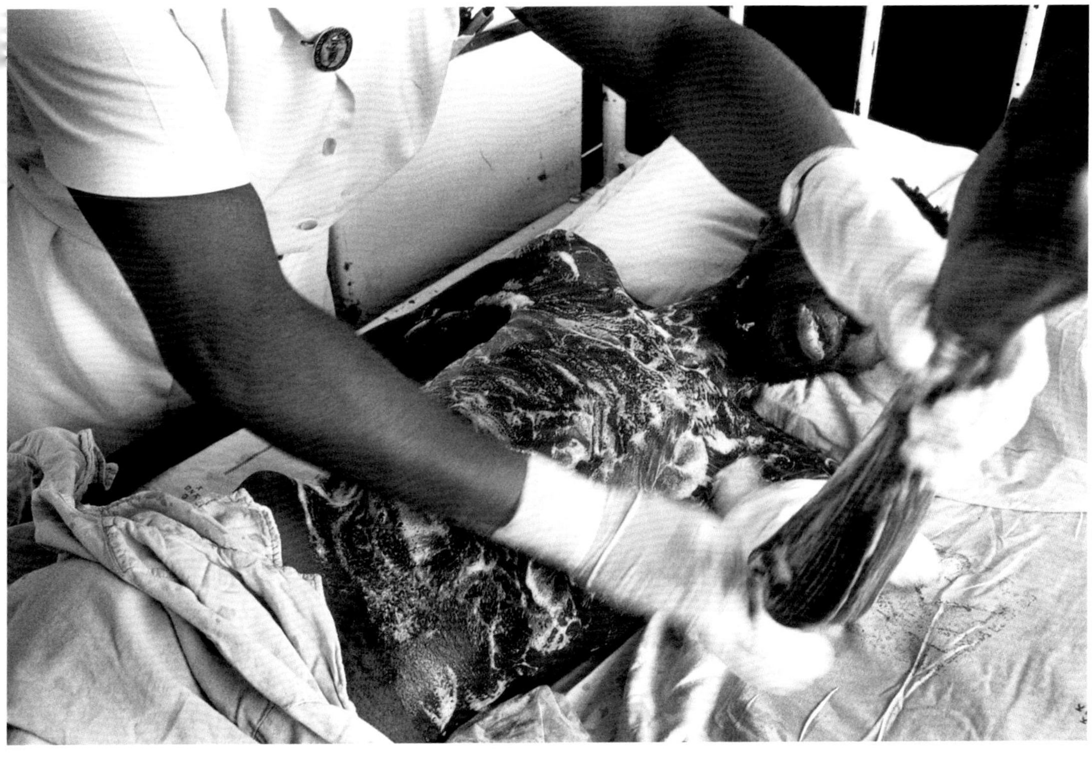

Hospital estatal. Nkhotakota, Malawi

Entre el 35 y el 40 % de las enfermedades de los pacientes del hospital están relacionadas con el sida.

Oficialmente tenemos camas para 110 pacientes, pero en la actualidad tenemos unos 130 internos y 250 externos cada día. Es difícil decir con exactitud cuántos han muerto de sida porque ya no tenemos reactivos para las pruebas y no hempos podido realizarlas. Tenemos cerca de una muerte por día.

No tenemos equipo, no tenemos personal, no tenemos medicinas. Ni siquiera tenemos esparadrapo, por lo que hemos de utilizar cinta adhesiva para sujetar las sondas de los sueros o entablillar los brazos de nuestros pacientes. Nos quedamos cortos en cualquier aspecto de esta epidemia.

DOCTOR MAURICE BONONGWE, DIRECTOR

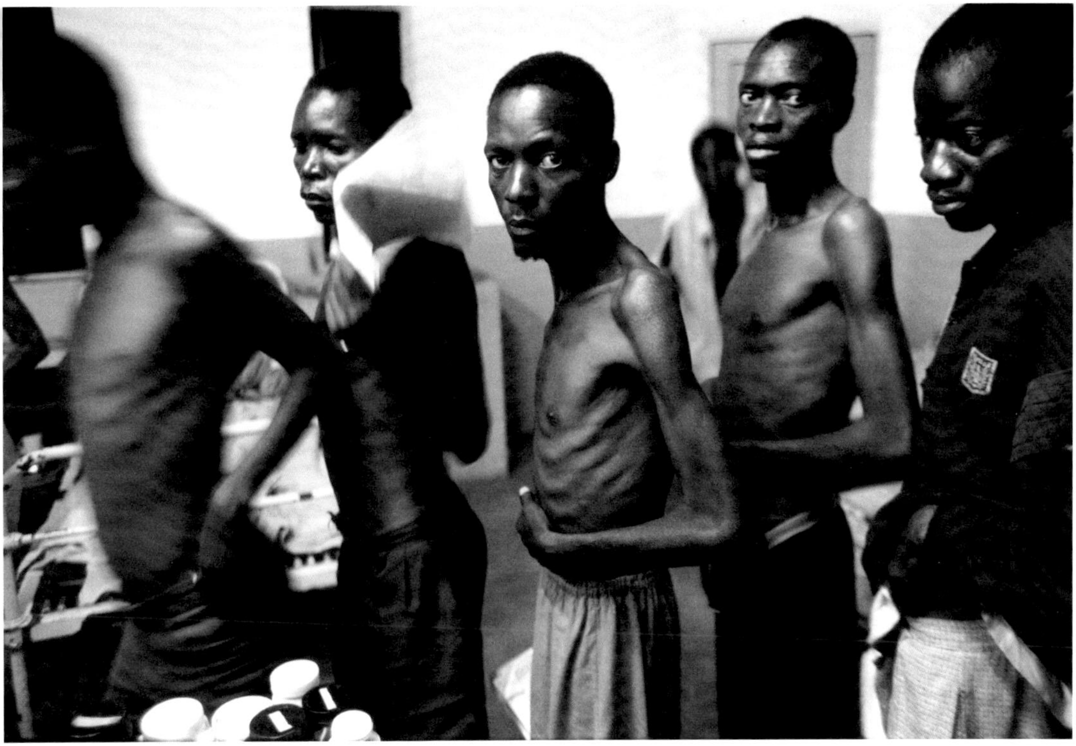

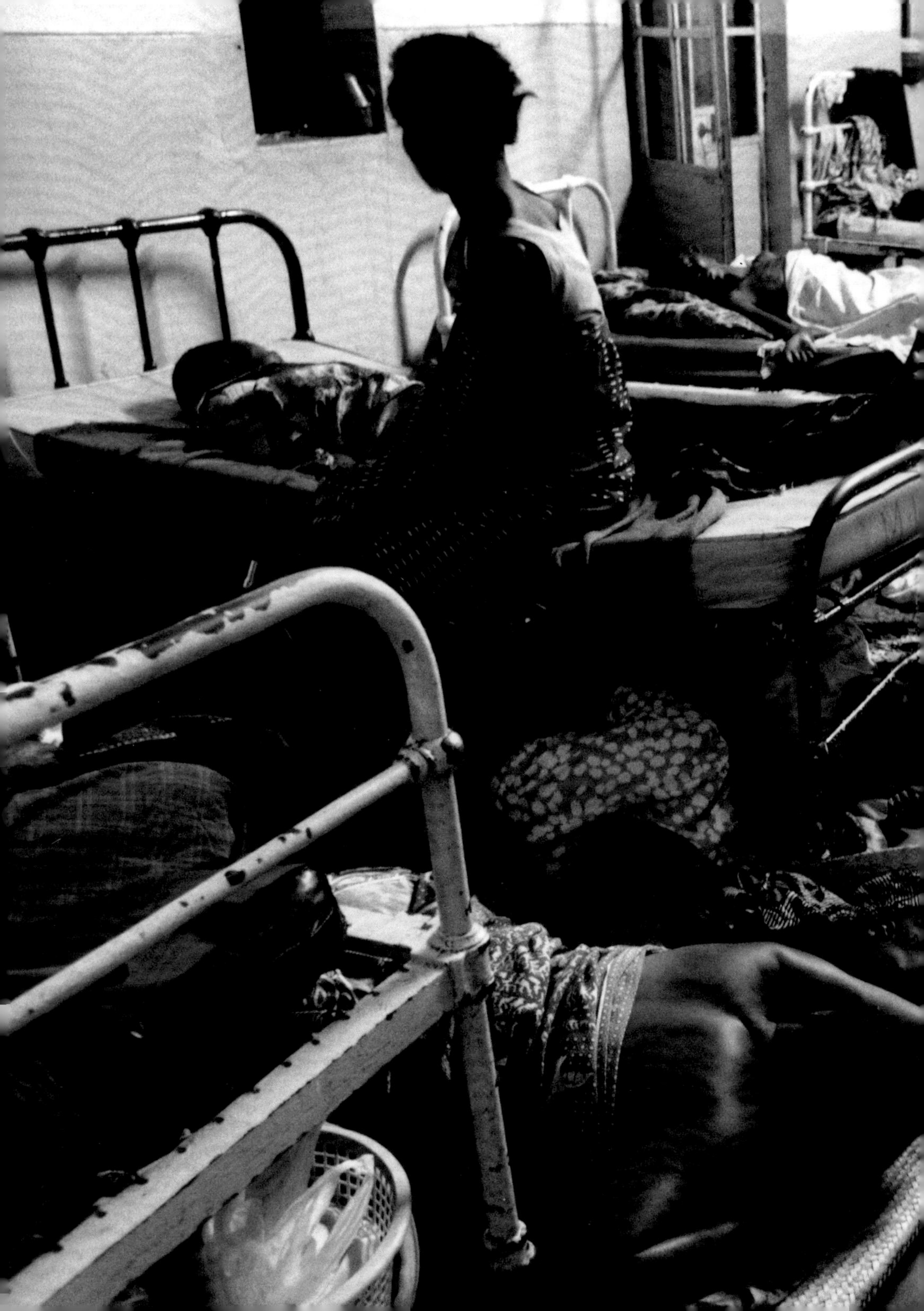

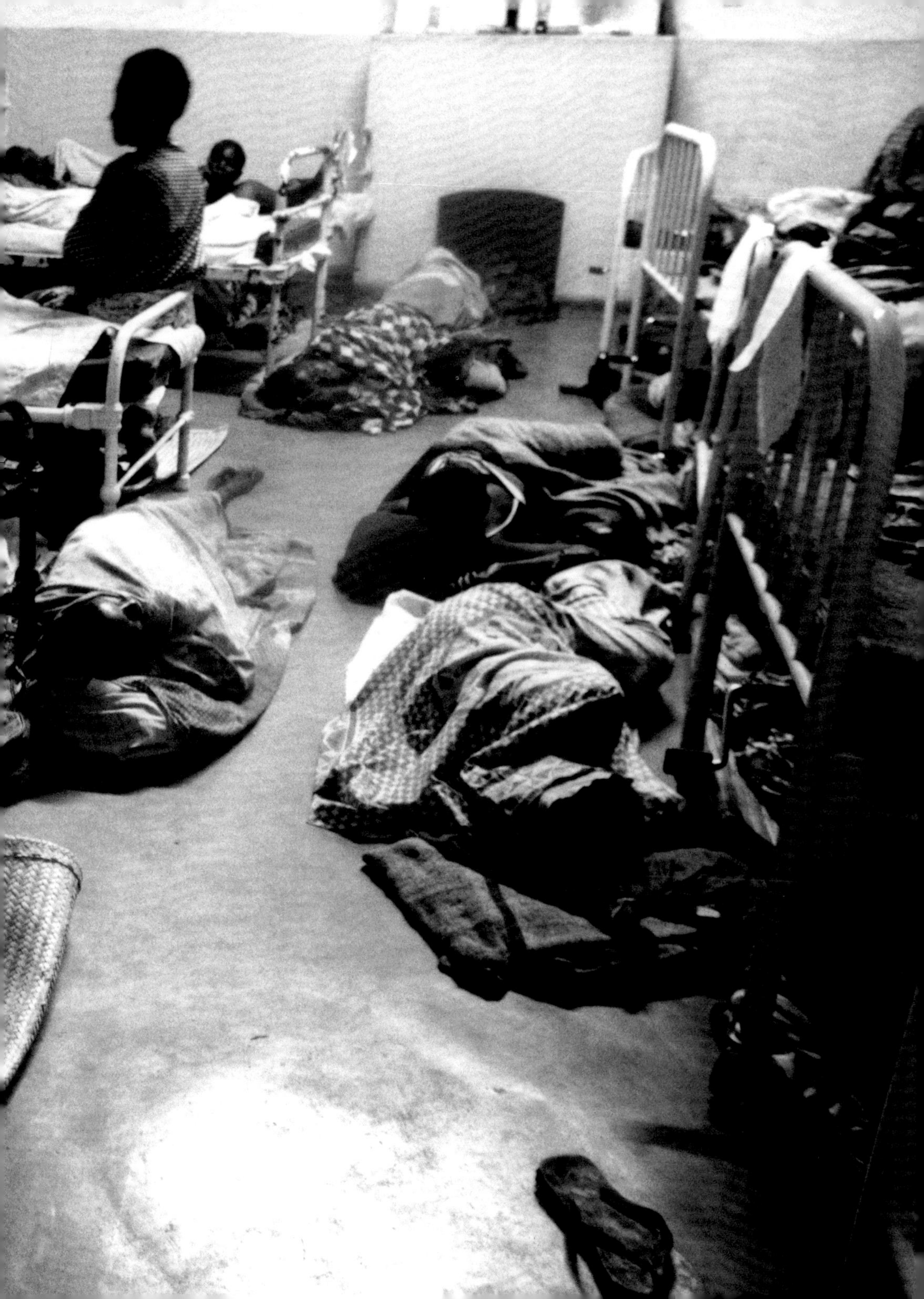

Eliza Myeni. Nkhotakota, Malawi

Eliza, una estudiante de 19 años, murió en el hospital de Nkhotakota en noviembre del año 2000.

Eliza era la hija de mi hermana menor Efrieda, que murió en 1994. Su padre, James, que era empleado en una tienda, murió el año pasado. Recibía formación en un colegio de Mzuzu, pero cuando enfermó fue a vivir a casa de su abuelo en Selemani Village, ya que no tenía padres que pudieran cuidar de ella. Se puso muy grave y estuvo en el hospital durante un mes. Me quedé allí con mis otras hermanas cuidándola lo mejor que pudimos. Vomitaba constantemente y perdió mucho peso, y sólo conseguíamos que comiera un poco de gachas de maíz. Tosía mucho y murió a la medianoche. Yo estaba con ella y sólo sujetó mi mano con fuerza. Dios la protegía y la llamó a su lado.

JANE CHIWRA

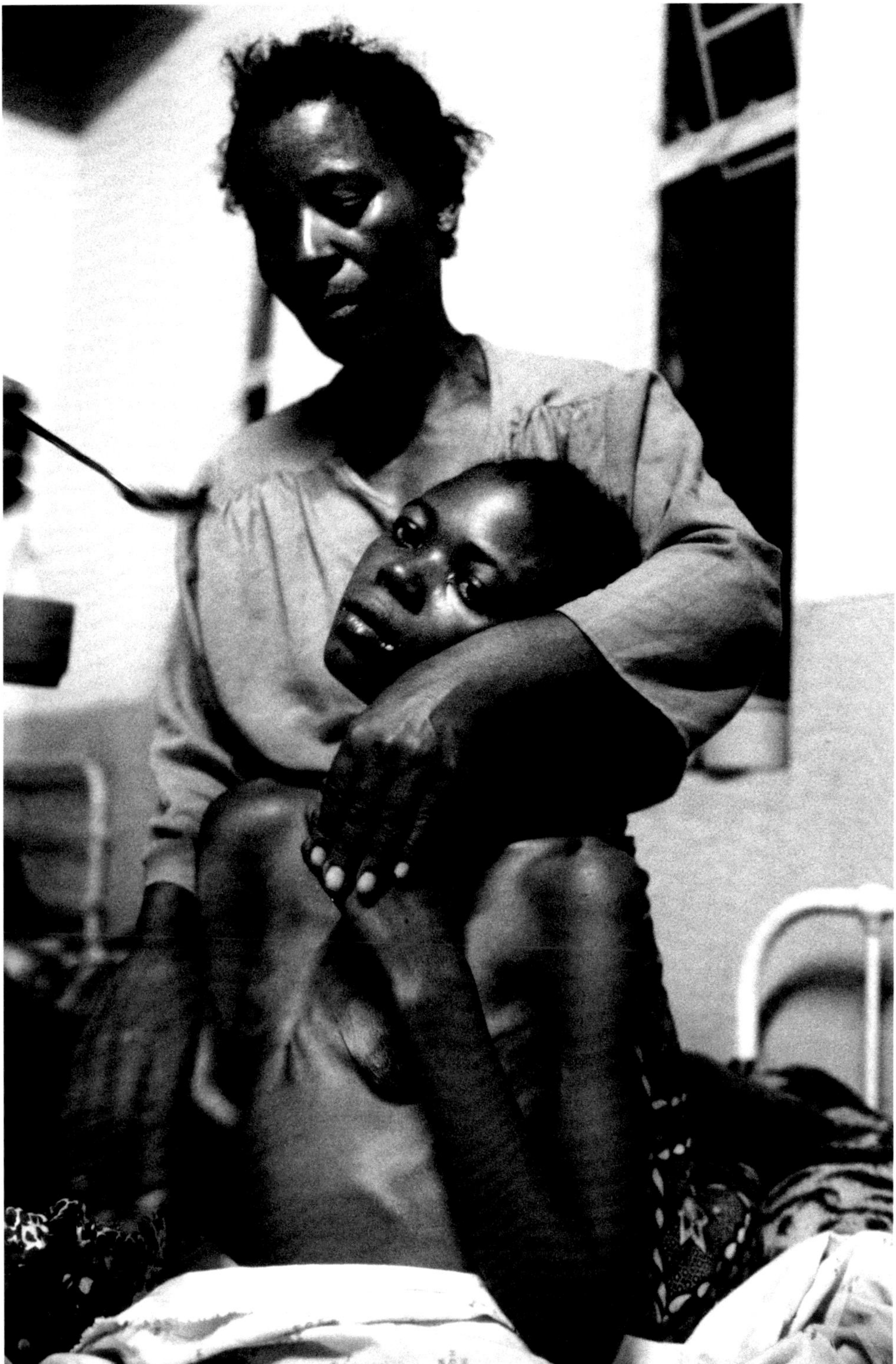

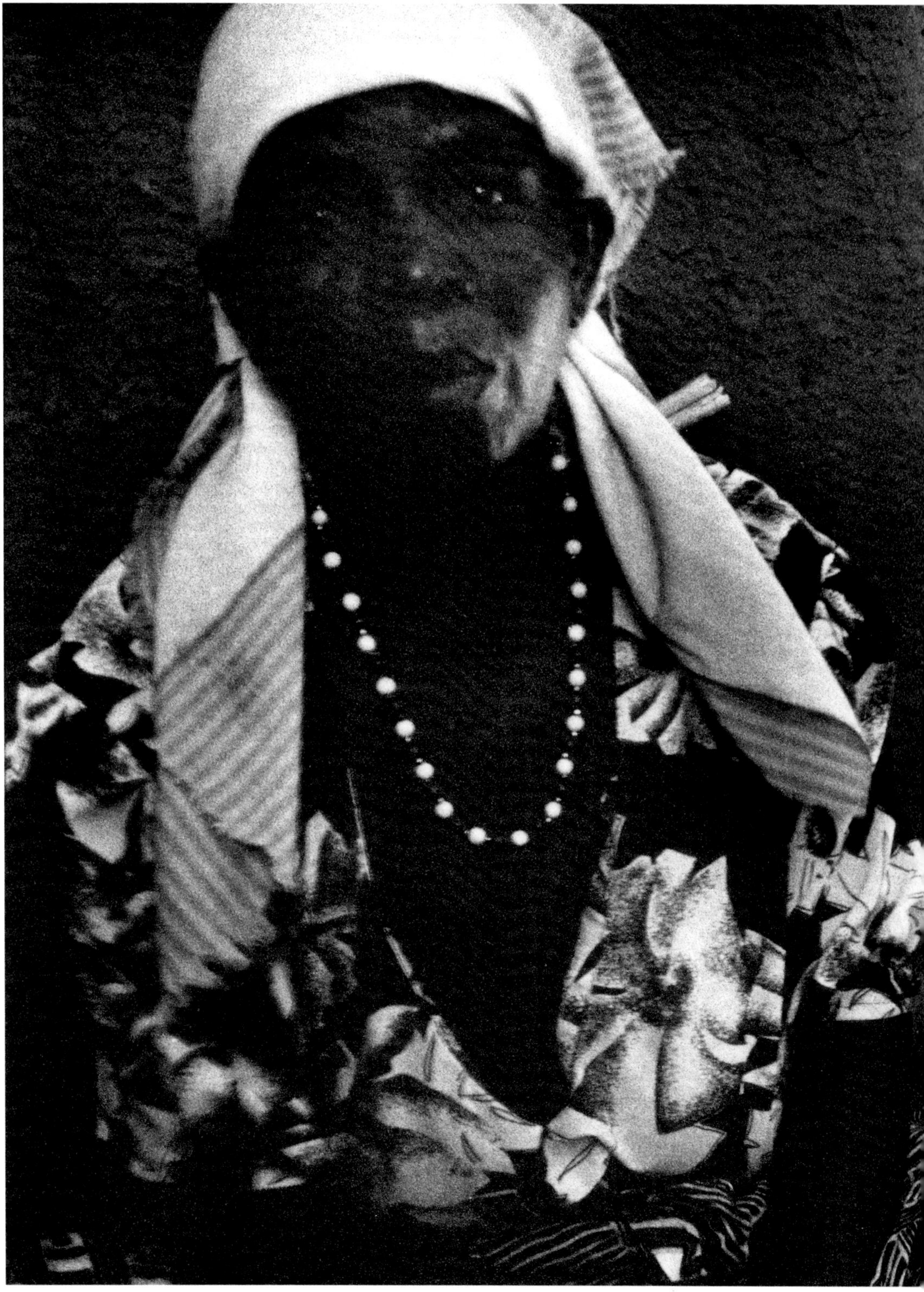

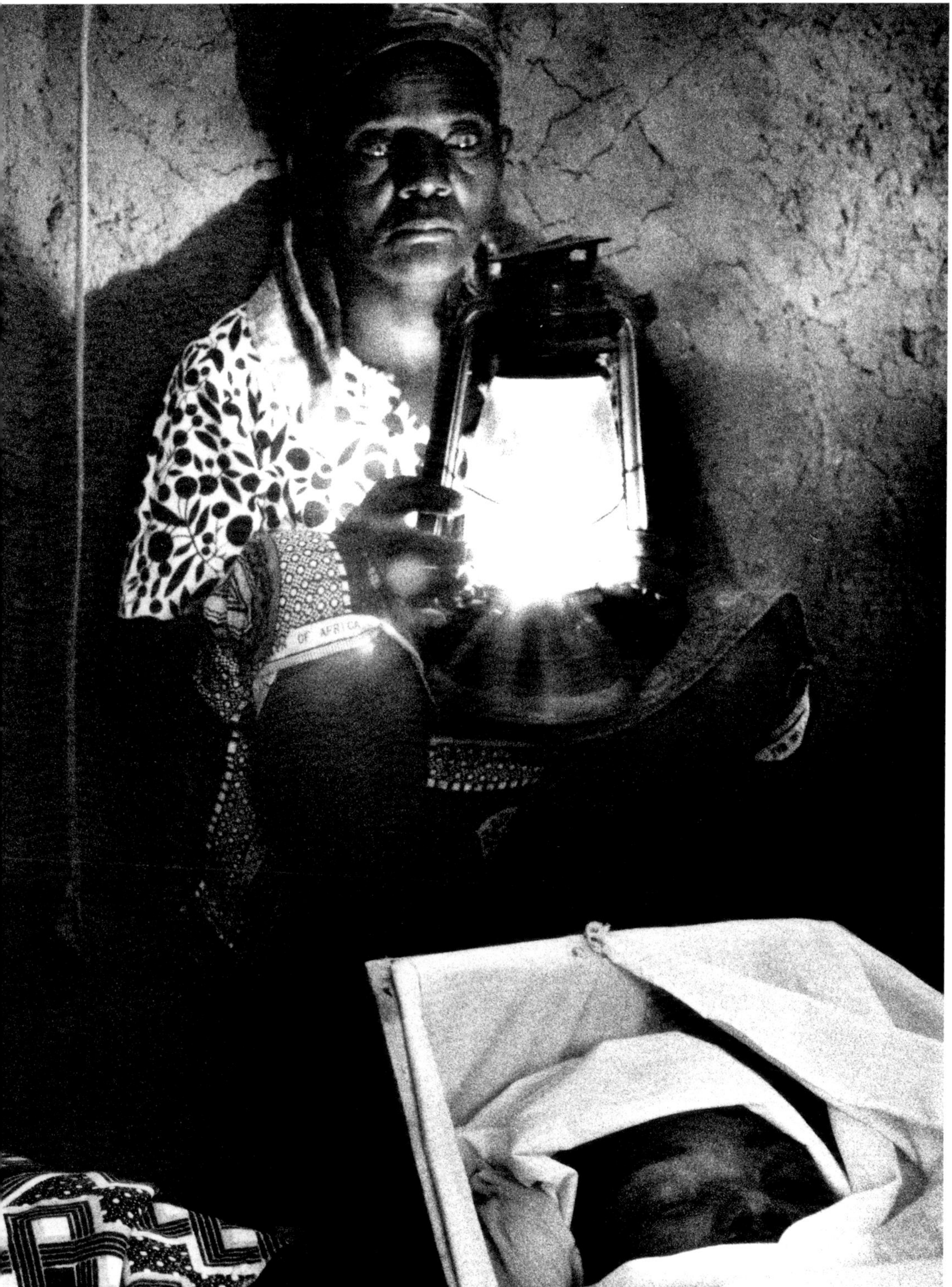
OF AFRICA

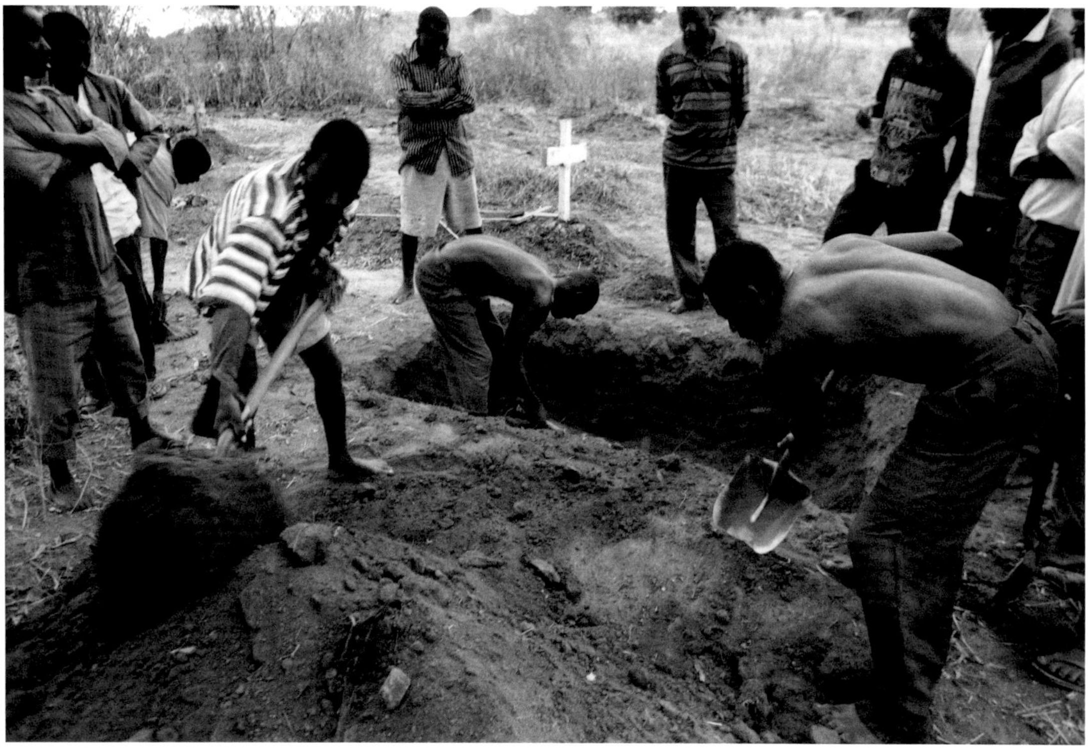

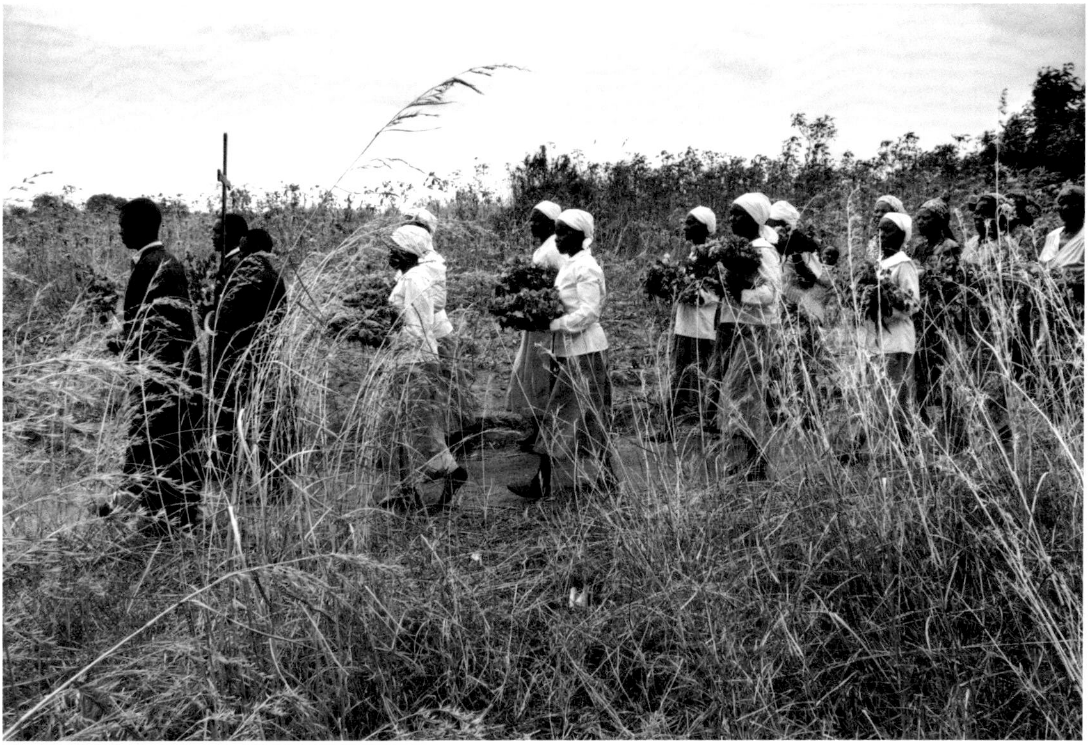

Miriam Mbwana. Nkhotakota, Malawi

Miriam cuidaba de su hija Mary. Ésta murió en noviembre del año 2000.

En mi vida he tenido 11 hijos, ocho chicas y tres chicos. Siete de ellos ya han muerto. El primero, Lawrence, murió en 1993, y desde entonces ha muerto uno de mis hijos cada año. Seis de mis nietos también han muerto. El sida se ha llevado a mi familia como una riada.

Cuido a 16 de los hijos de mis hijos. Mi nieta Madrin está en el hospital con su hijo John, y ambos están muy débiles. Ya ha perdido a tres hijos. Mi hija Mary está muy enferma. Estamos muy unidas. Ella es mi mejor amiga.

¿Qué hemos hecho para merecer esto? Mi padre solía decir: «Cuando llegue la muerte, pasa por el otro lado». Pero eso ahora no es posible. La muerte está en todas partes.

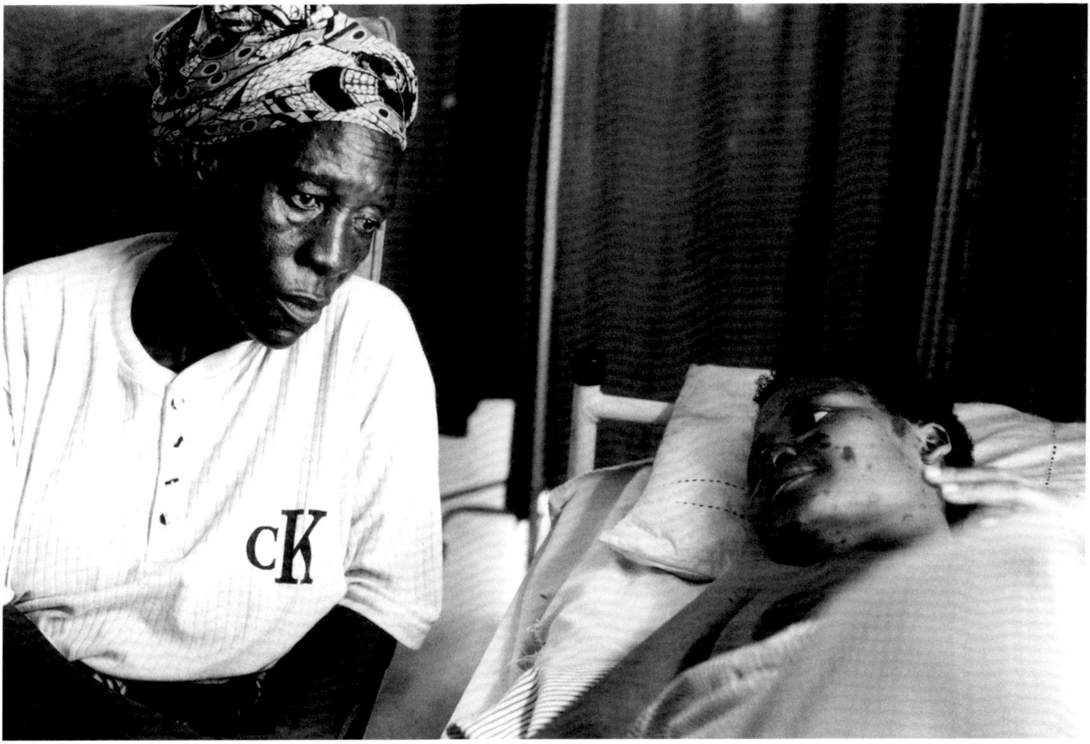
cK

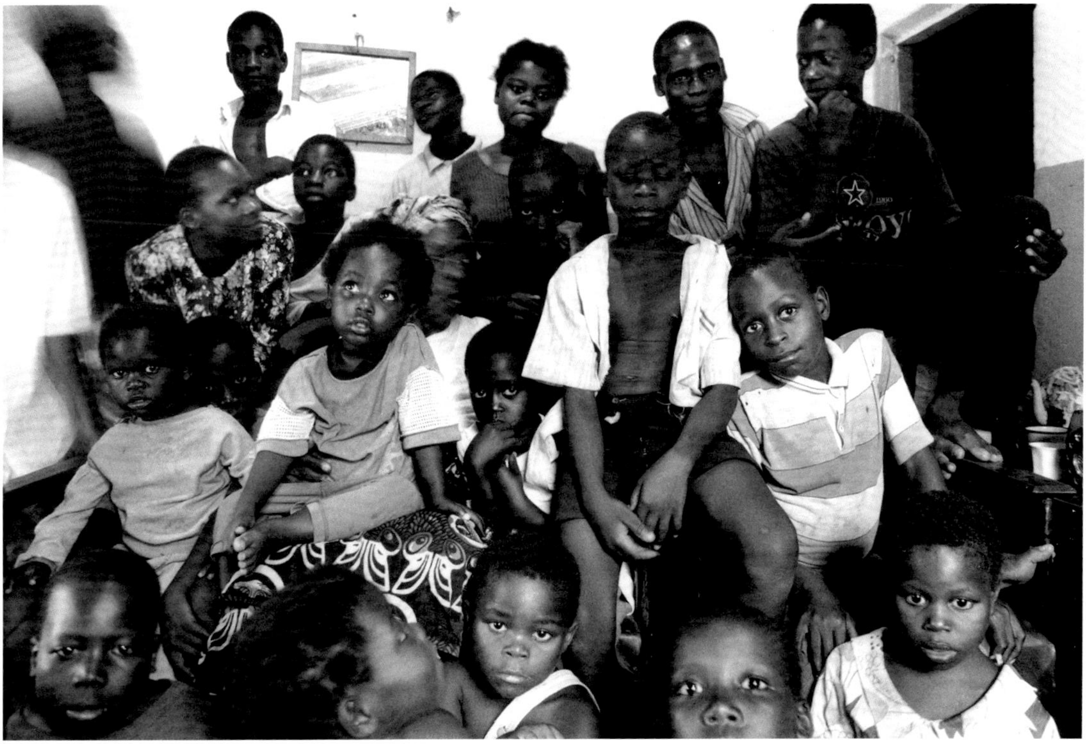

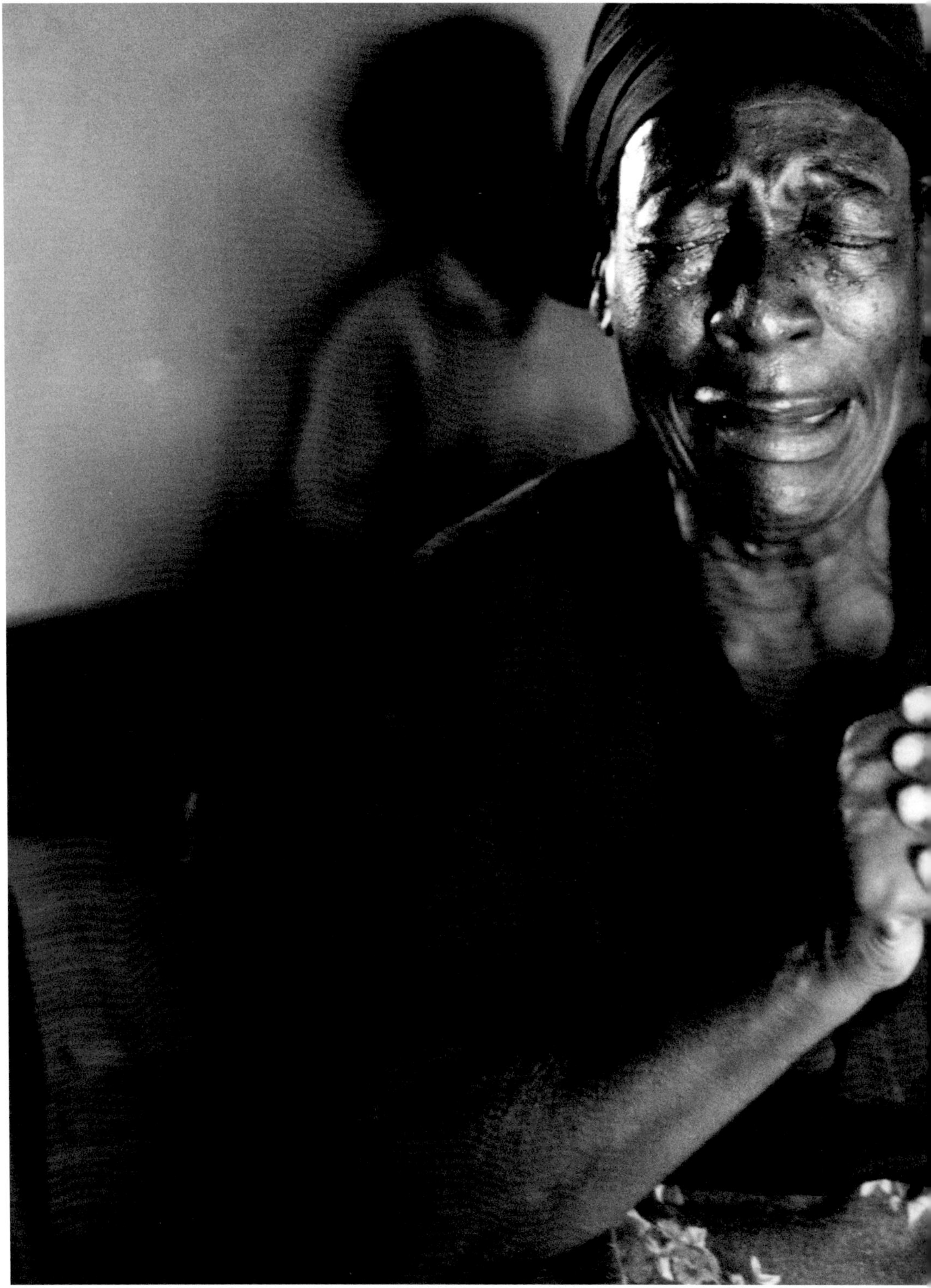

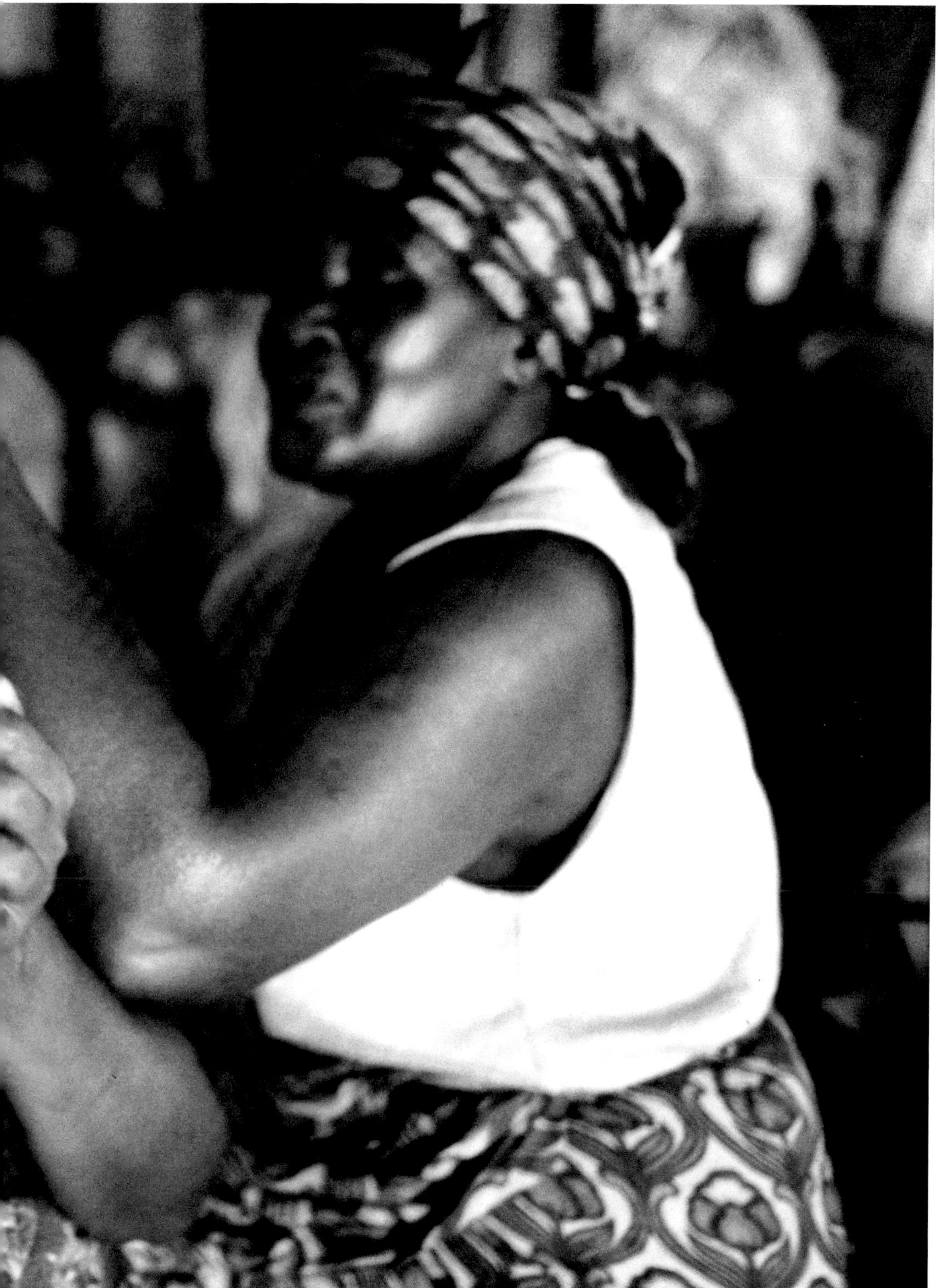

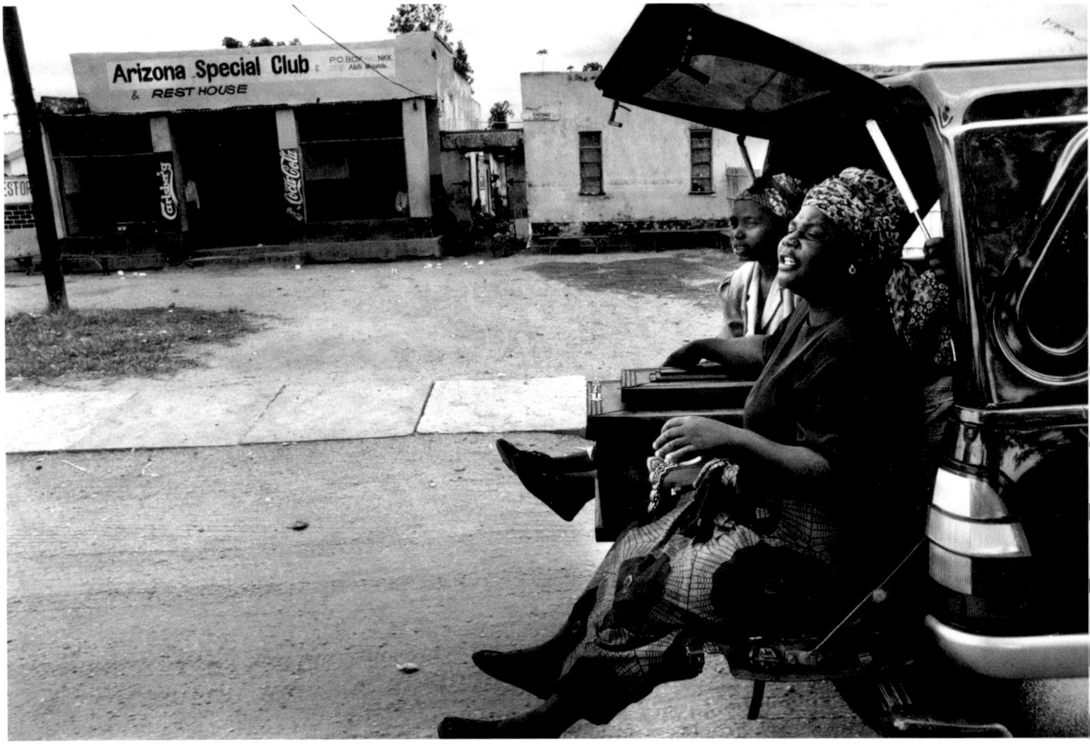
Arizona Special Club
& REST HOUSE
Carlsberg
Coca-Cola

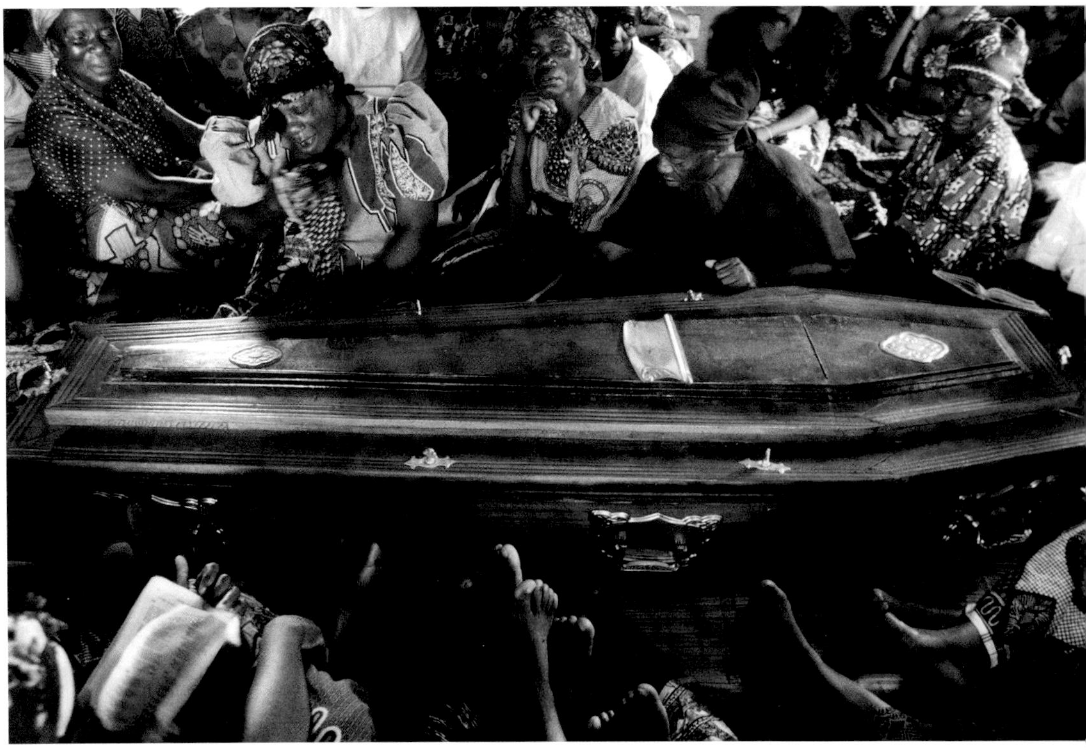

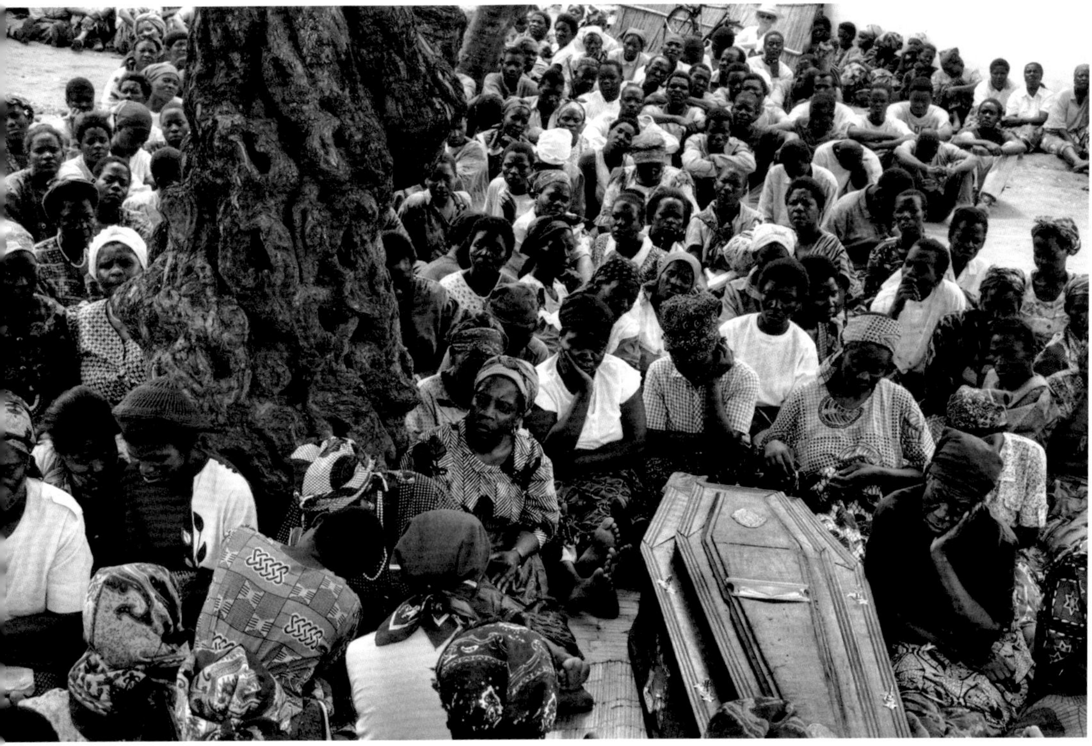

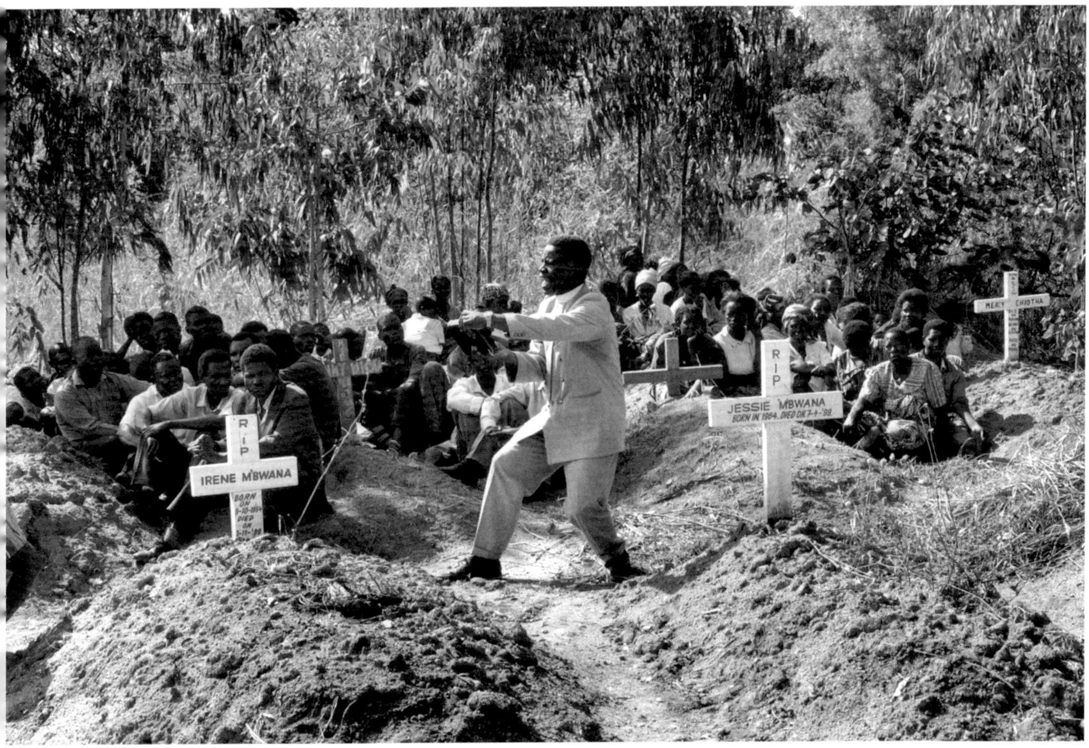
RIP
IRENE M'BWANA
BORN ON
DIED ON
RIP
JESSIE M'BWANA
BORN IN 1964. DIED ON 7-4-'99
MERCY CHIOTHA

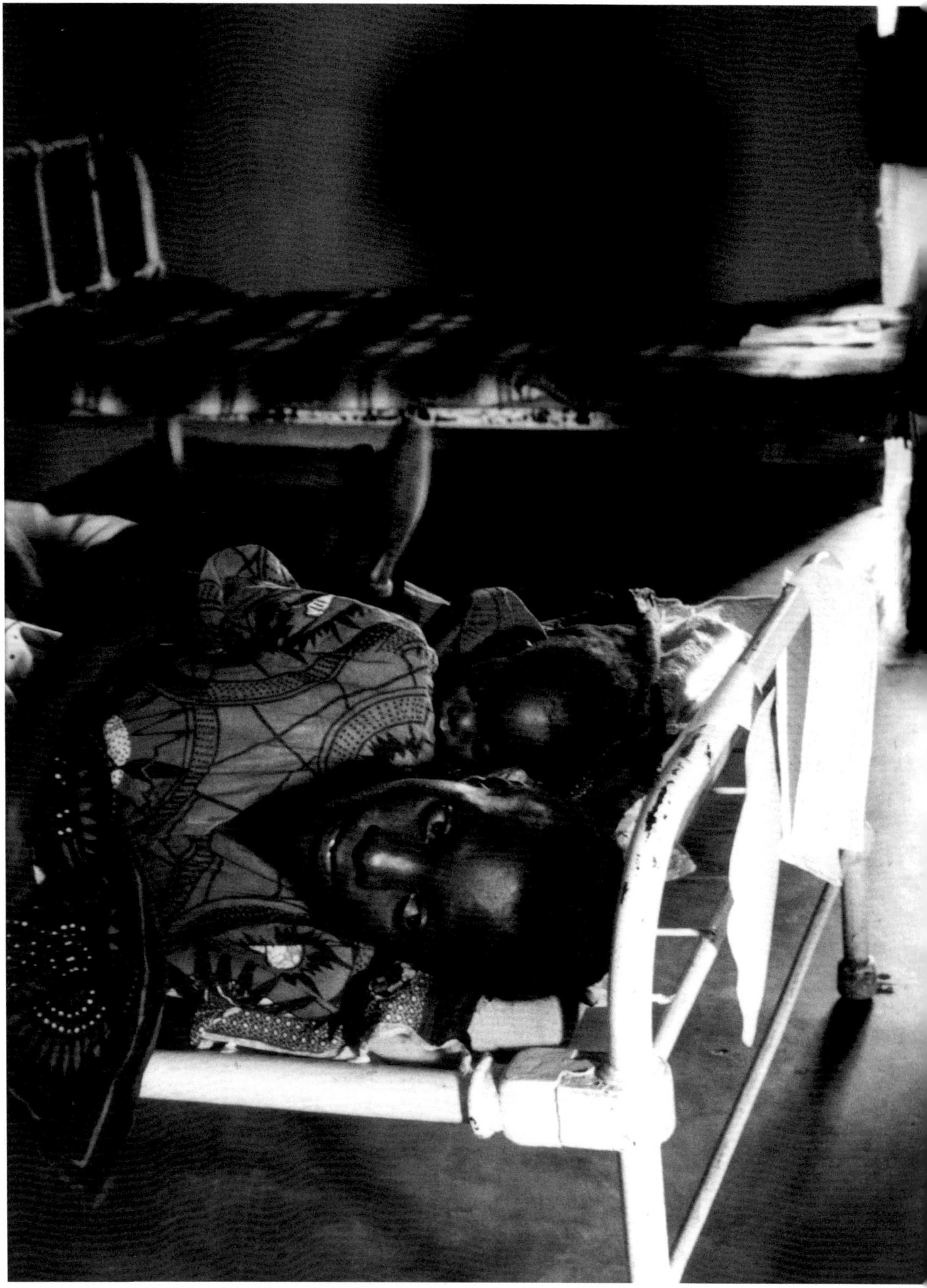

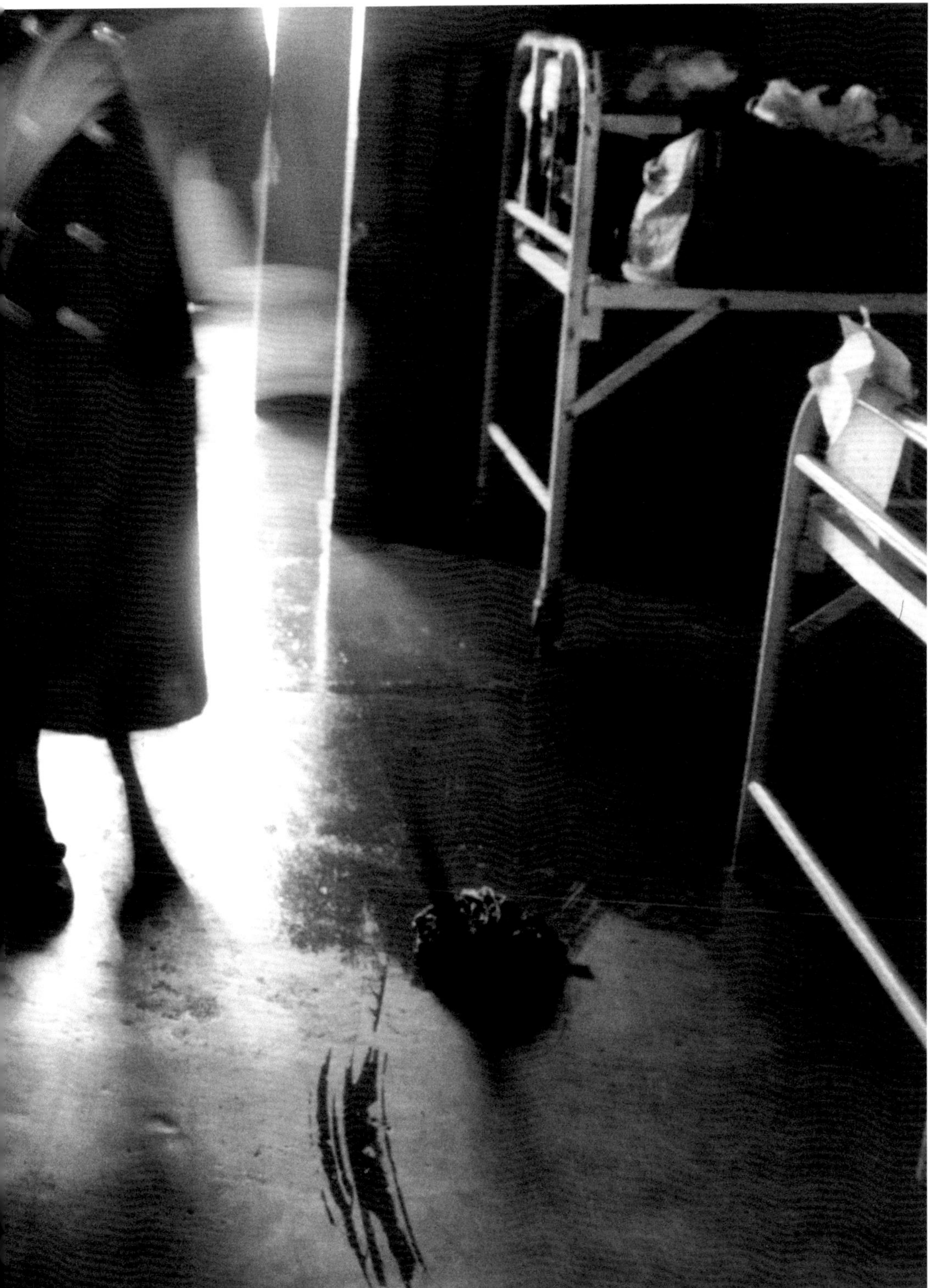

Mzokhona Malevu. Municipio de Enseleni, Sudáfrica

La familia de Mzokhona vive en un campo de refugiados, después de perder su casa durante los enfrentamientos políticos de 1992.

He estado enfermo durante mucho tiempo, pero hasta 1998 no descubrí que era seropositivo. El primer refuerzo vino de mi asesora, que me preguntó: «¿Con quién compartirás esta información?». Primero pensé que únicamente debería decírselo a mi madre, pero luego me dí cuenta de que esto supondría una gran frustración para mí. Decidí compartirlo con toda la comunidad.

El domingo pasado celebramos un servicio religioso y encendimos velas por aquellos que han muerto y por los que viven con el VIH o el sida, así como por los huérfanos y sus cuidadores. Cientos de personas estaban allí y por primera vez hablé en público. Me presenté a todas las personas y le expliqué a los jóvenes que deben protegerse. Yo me contagié con el VIH por no tener cuidado. La gente me aceptó después de lo que dije. Las personas deben recibir la información, y yo no tengo miedo de hablar en mi comunidad.

Espero que mi muerte sirva para ayudar a educar a mi familia y a mi comunidad. Mi funeral será un mensaje educativo sobre el sida: la información se extenderá a lo largo y ancho de mi entorno.

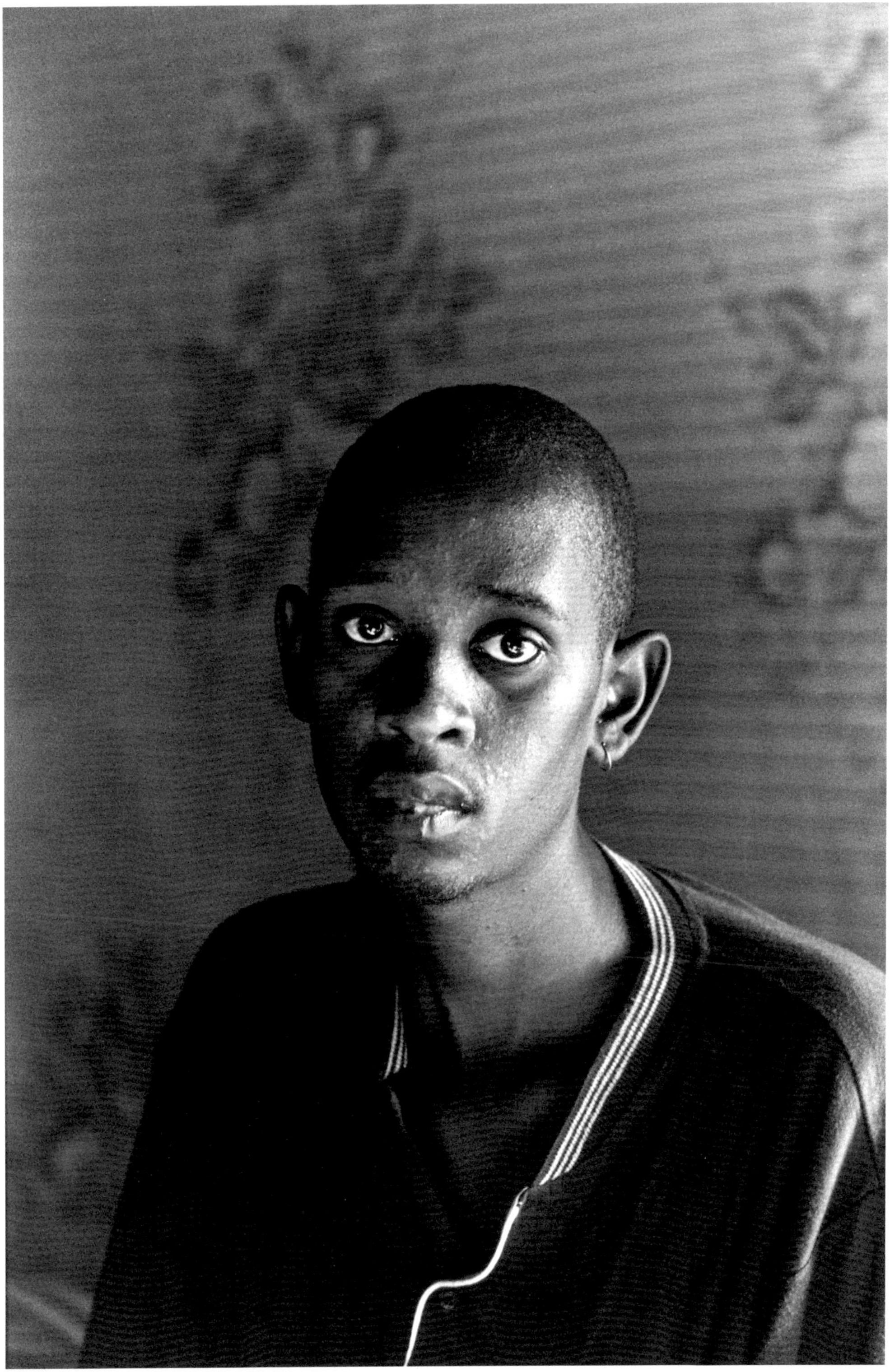

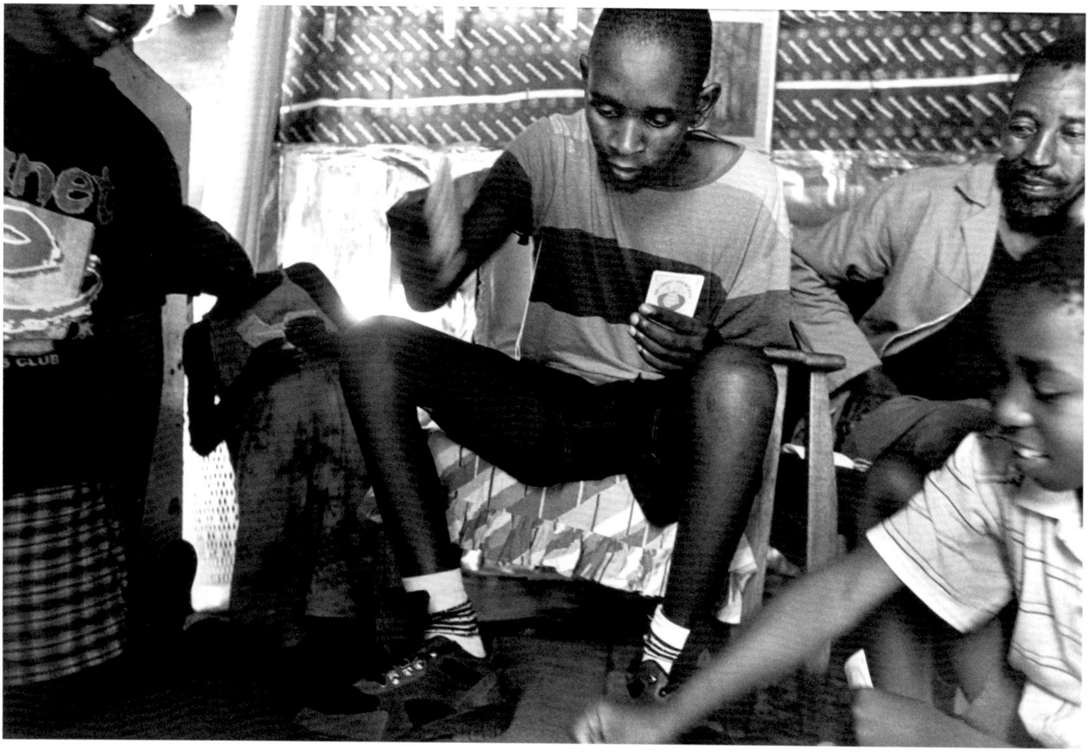

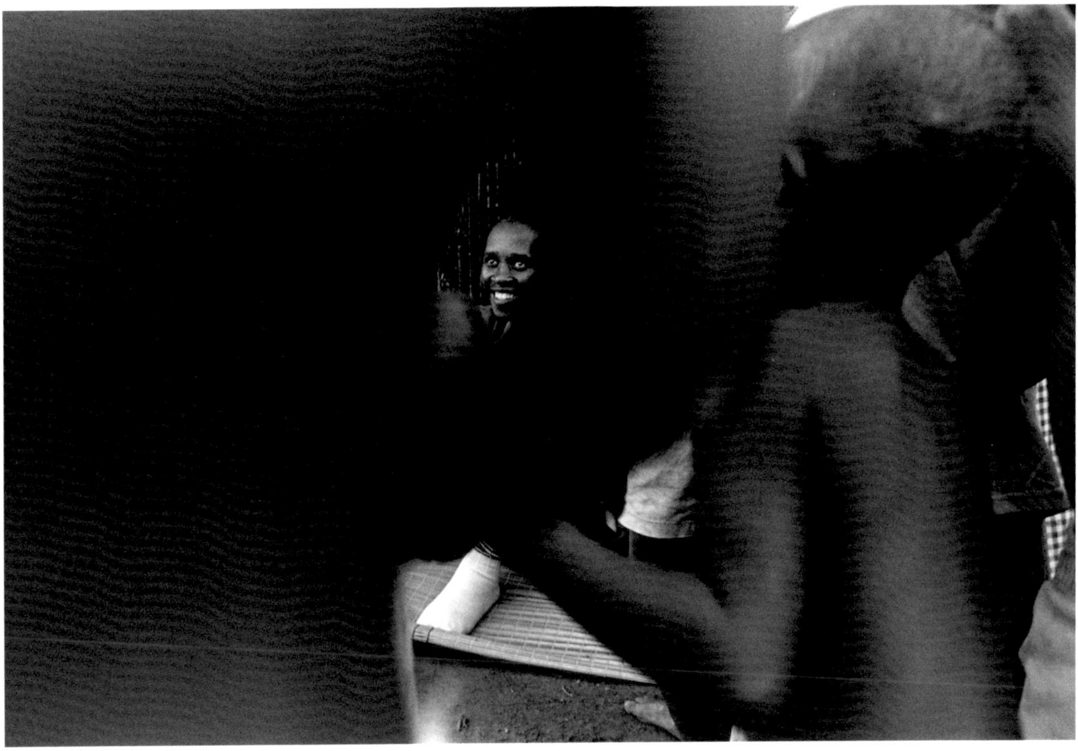

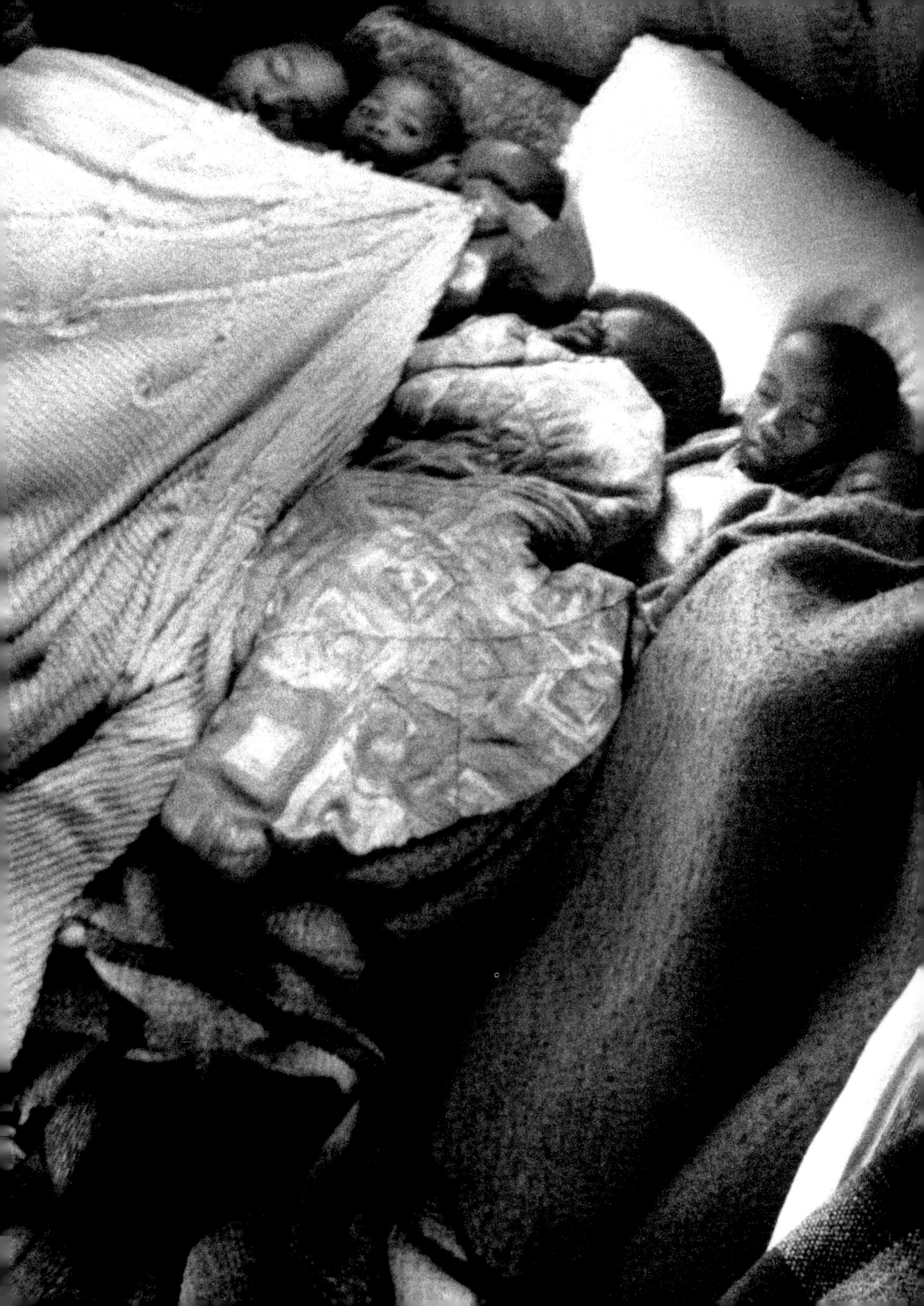

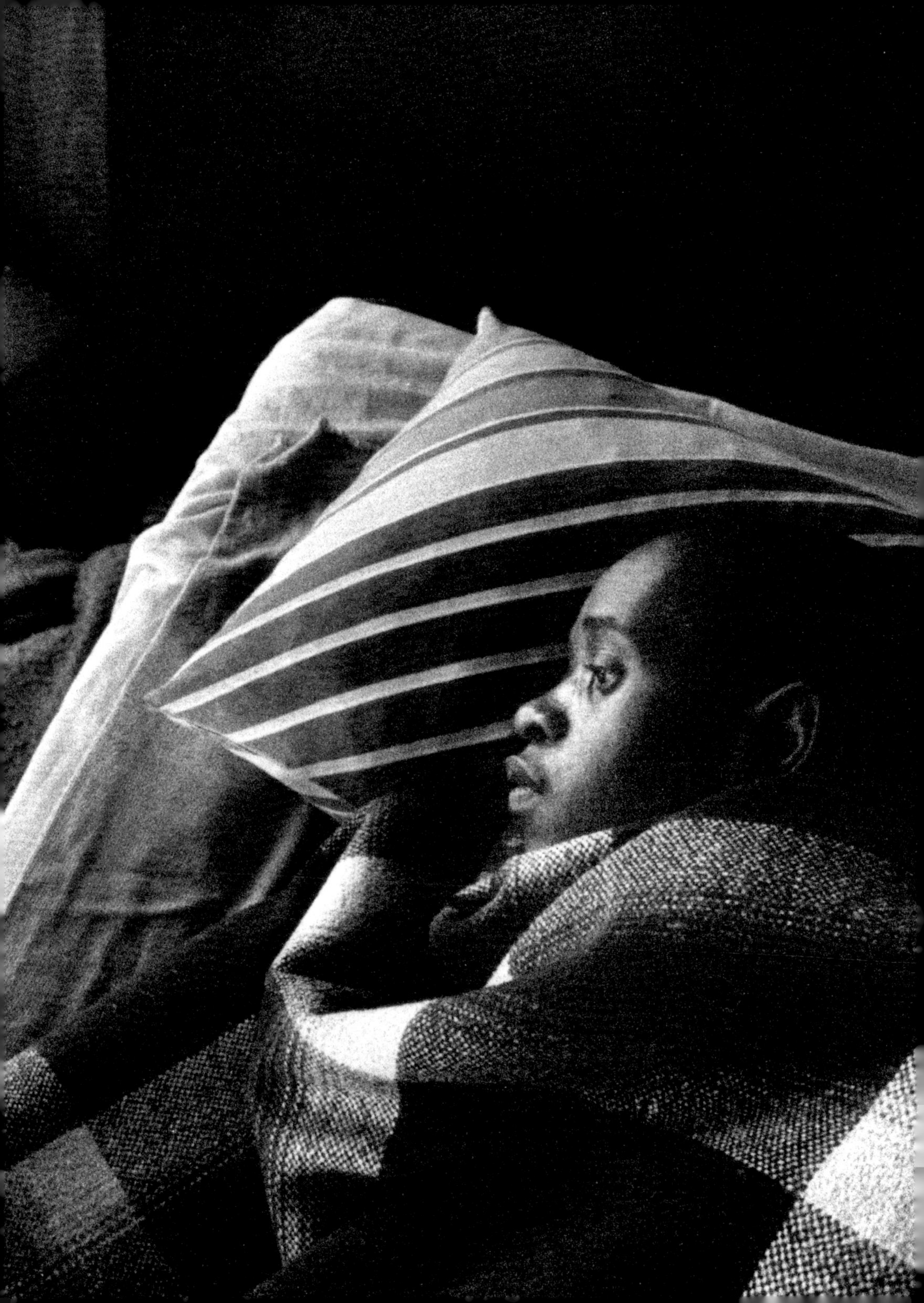

Mi padre tiene empleos temporales en la construcción y mi madre trabaja como empleada doméstica de unos blancos en Richard's Bay. Vivimos aquí en dos habitaciones. Somos mi madre, mi padre, sus ocho hijos y once nietos, veintiuna personas en total. Duermo en la cama aquí, en esta habitación, y siete niños duermen en el suelo a mi lado. A veces no tenemos suficiente comida para todos. Mi padre suele bañarme temprano por la mañana antes de ir a trabajar, pero desde hace poco tenemos que pagar por el agua que vamos a buscar en la fuente comunitaria. Vale 60 céntimos por cubo. Ahora incluso el agua para lavarse puede convertirse en un problema.

Me dieron unos medicamentos que me hacían sentir mucho mejor, pero ahora ya no puedo pagarlos. He oído que en otros países el gobierno da medicación y alimentación gratuitas a los enfermos de sida, pero aquí en Sudáfrica no es así. En la clínica dicen que no pueden hacer nada. Uno se ha de ir a casa. No es justo. La gente en el extranjero se recupera con los medicamentos que recibe; en Sudáfrica tenemos que morir.

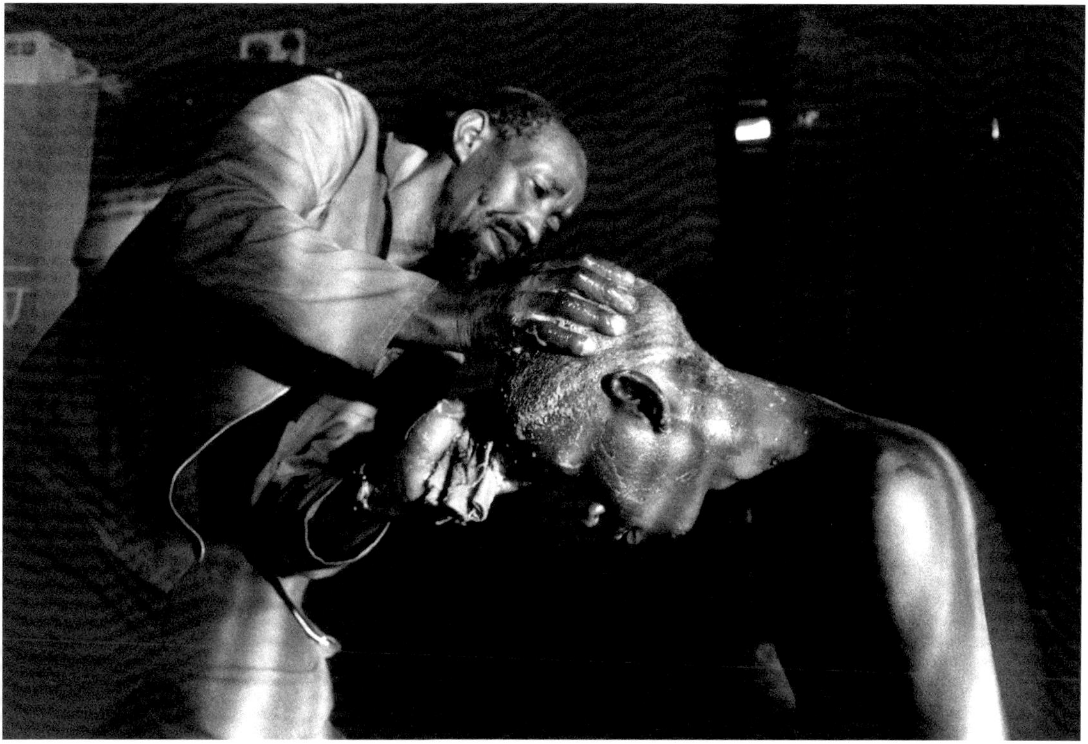

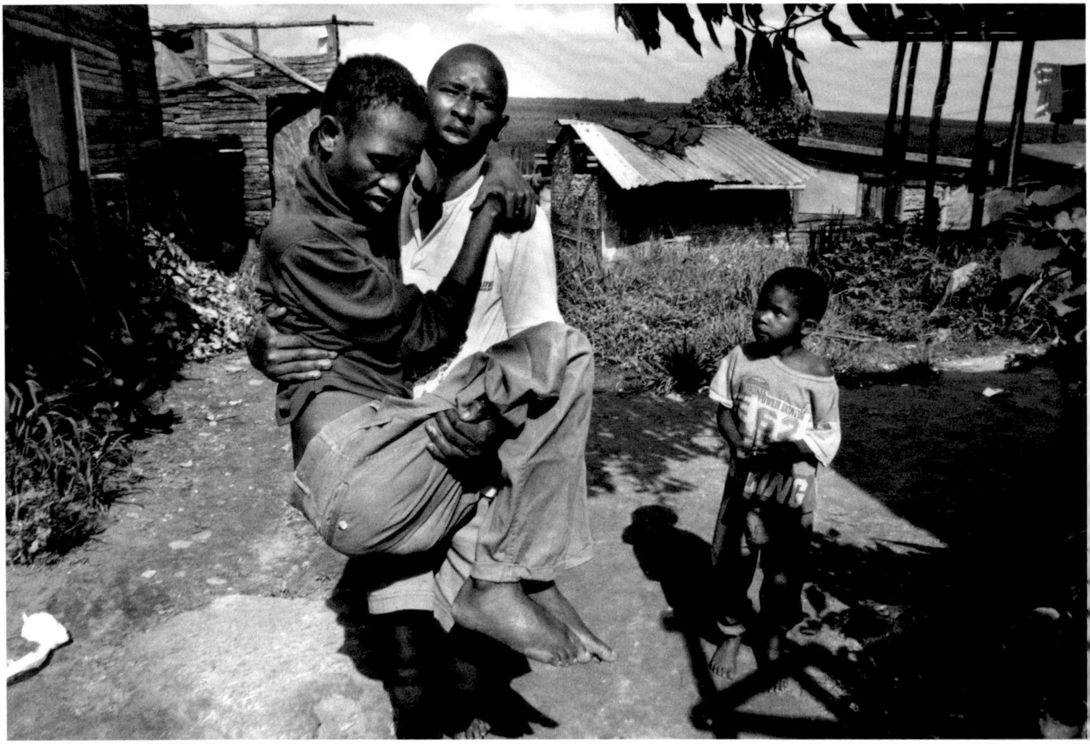

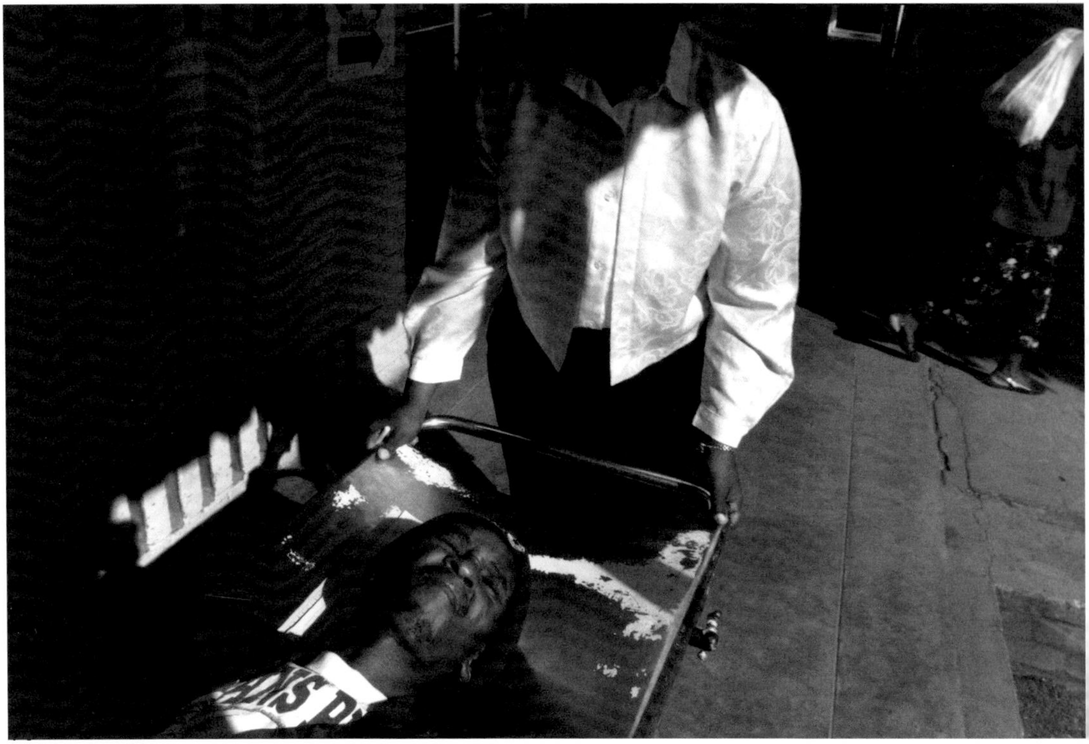

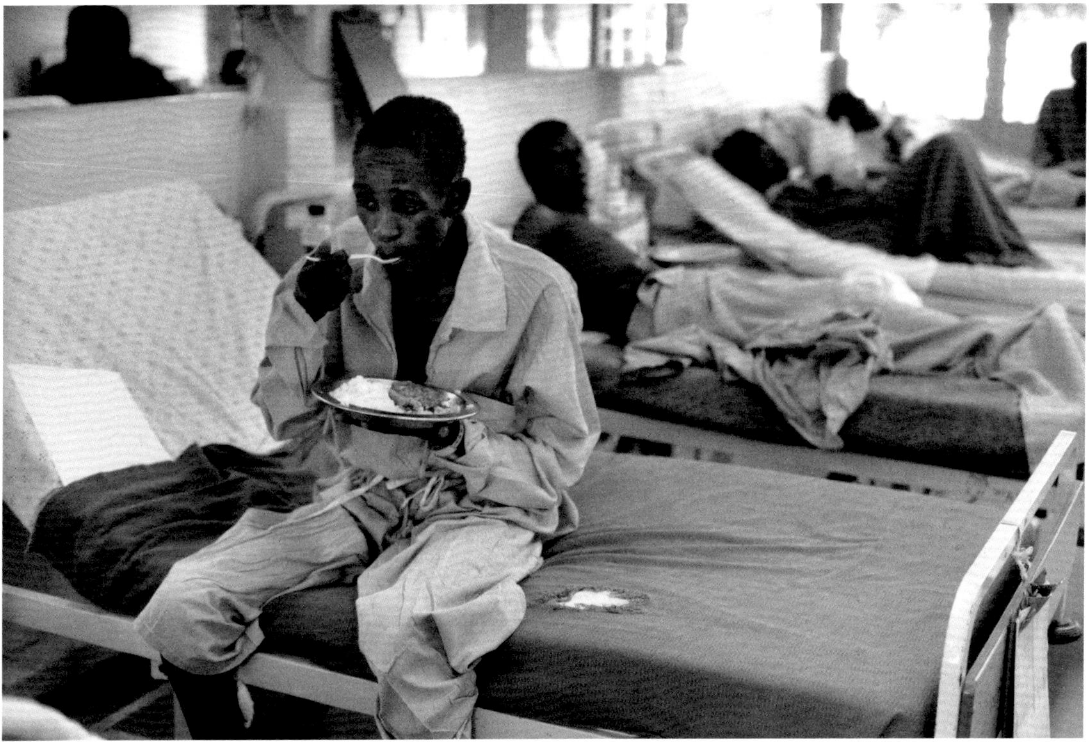

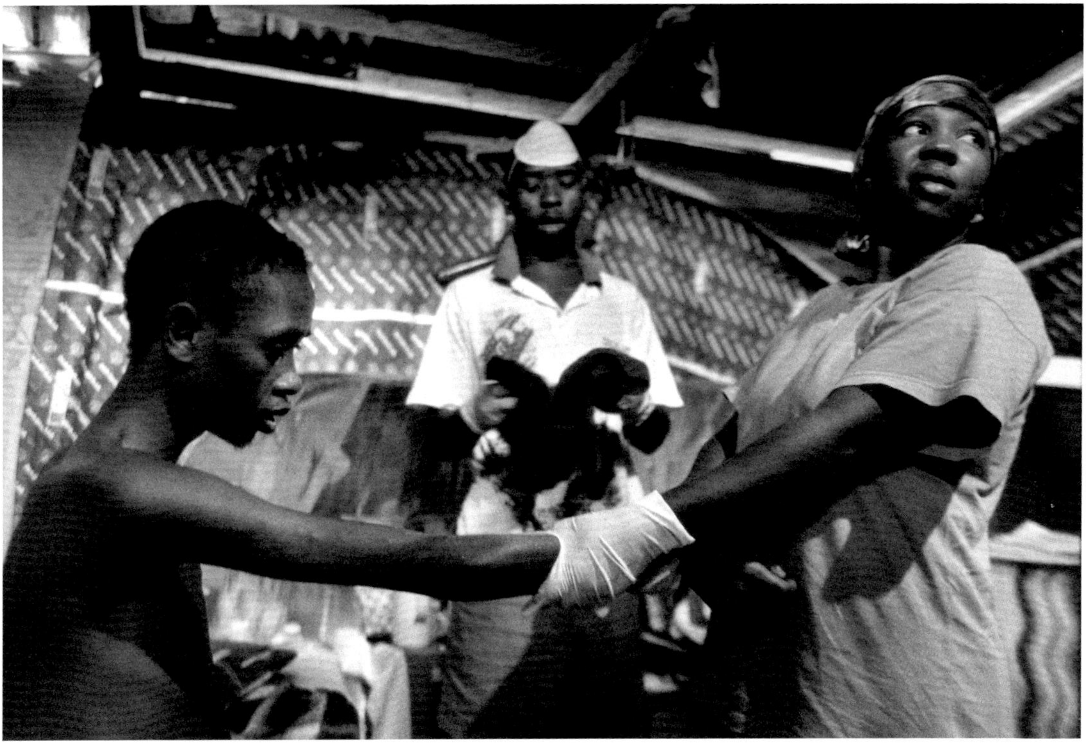

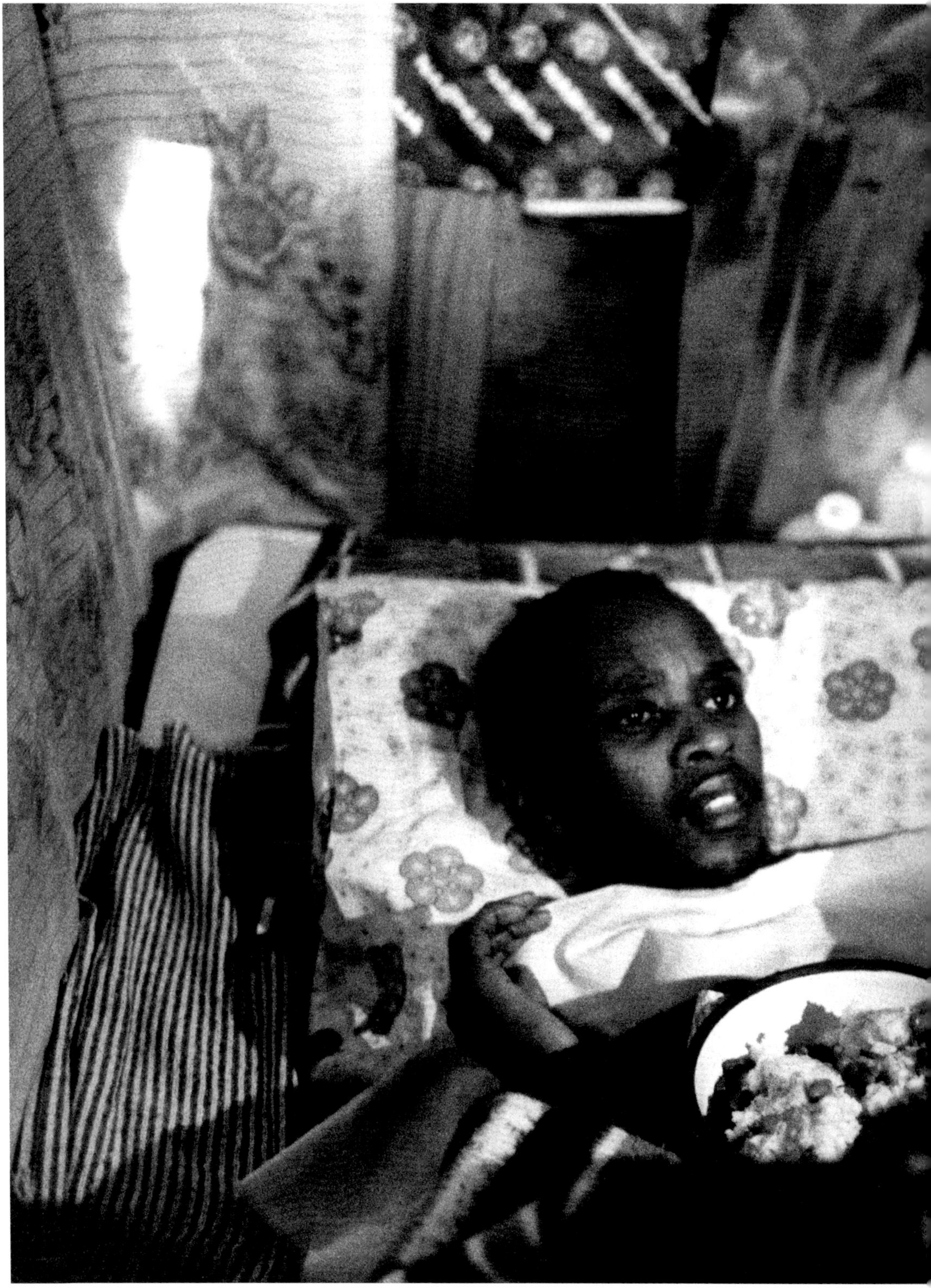

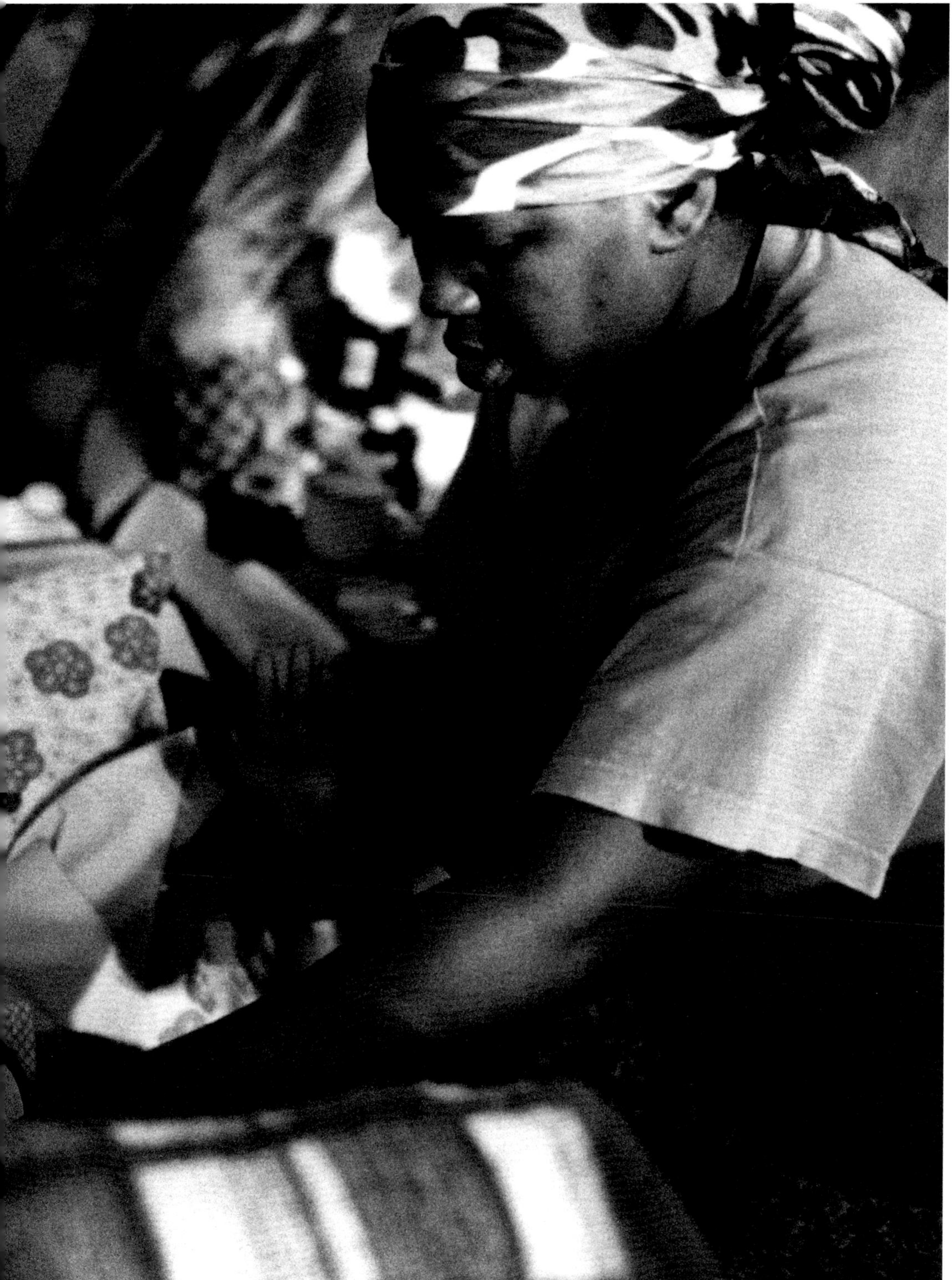

Mzokhona falleció el 2 de septiembre de 2000 a los 29 años. Murió en casa. Su padre entró en la habitación por la mañana y dijo que estaba lleno de llagas y decidió llevarlo a la clínica. Por la tarde decidí ir a su casa y ver cómo estaba. En los últimos meses de vida se había quedado ciego; hubo una época en la que no dejaba de gritar. Desde el miércoles había dejado de comer y no quería beber. Su familia tenía que obligarle. Él sabía que estaba a punto de morir. El sábado llamó a la hermana y al hermano que le habían cuidado con mayor dedicación y les dijo de todo corazón: «Muchas, muchas gracias». Cuando le ví en su habitación, parecía dormir, pero su hermana me dijo que ya se había ido.

El cuerpo fue recogido esa noche, cerca de las 20.00 horas. Todos los hijos y los hermanos, y todos los que vivían en esa casa, estaban muy tristes, y lloraban o sollozaban. A veces hablaba en las iglesias. Pedimos prestada una silla de ruedas para que pudiera ir a la iglesia a hablar. Le ví enfermar cada vez más, hasta el final. Él era una de mis personas favoritas.

NANCY KHUSWAYO, ASESORA DE ENFERMOS DE SIDA

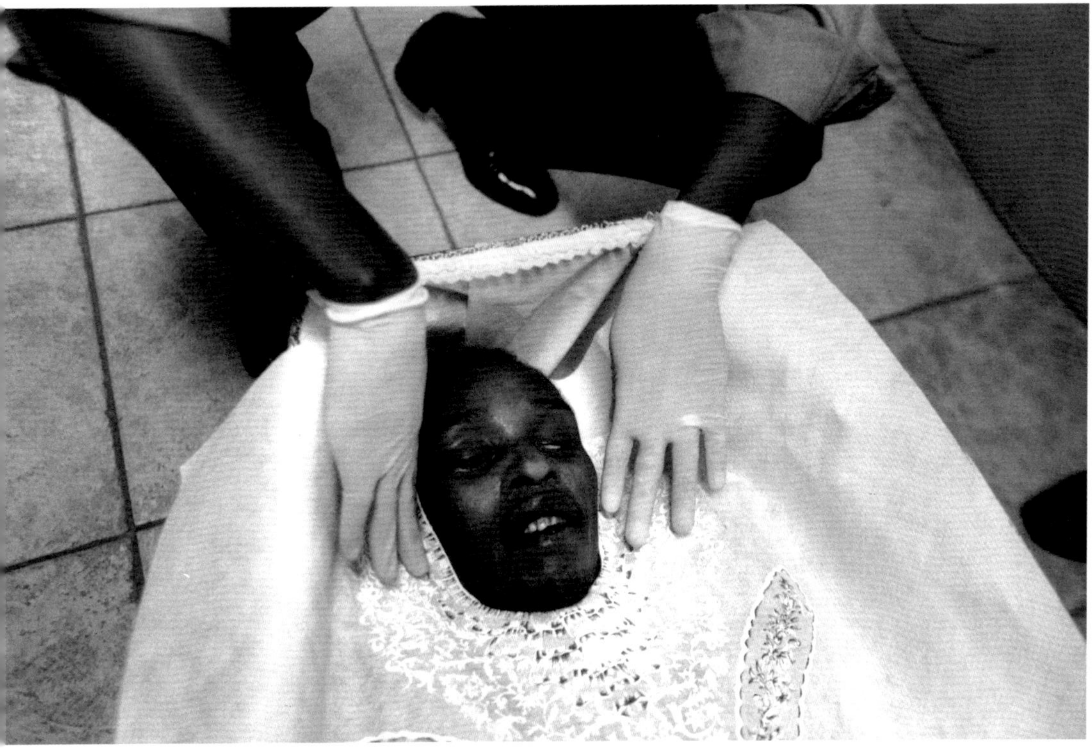

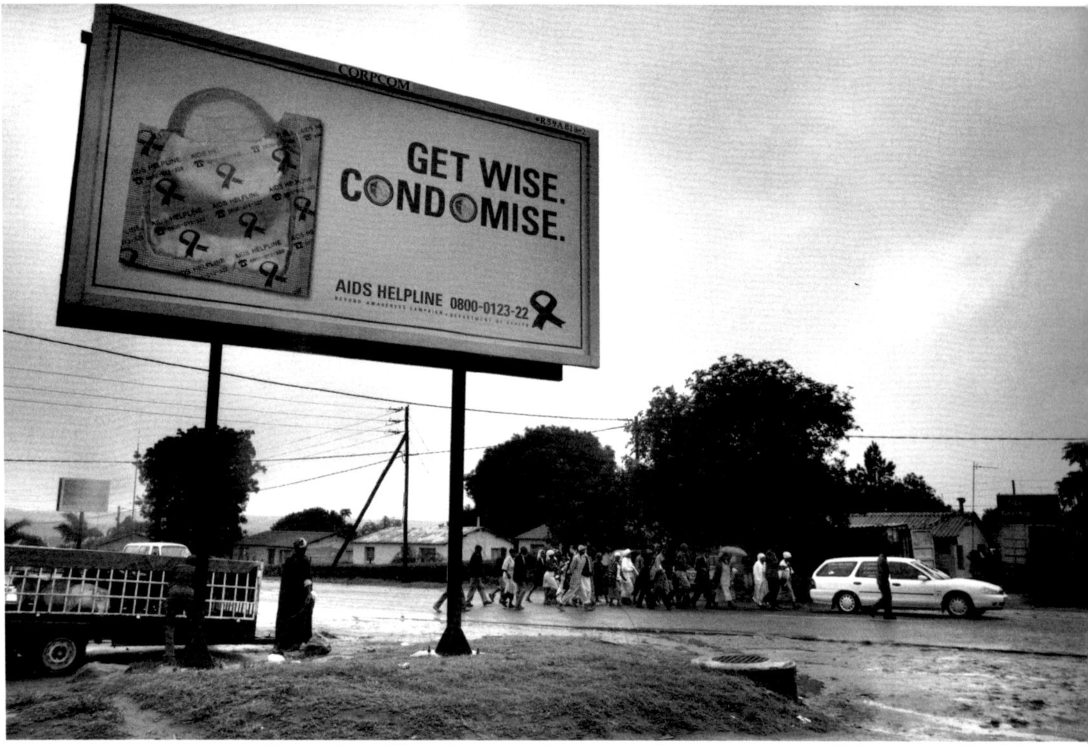
CORPCOM
GET WISE.
CONDOMISE.
AIDS HELPLINE 0800-0123-22

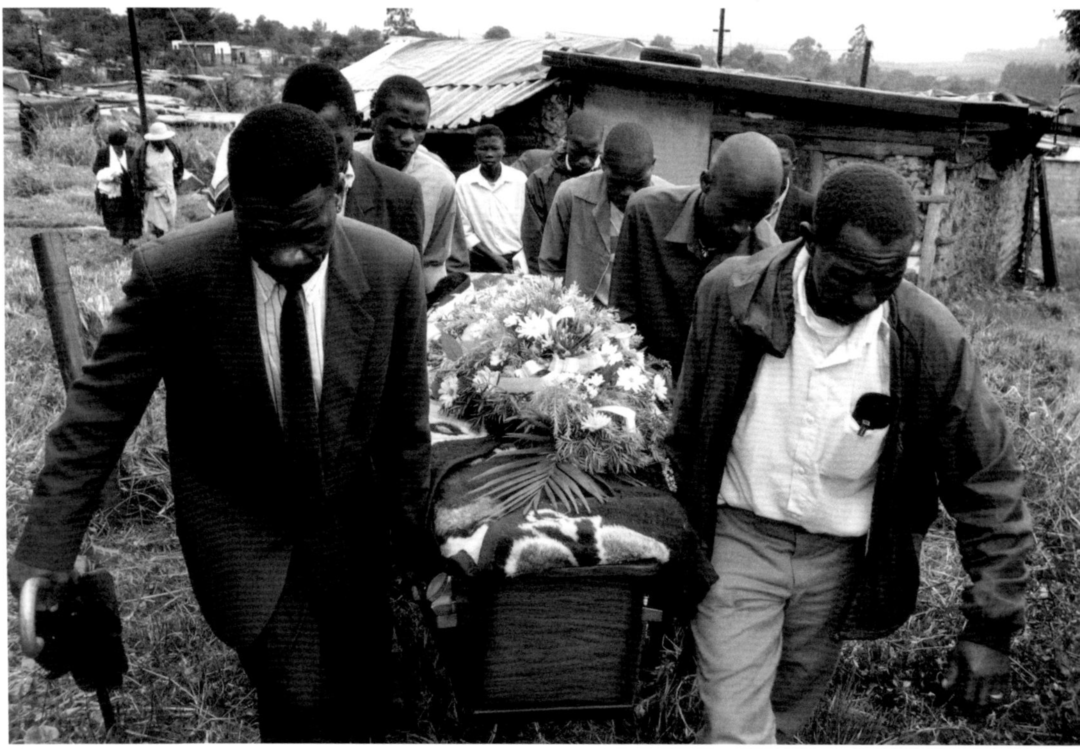

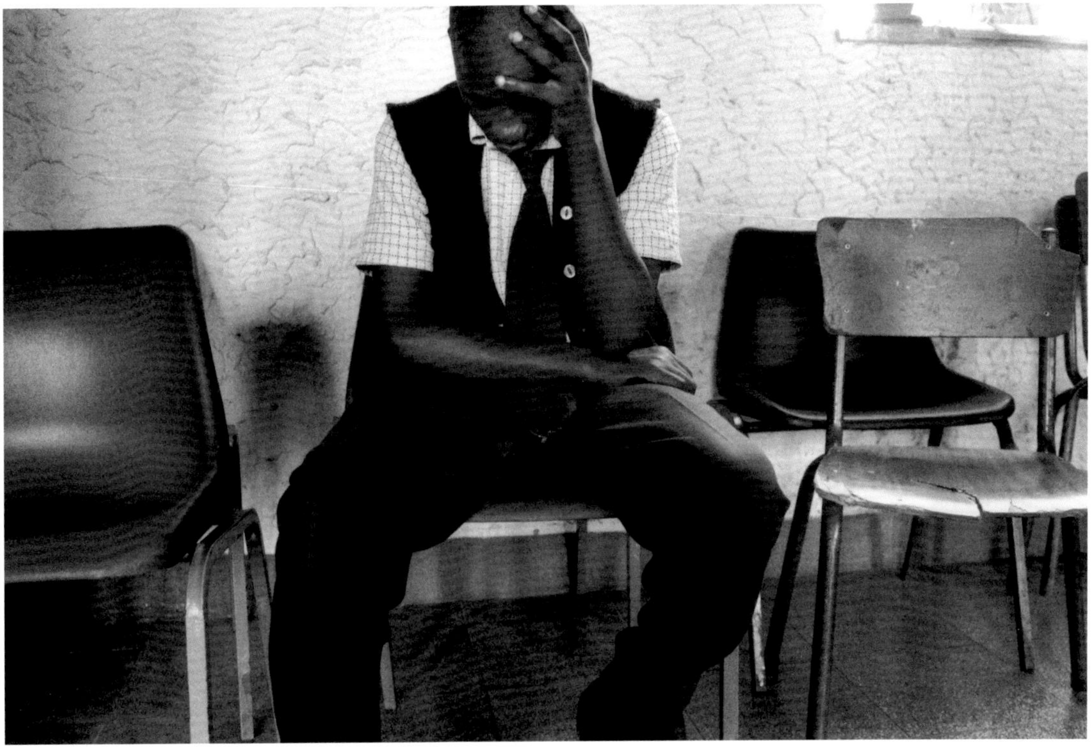

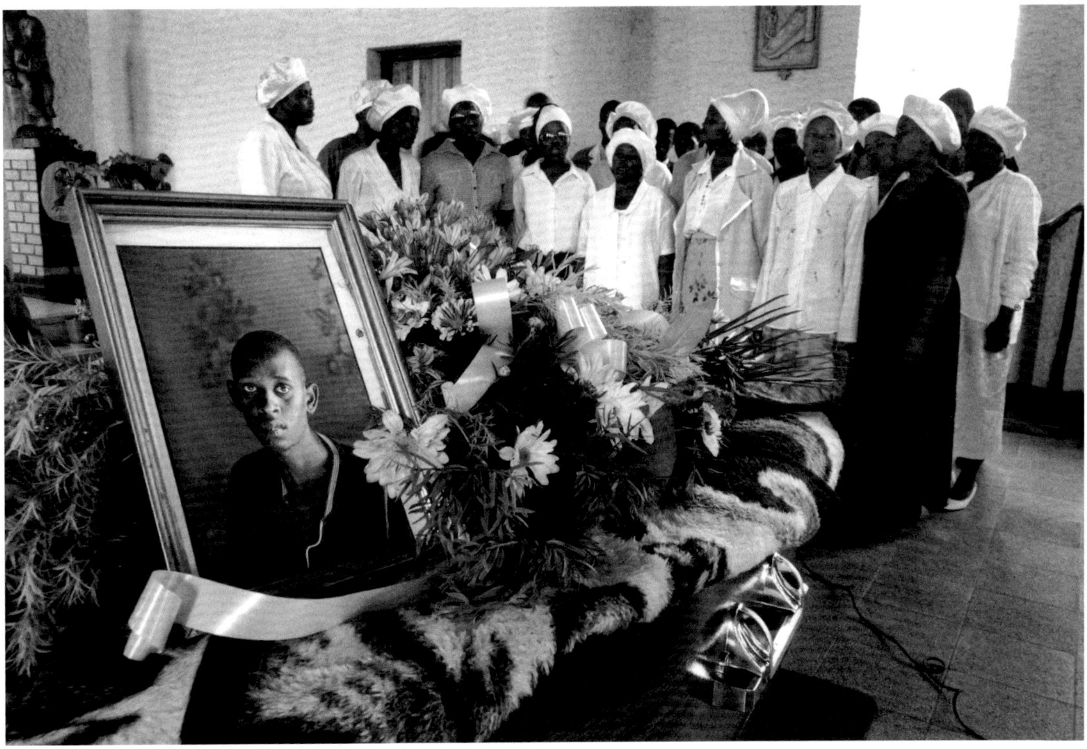

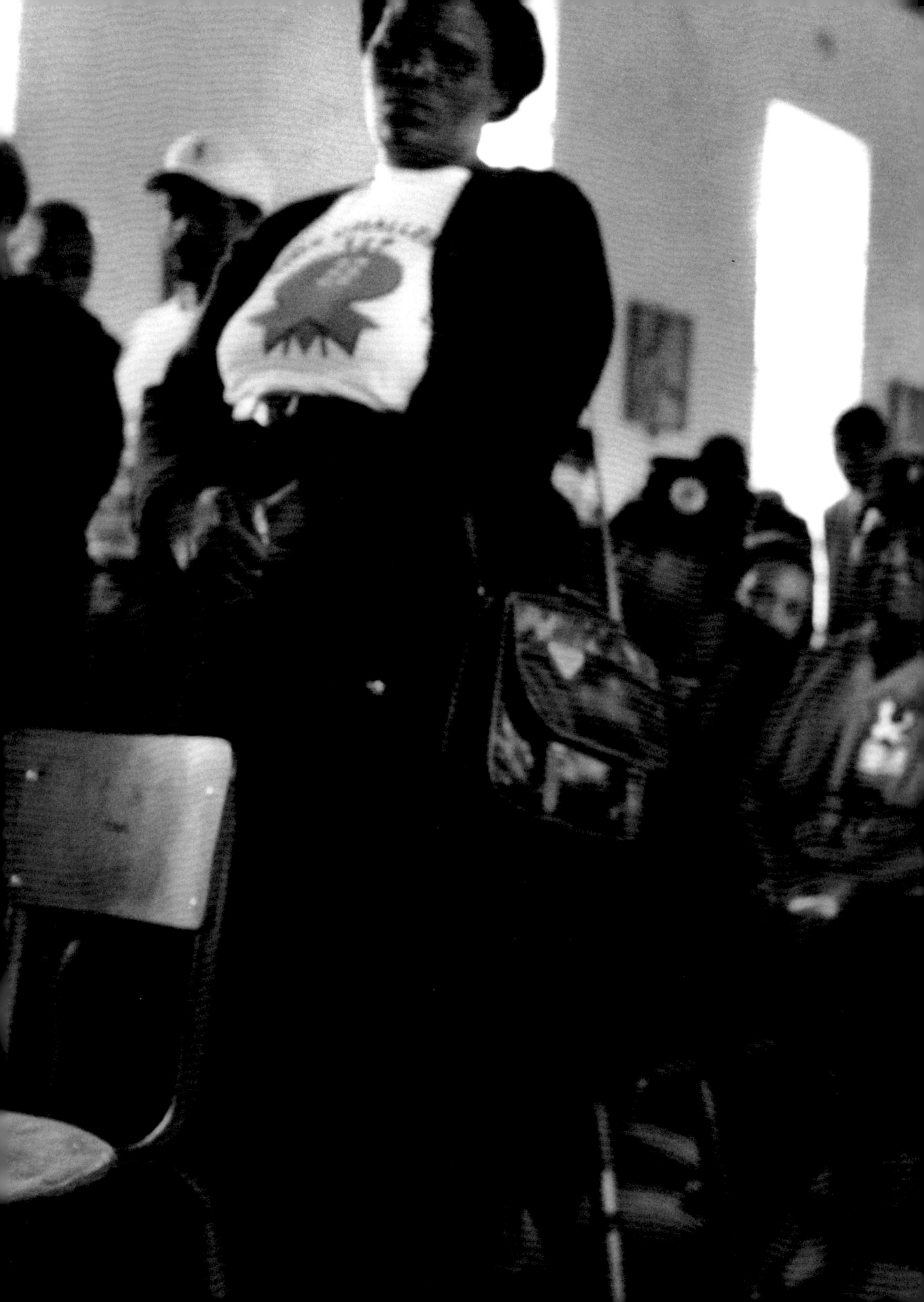

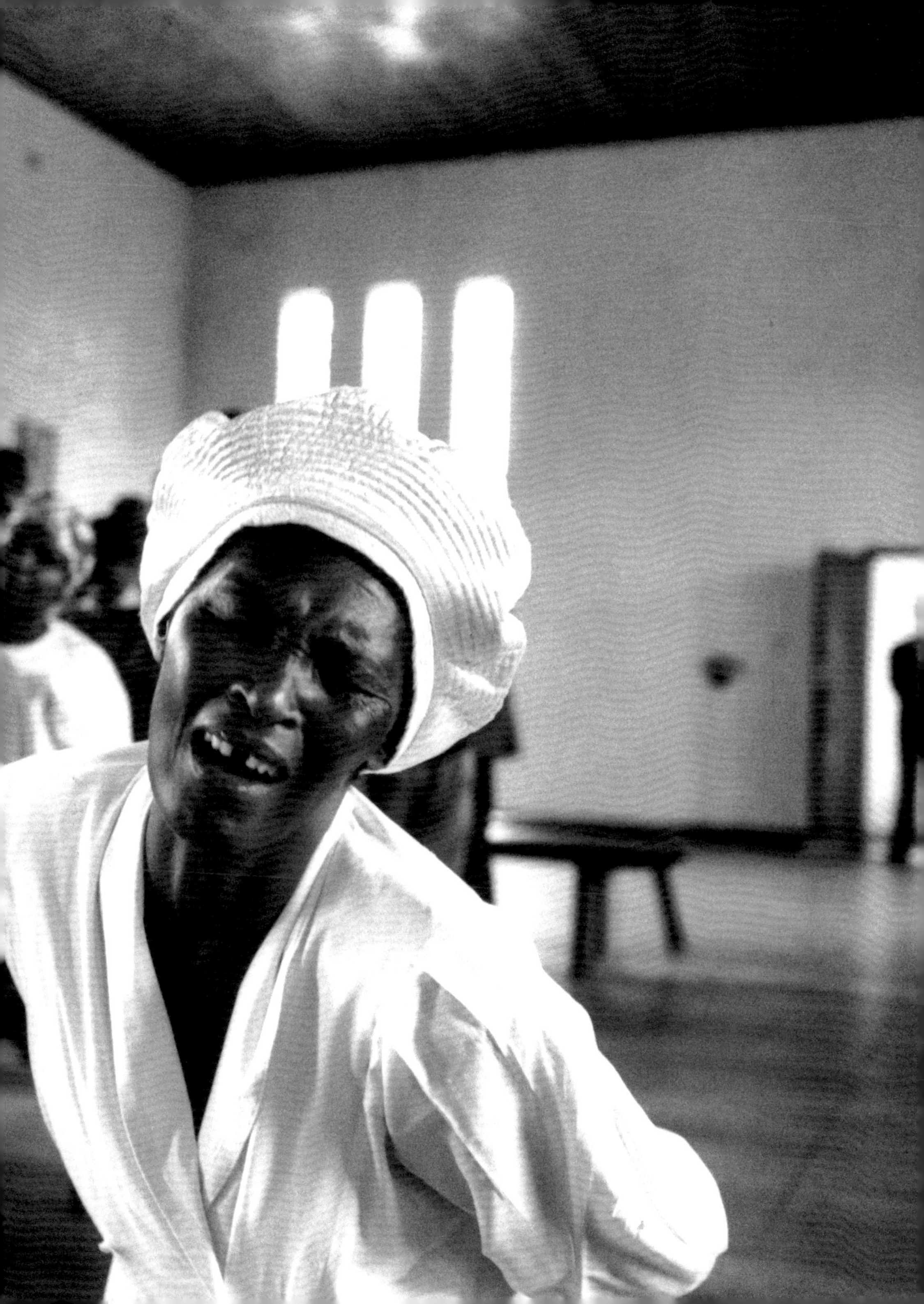

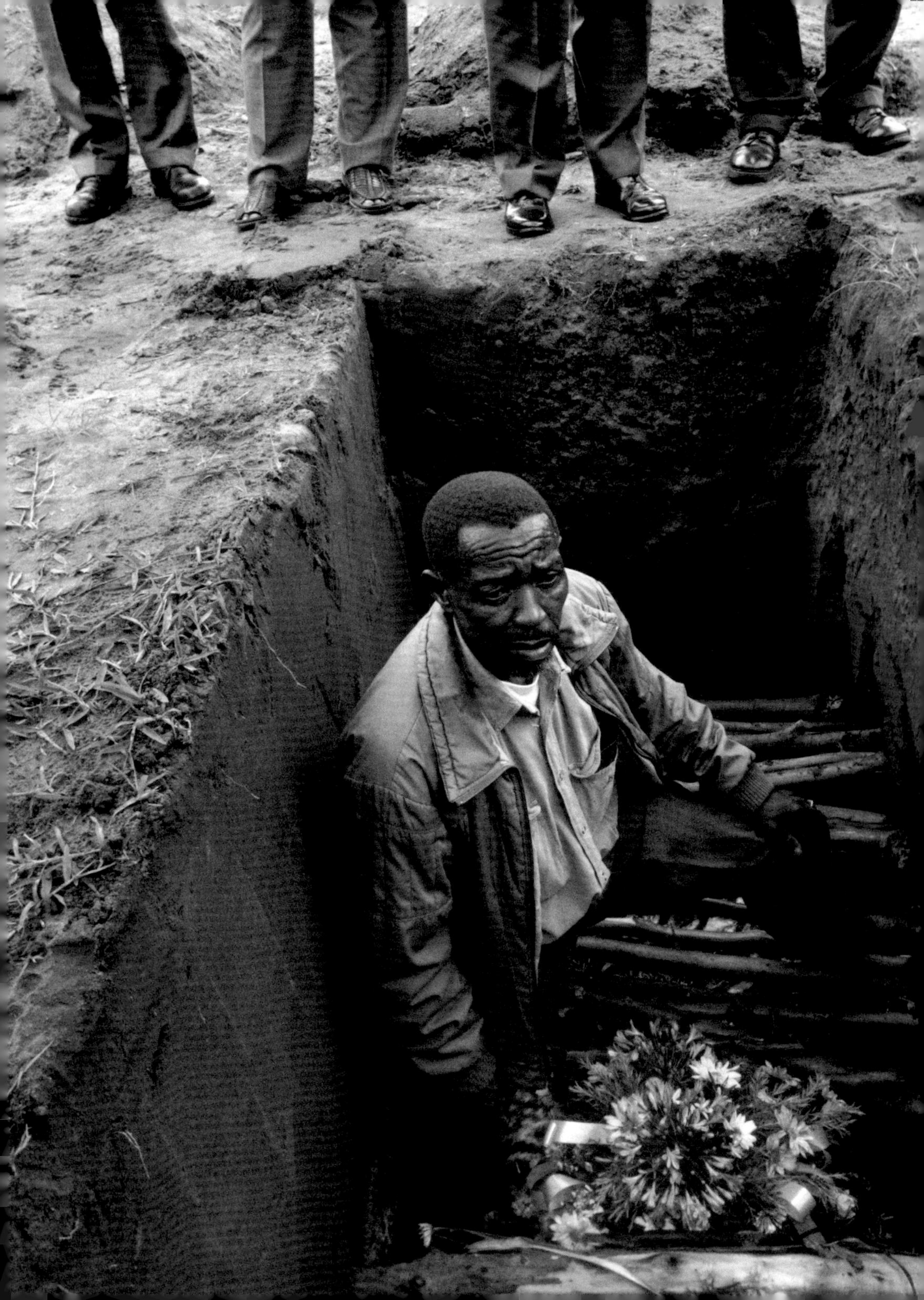

Sipho Gamede. Municipio de Enseleni, Sudáfrica

Sipho murió en abril del año 2000. Quiso tener un funeral educativo para los afectados de sida.

Era un héroe. Quería que las personas supieran que el sida mata. Hemos reunido a todos para que la gente sepa que es un héroe por el hecho de decidir revelar su situación a su novia para que todos pudieran beneficiarse. Su familia tuvo la oportunidad de aprender. Su familia pudo hacerse a la idea.

Damas y caballeros, esta persona amaba a su comunidad y a su país. Él pidió que les trajéramos a todos ustedes aquí para que pudieran aprender y extender el mensaje de que esta enfermedad es como una serpiente, que mata a todos; que aunque se niegue que se tiene el sida, uno se muere igual. Toda persona debe saberlo. Esta información que les doy es muy importante. Vayamos todos a hacernos una prueba para poder saber.

SHUMI KHUMALO, LÍDER JUVENIL (EXTRACTO DE SU DISCURSO FUNERARIO)

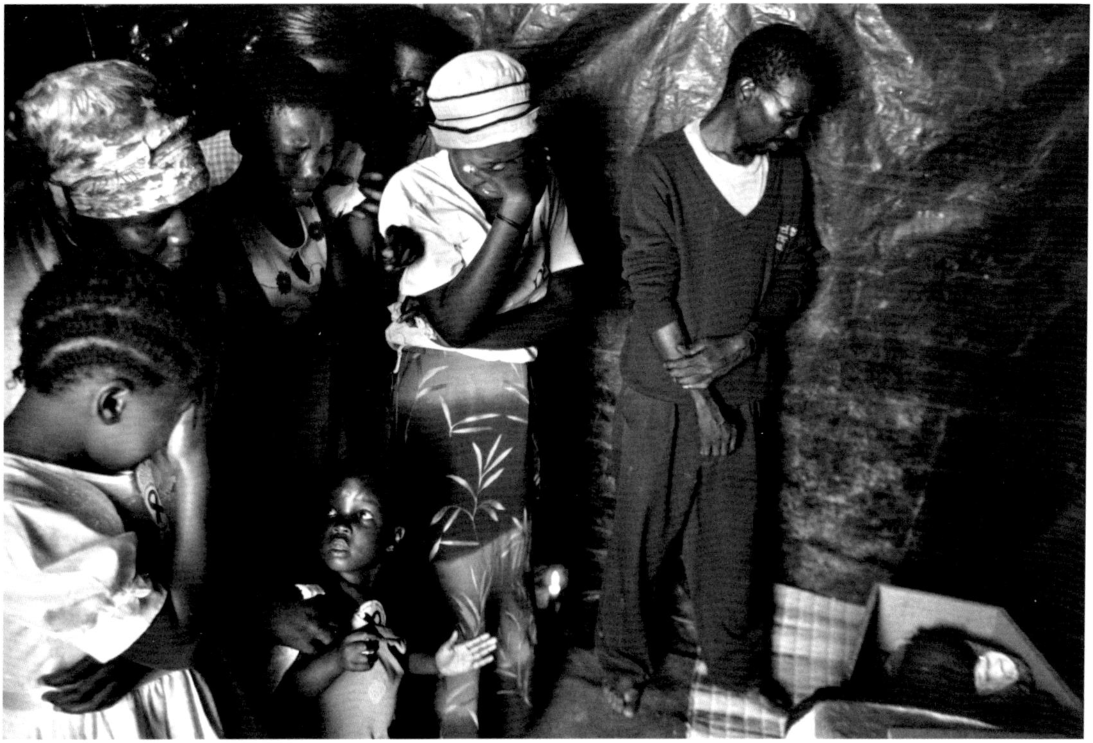

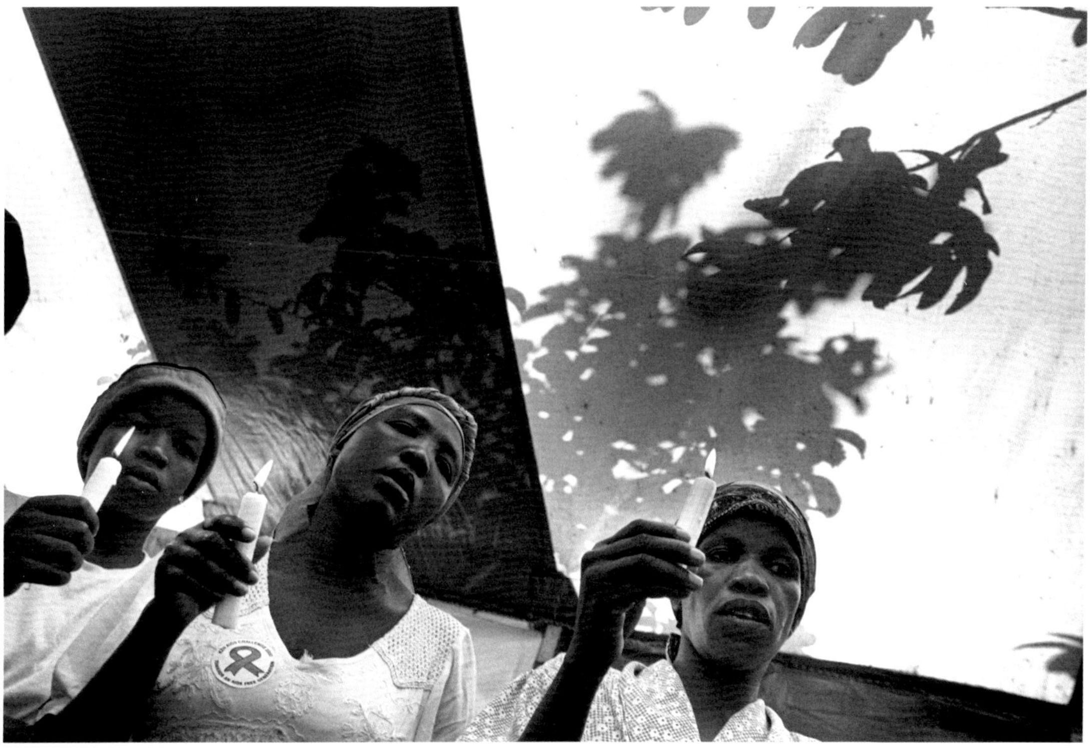

Funerales. Copperbelt, Zambia

Para las familias urbanas de la región minera de Zambia, como en cualquier otro lugar, el coste de un funeral complica los problemas de la pérdida del sostenimiento familiar.

La primera preocupación cuando alguien muere es el dinero. Si muere en casa resulta caro alquilar un coche para llevar el cuerpo al tanatorio. La mayor parte de las personas de aquí no pueden pagarlo, por lo que si les parece que su familiar morirá pronto les ayudamos a llevarlo al hospital para que muera allí. Es posible conseguir un coche para llevar a una persona viva al hospital por 5.000 kwacha, pero en cambio cuesta 15.000 llevar un cadáver al tanatorio. Un cuerpo se puede quedar en casa hasta una semana por falta de dinero.

No llevamos la cuenta de los funerales a los que vamos. Hoy en día se les entierra en los cementerios como si se tratara de una competición. En ocasiones coinciden hasta seis funerales a un tiempo. Mientras uno de los grupos reza, el otro cava el suelo y el otro coloca el ataúd en la sepultura. Casi cada día se celebra un funeral por algún conocido. En Ndola y Kitwe, los asistentes al entierro tienen que ayudar a cavar las tumbas. Como hace ya cuatro meses que no les pagan, los sepultureros están en huelga.

VIOLET MUKOSHA, VOLUNTARIA DE ASISTENCIA DOMICILIARIA EN NDOLA

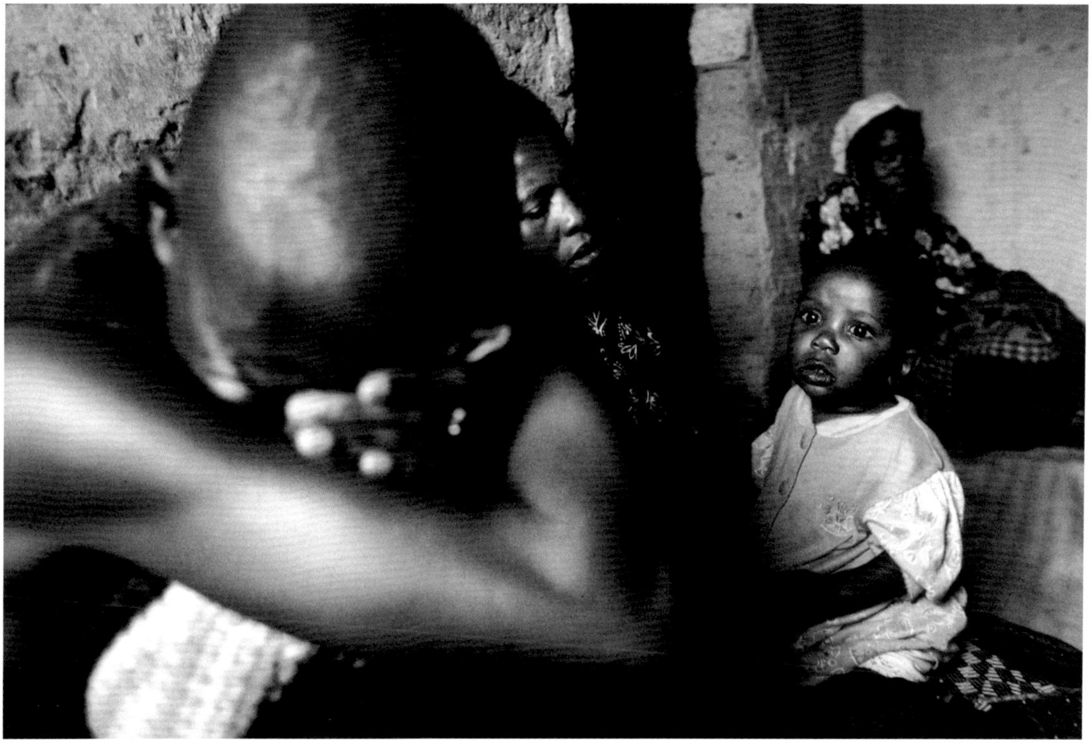

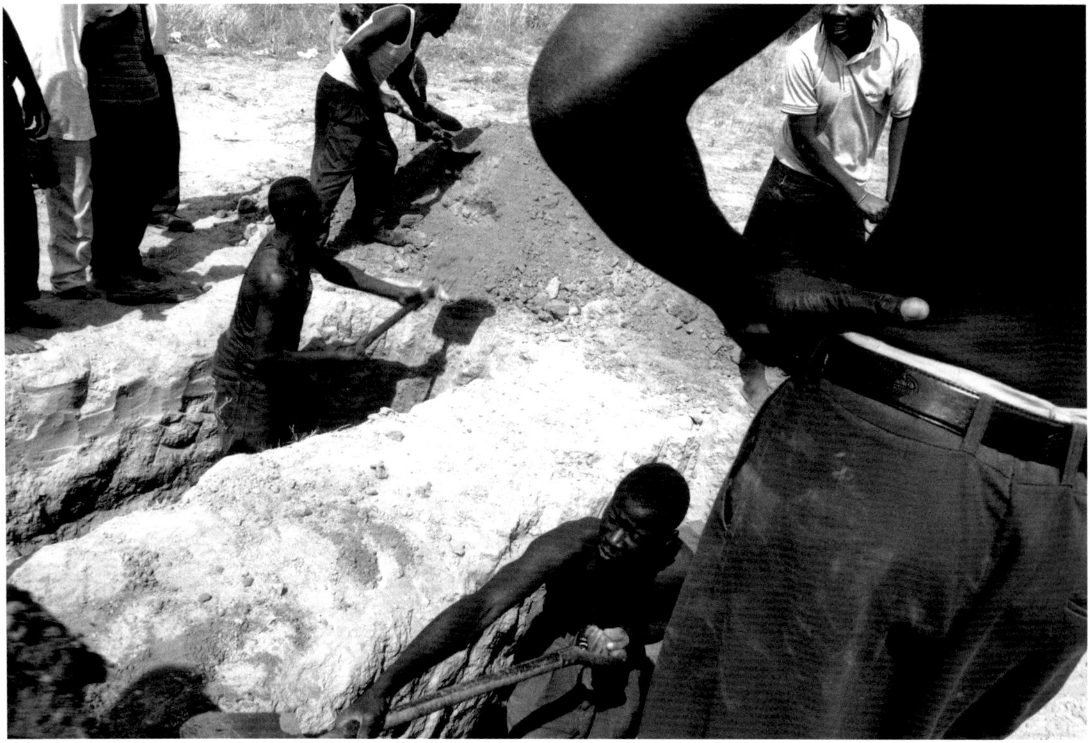

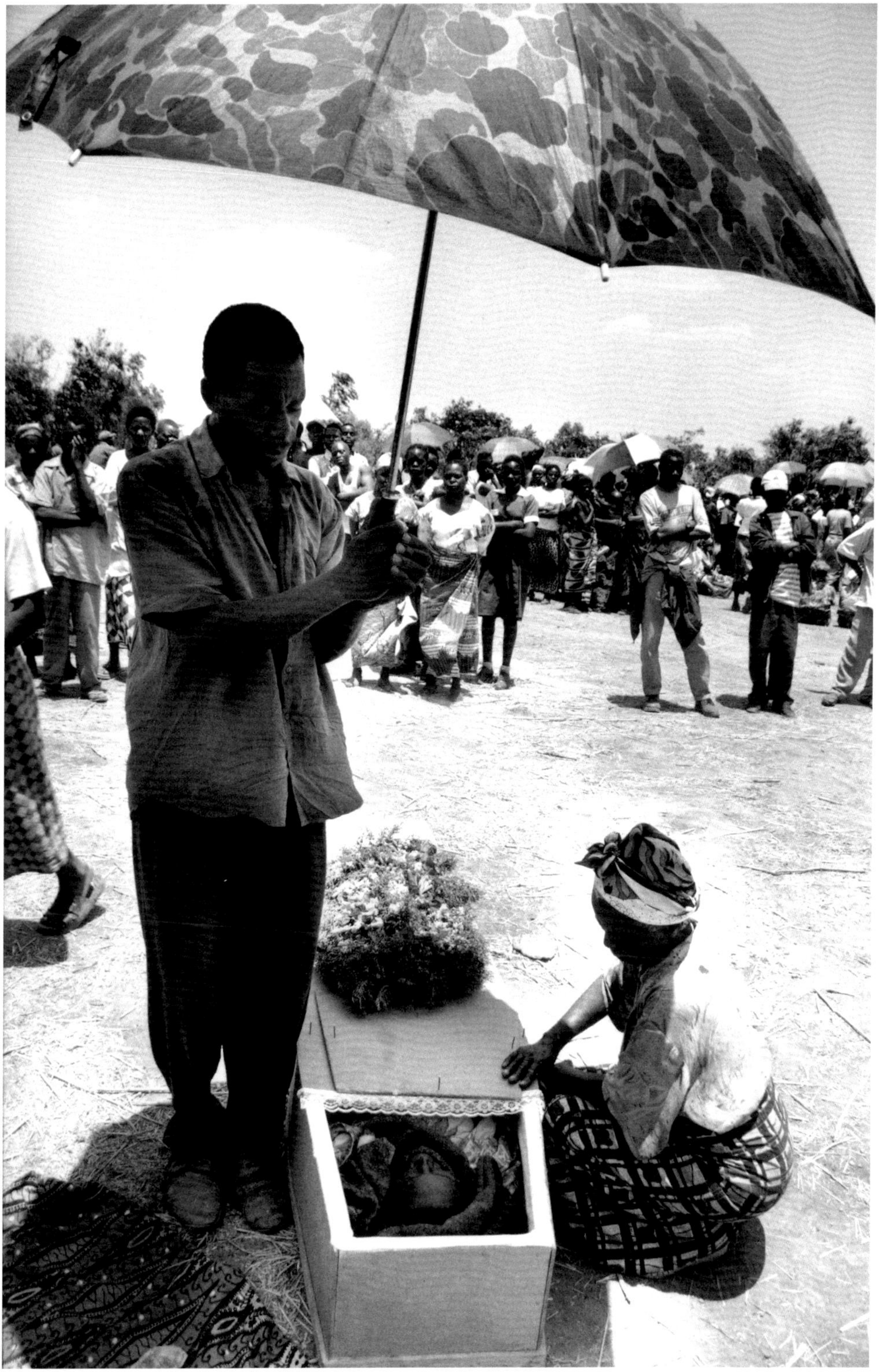

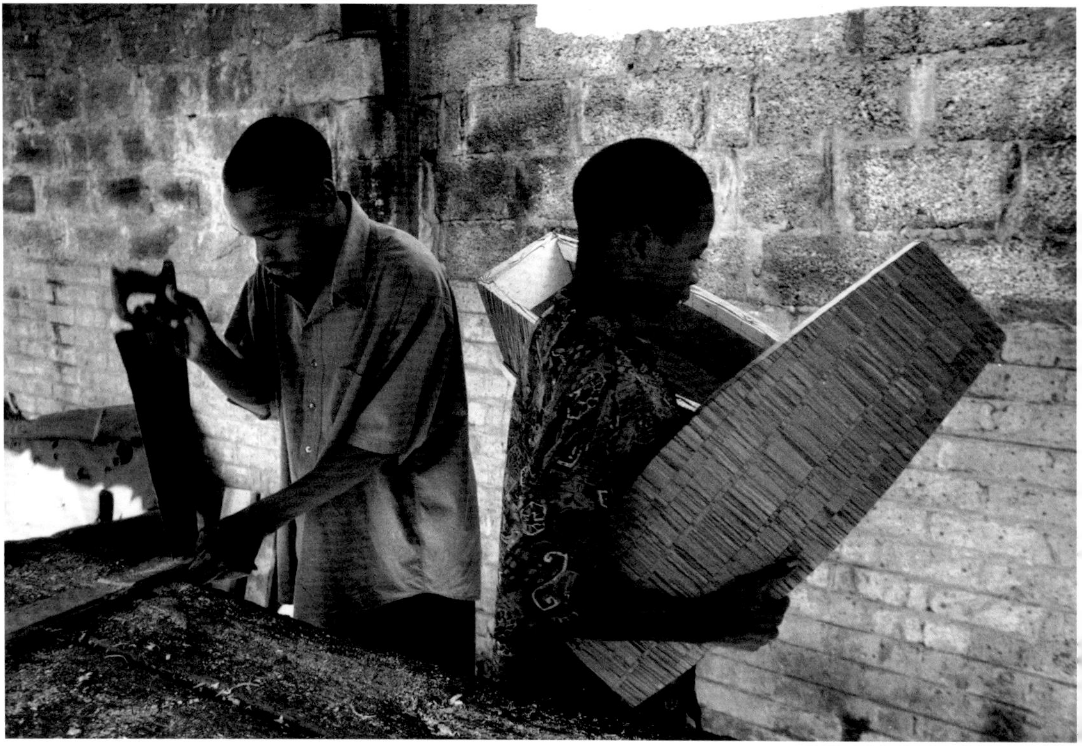

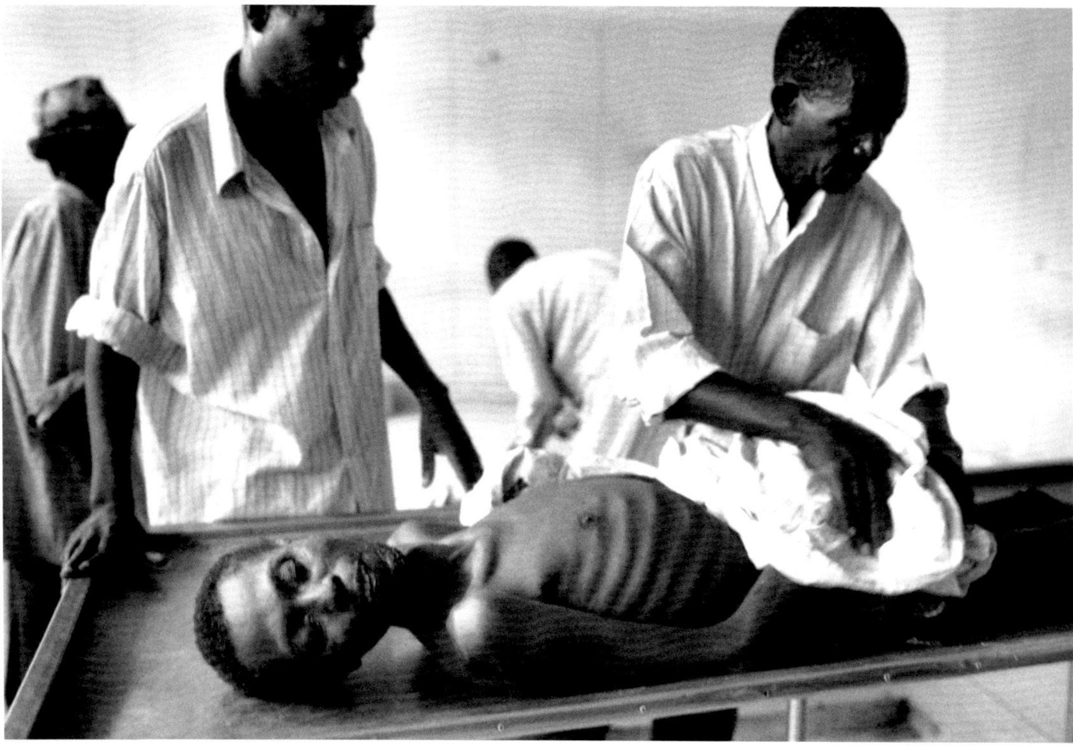

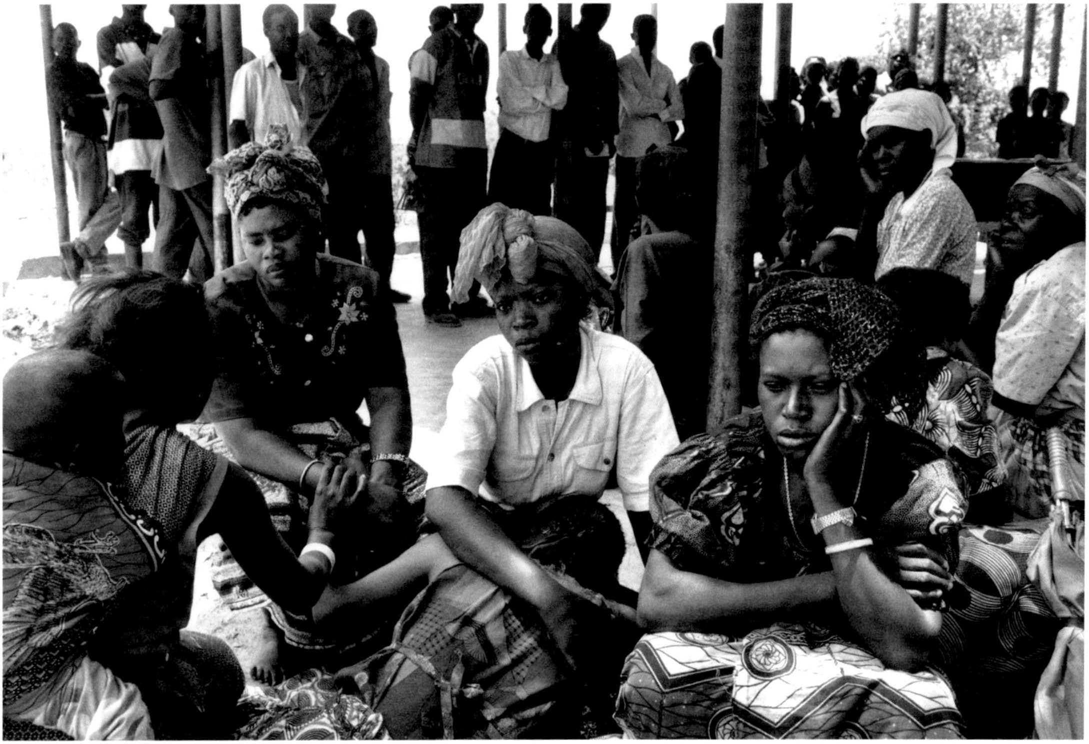

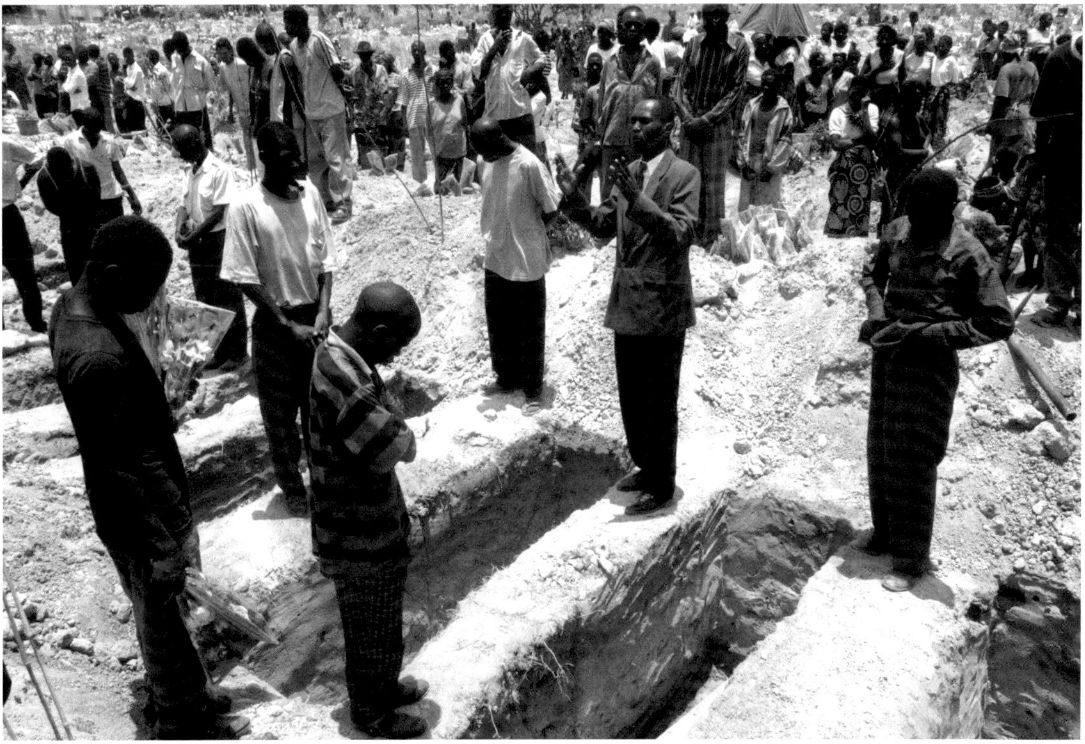

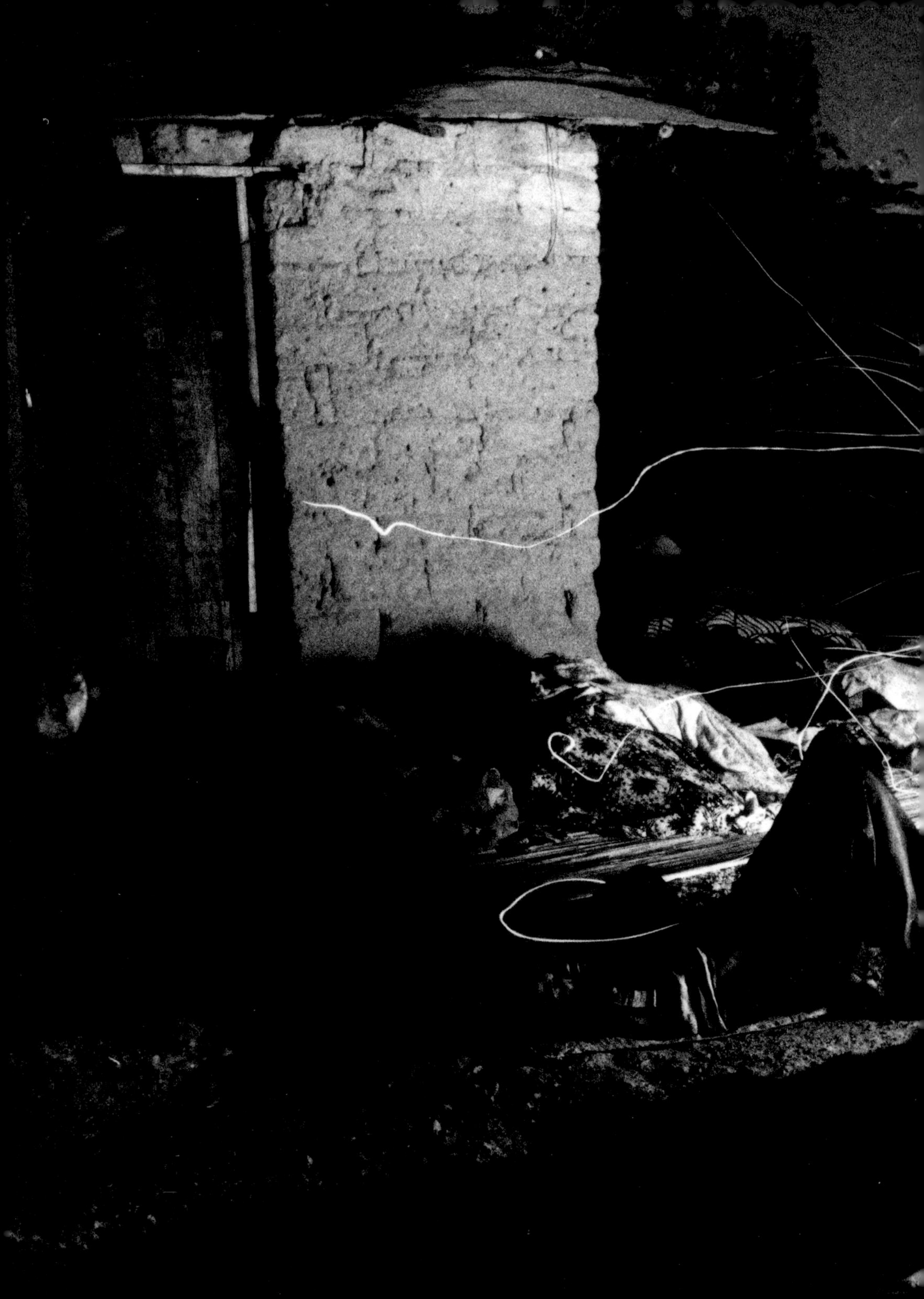

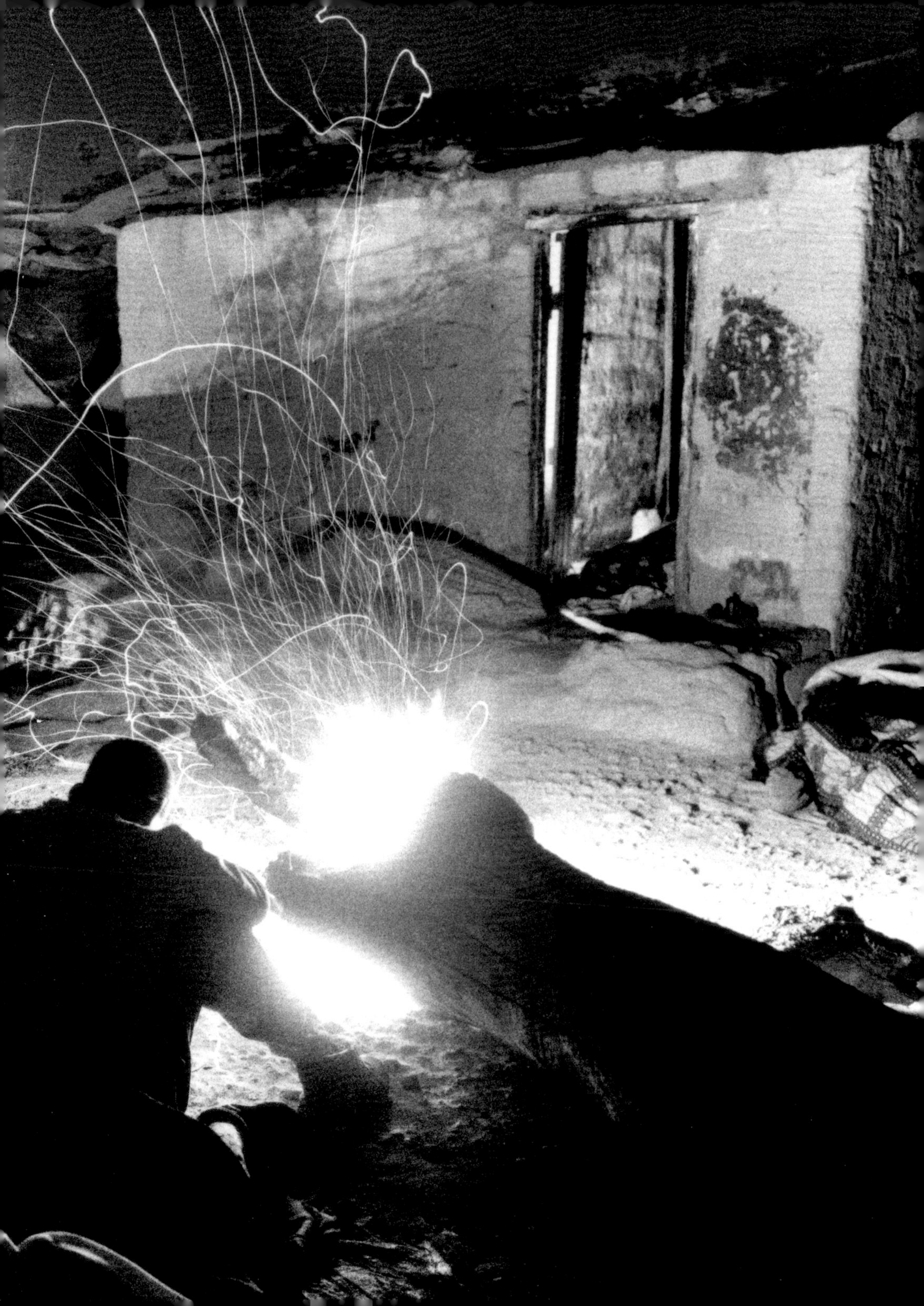

Asistencia domiciliaria. Chikankata, Zambia

El camión del hospital de Chikantaka visita los pueblos vecinos como parte de su programa de asistencia domiciliaria.

Aquí resulta difícil pagar una aspirina. Nuestro país dedica 6,50 USD per cápita a la sanidad al año, y tenemos más de 300.000 personas con sida, por lo que es imposible cuidarlas a todas en hospitales u hospicios. Así que el equipo de sida en nuestro hospital comenzó a buscar soluciones alternativas. Decidimos intentar ayudar a las personas en sus casas, donde podemos visitarlas, darles consejo y enseñar a sus familias a cuidarlas.

Funciona. La asistencia a domicilio utiliza la fuerza de la familia y satisface el deseo de la mayoría de los zambianos de morir en casa. También descubrimos que los vecinos sentían curiosidad. Aprendimos que el paciente era el punto de entrada no sólo a la familia, sino a toda la comunidad. Nos dimos cuenta de que el cuidado y la prevención están muy relacionados.

También hemos entrenado en cuidados y prevención a voluntarios de 16 pueblos. Nuestra filosofía de «asesorar a la comunidad» desafía a las personas a cuestionar sus costumbres. En muchos casos éste ha sido el motivo del cambio.

DAPHETONE SIAME, DIRECTOR DE GESTIÓN SOBRE EL SIDA

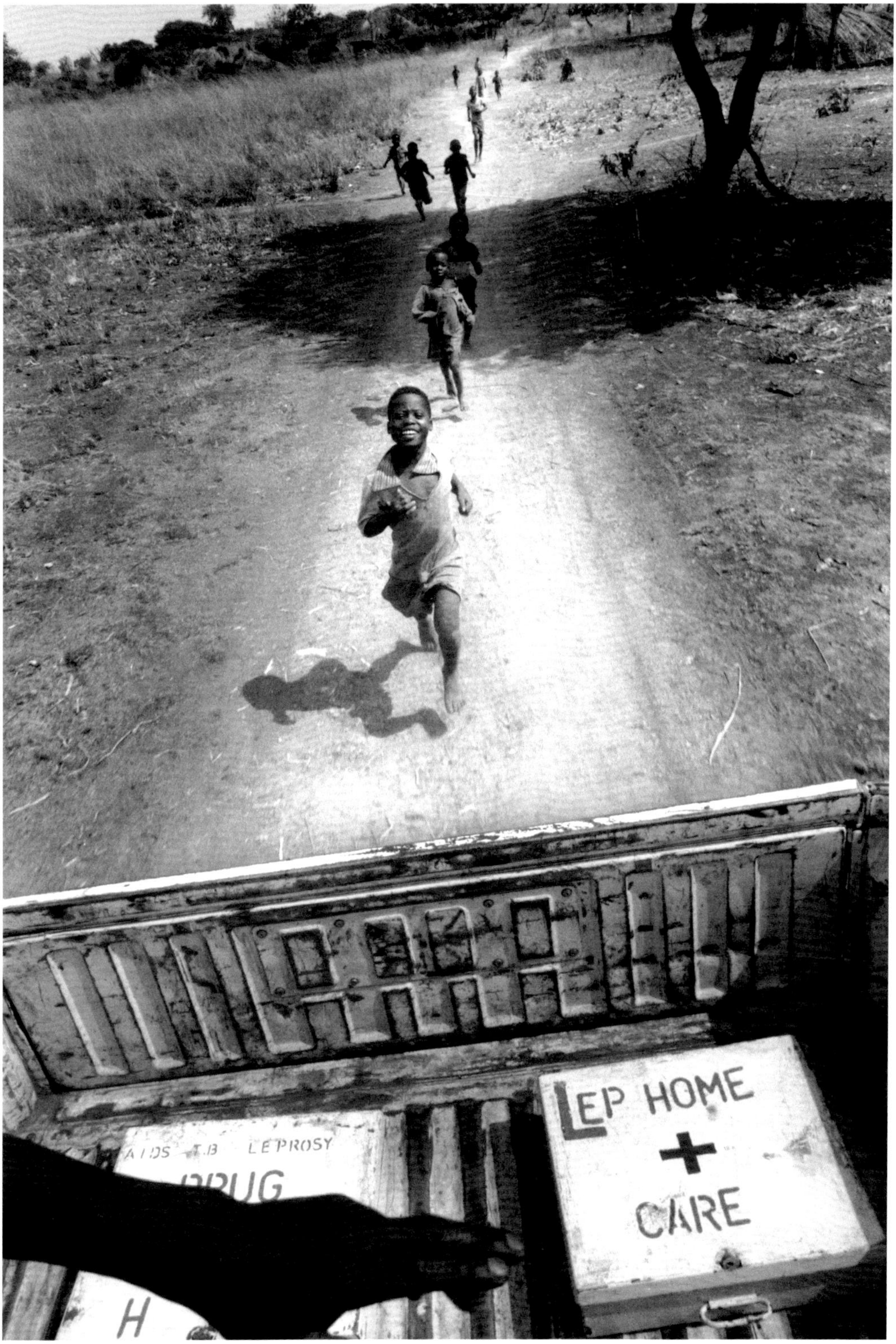
AIDS T.B LEPROSY
LEP HOME
+
CARE

Asistencia domiciliaria. Ndola, Zambia

Los voluntarios son un elemento clave en el programa de asistencia domiciliaria de Ndola.

Cuando eres pobre y enfermas, es aún peor. En nuestra zona muchas personas con la enfermedad mueren rápidamente a causa del hambre y la falta de medicamentos. La pobreza y esta enfermedad trabajan mano a mano, en especial cuando el enfermo es la base del sostenimiento económico familiar. Si están enfermos el lunes, la familia no come el martes. Si la base económica de la familia está enferma, toda la familia lo está.

Como voluntarios intentamos cuidar a nuestros pacientes atendiendo a sus necesidades. Nuestro trabajo consiste en ayudarles a tomar su medicación, en especial en los casos de tuberculosis, en los que resulta primordial tomarla cada día. En los casos en los que los padres no son capaces, nosotros debemos hacerlo todo: bañar a los niños, fregar el suelo, lavar los platos, lavar la ropa y cocinar para la familia. Cuando están muy enfermos, lo que podemos ofrecerles es nuestra compañía y apoyo. Y si son creyentes, rezamos con ellos.

VIOLET MWINUKA, VOLUNTARIA DE ASISTENCIA DOMICILIARIA

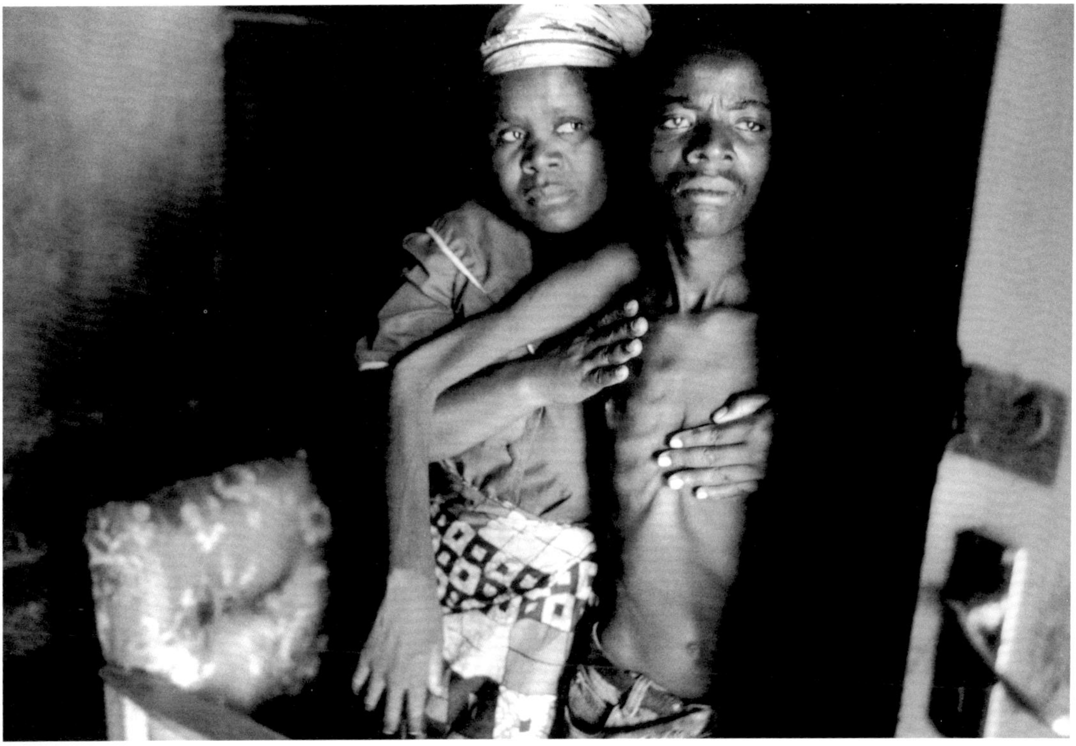

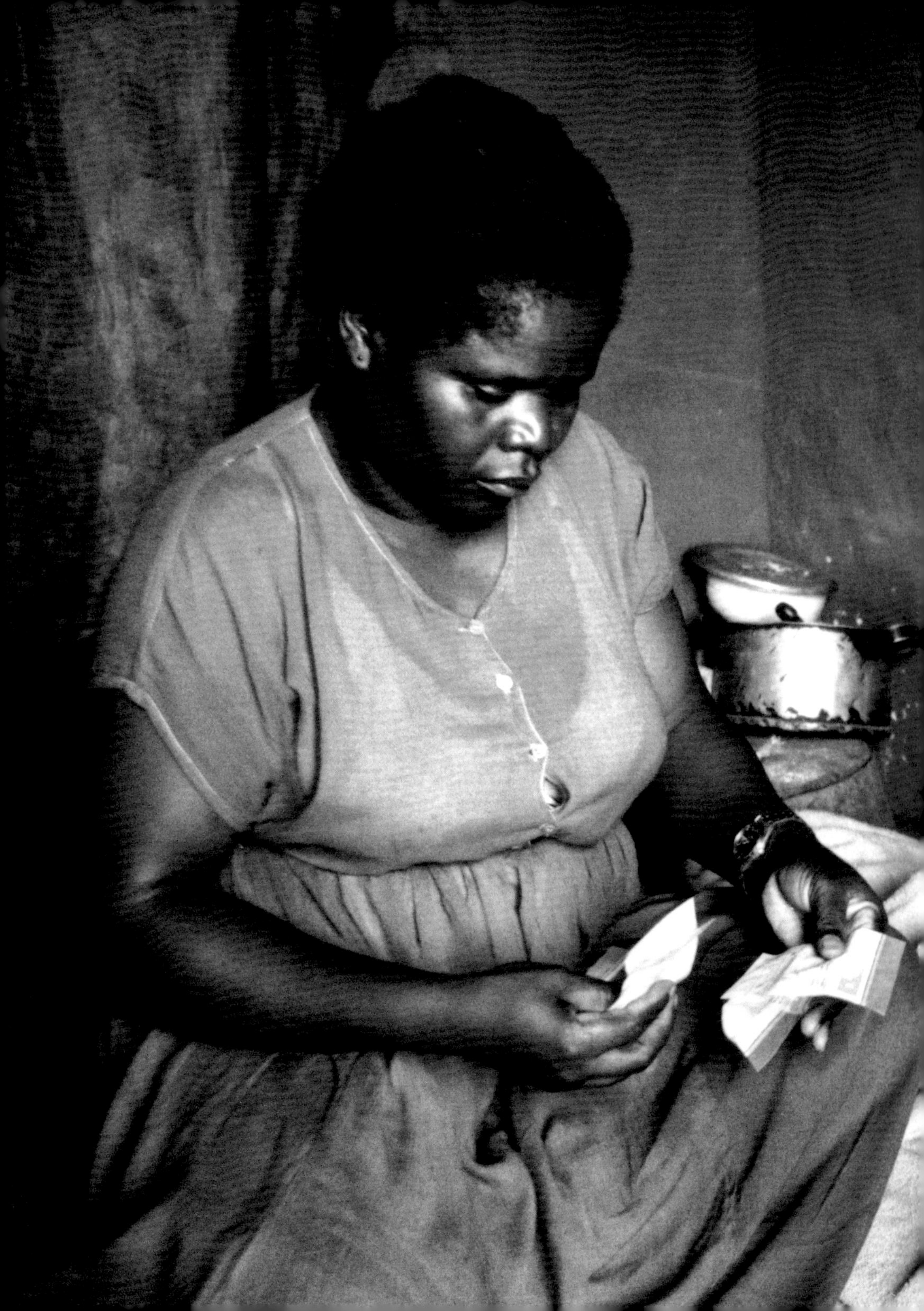

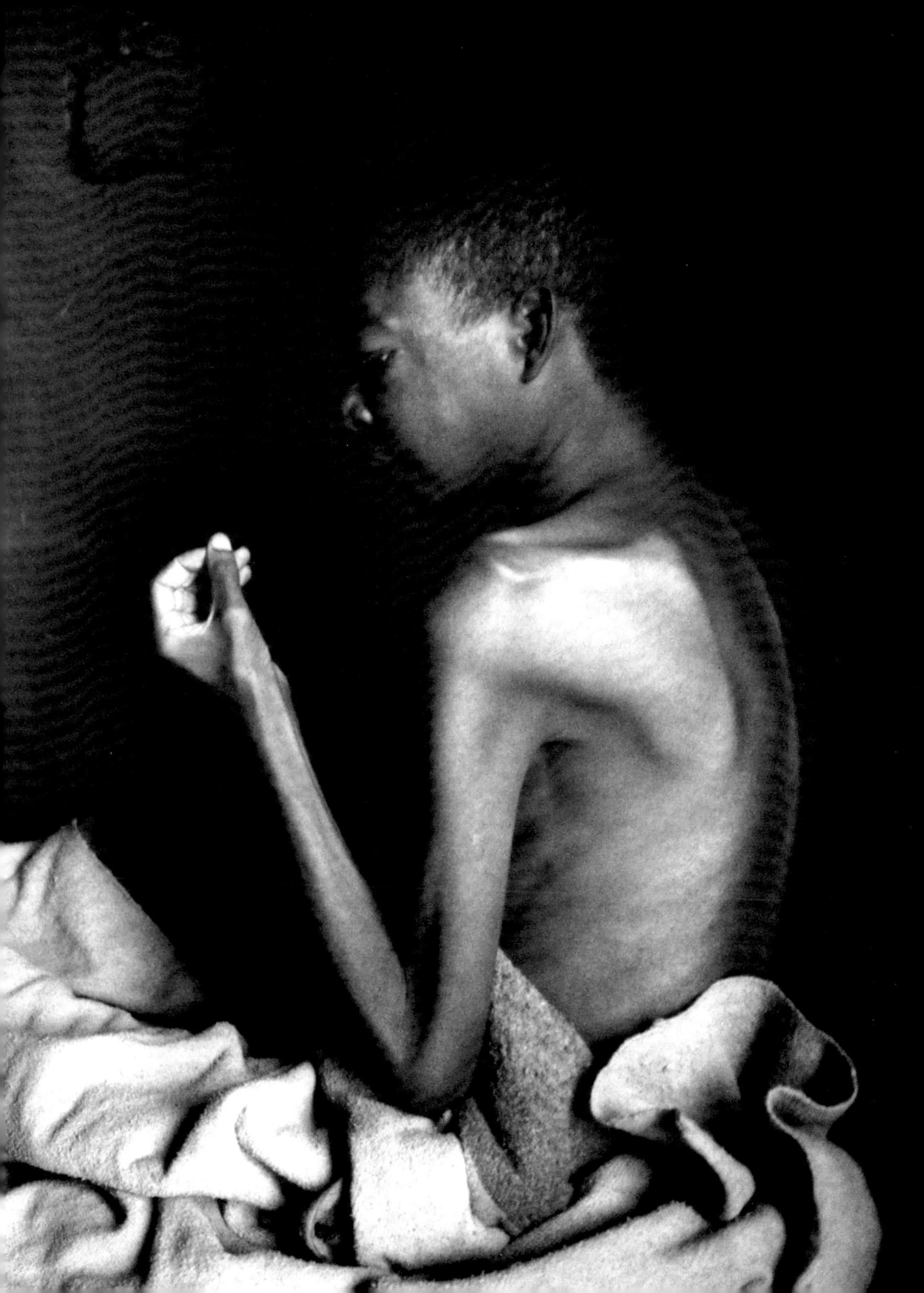

Mi trabajo incluye la visita a los pacientes en sus casas, donde les doy los medicamentos adecuados y aconsejo a los pacientes y a sus familias sobre las medidas que deben adoptar para mantener o mejorar su salud. Visito a los voluntarios que viven en las comunidades locales y que comprueban con regularidad los progresos de los pacientes. Ellos son los que mejor identifican a los que tienen las necesidades más imperiosas.

Sabemos que las madres infectadas con el VIH pueden transmitir el virus a sus bebés al darles el pecho, pero en estos lugares la gente es tan pobre que les resulta imposible proporcionar una dieta equilibrada a los bebés sin leche materna. Si no la reciben, no pueden luchar contra las infecciones y mueren a causa de desnutrición o diarrea, u otras infecciones como las paperas o la malaria.

El sida afecta a nuestra comunidad a todos los niveles. La mayoría de nuestros clientes no son capaces de trabajar, por lo que la economía se resiente. Se aproxima la época de la plantación, pero los pacientes crónicos no pueden trabajar. El resultado es la pobreza más absoluta.

JESSIE KAMOLONDO, ENFERMERA TITULADA

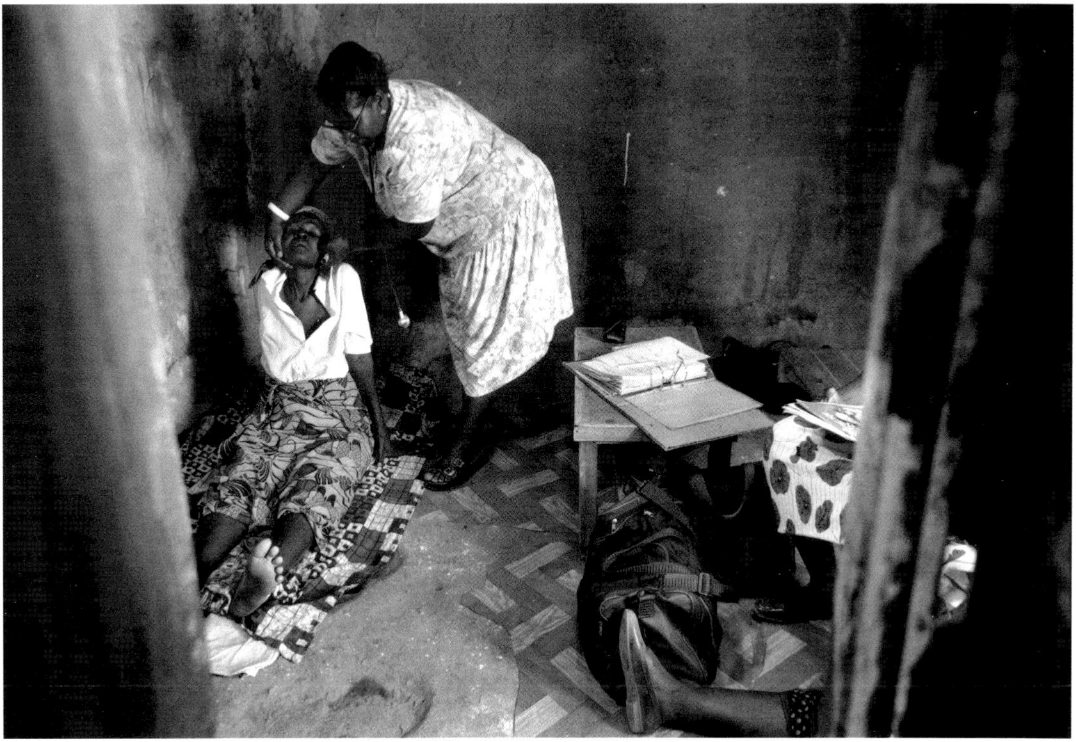

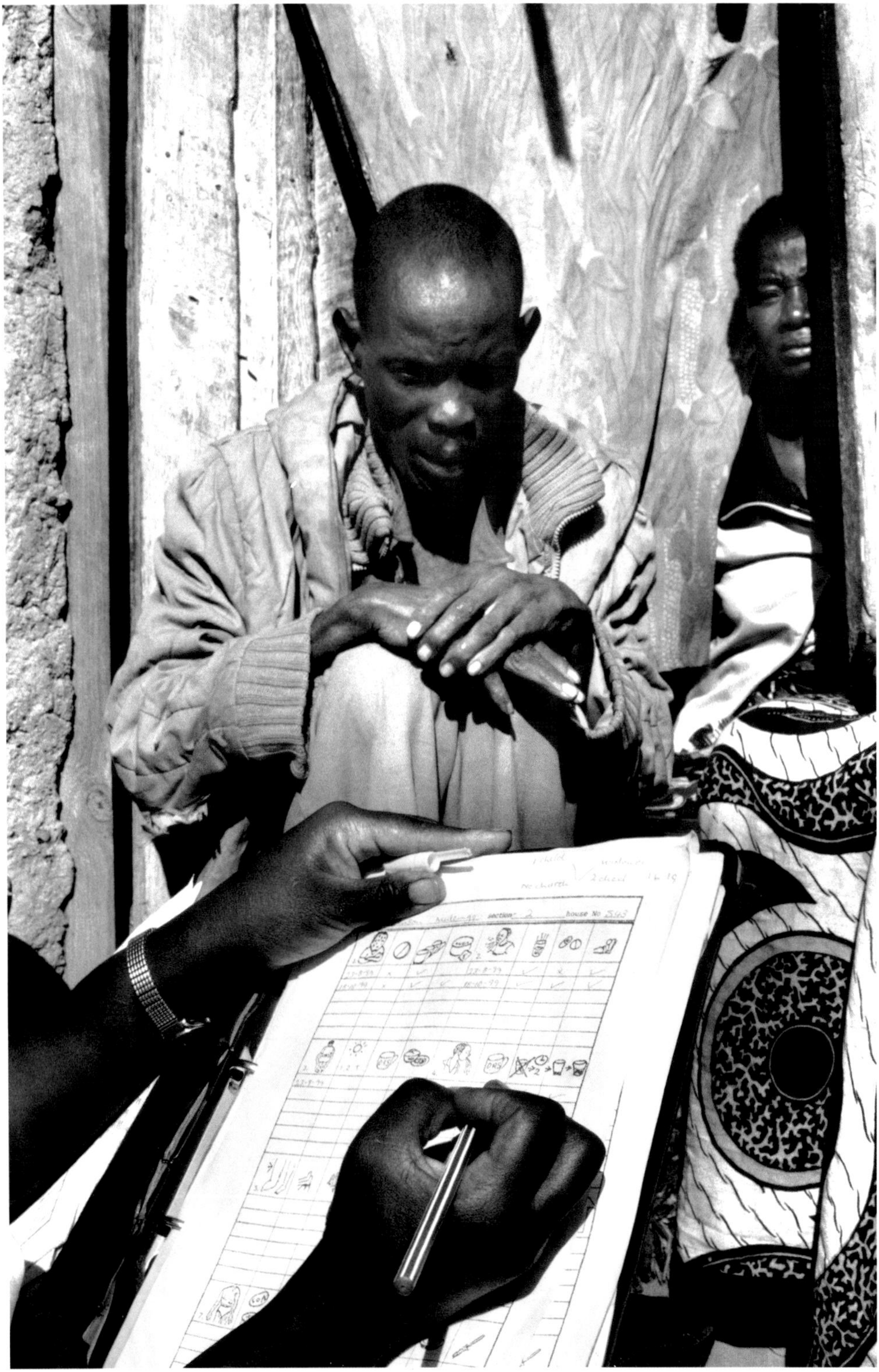

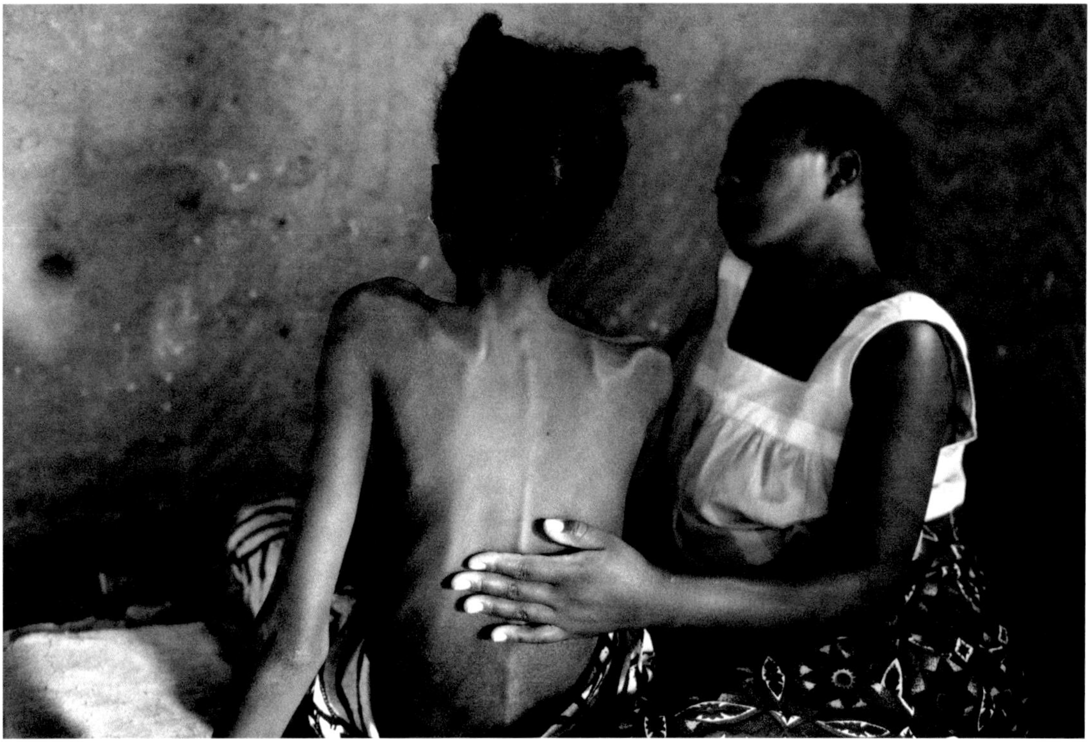

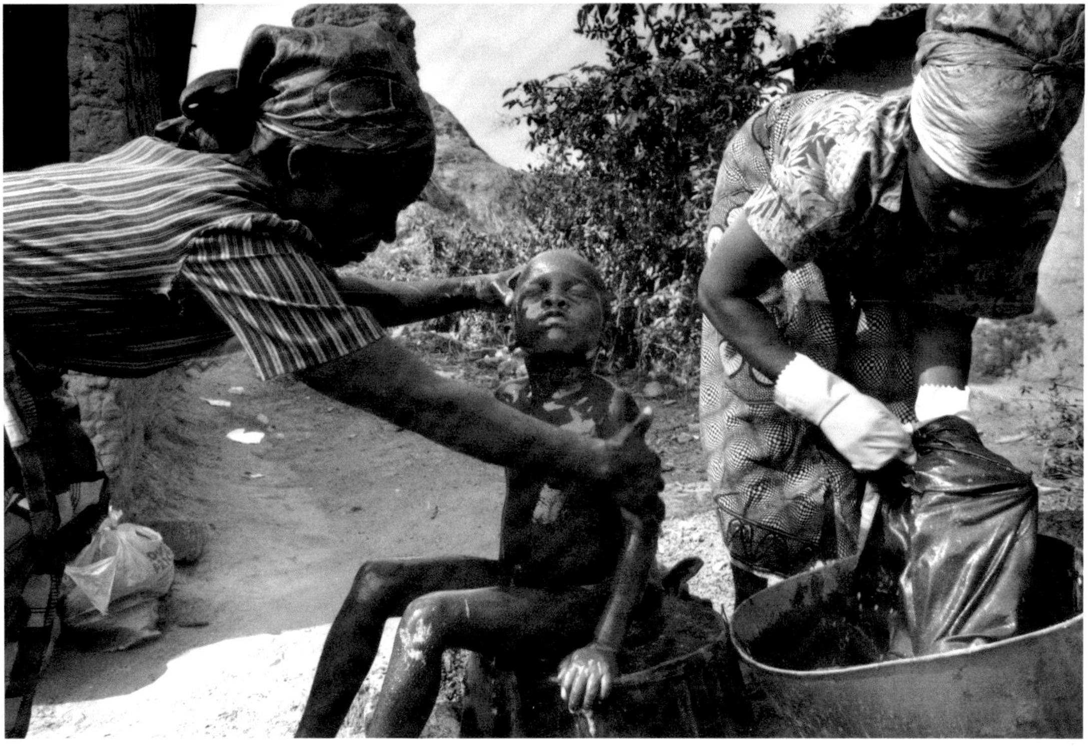

Janet es la principal paciente a mi cuidado. Vive cerca de mi casa, pero no la conocía antes de que enfermara. Limpio su casa, la lavo, la ayudo a tomar sus medicamentos, cocino para ella y le ayudo a comer. Janet está sola; no tiene familiares que la cuiden. Como ahora está muy enferma, intento entrar en contacto con sus familiares para que se la lleven con ellos o envíen a alguien para que se haga cargo de ella aquí. El problema son las noches, porque está sola. No hay nadie que le dé agua para beber o que la cuide.

EDWICE ZULU, VOLUNTARIA DE ASISTENCIA DOMICILIARIA

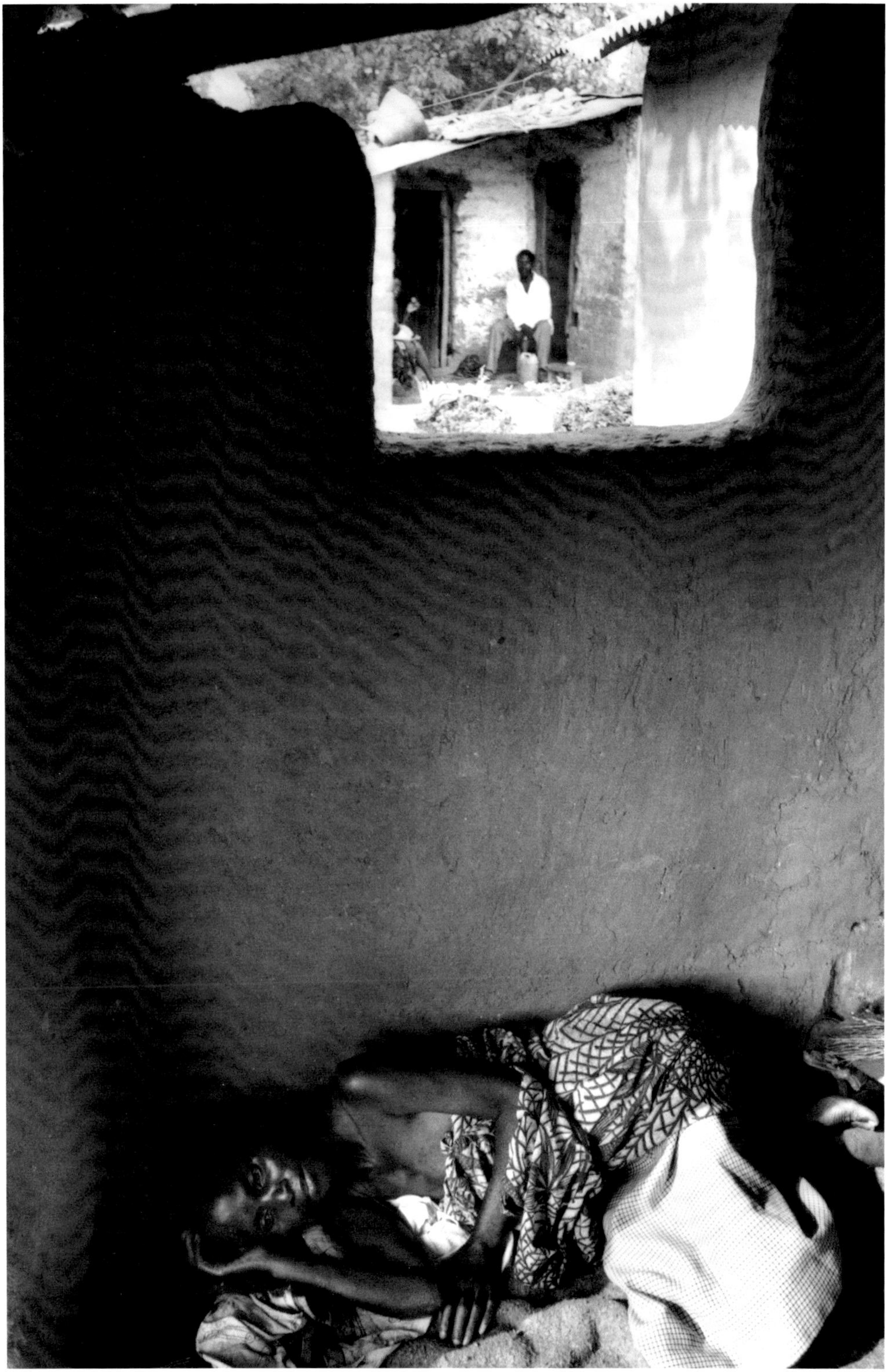

Asistencia domiciliaria. Bukoba, Tanzania

Un médico a domicilio visita a Faustina Nyauga en su casa para proporcionarle cuidados y ayudar a su familia a decidir lo mejor para los hijos de la enferma.

Vamos a pie o en bicicleta para comprobar cómo están nuestros pacientes. Sería un problema ir en coche, pues la gente sabría que se trata del coche del sida.

En la mayoría de los casos uno se da cuenta de que los niños no van al colegio y de que no tienen suficiente comida. La madre está enferma, los niños están enfermos. Pero ¿quién les ayuda? No hay nadie más.

Siempre se trata del mismo problema: la pobreza. Nos esforzamos al máximo, pero esta enfermedad es desmoralizadora. Para nosotros es demasiado ver tanto sufrimiento cada día. Pero vemos que los pacientes necesitan nuestra ayuda y que somos felices al ayudarles, y que ellos son felices porque hablamos con ellos; bromeamos con ellos y les consolamos. Los pacientes sienten que «aún son seres humanos» cuando les toco.

AMRI BYARUGABA, ASISTENTE MÉDICO

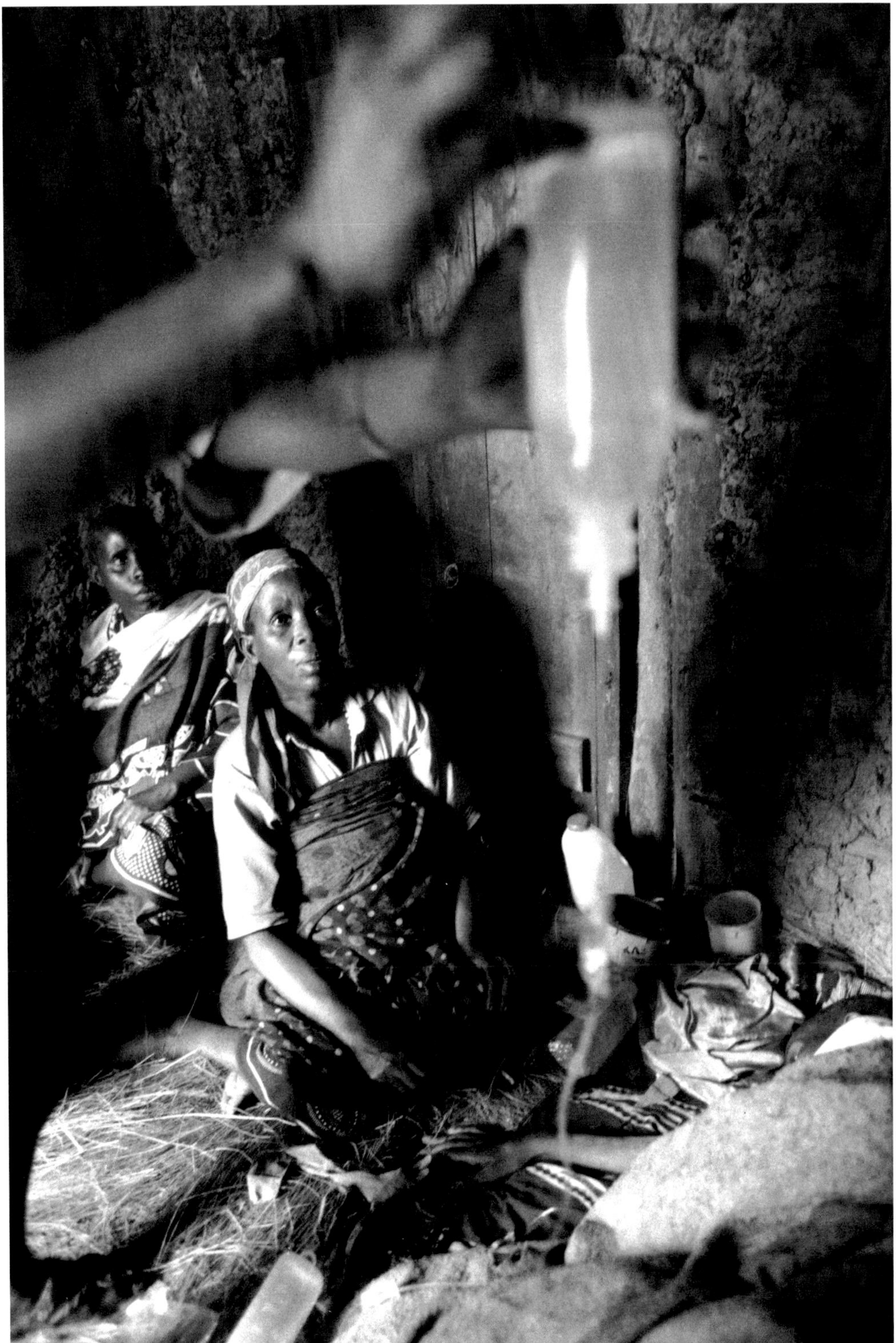

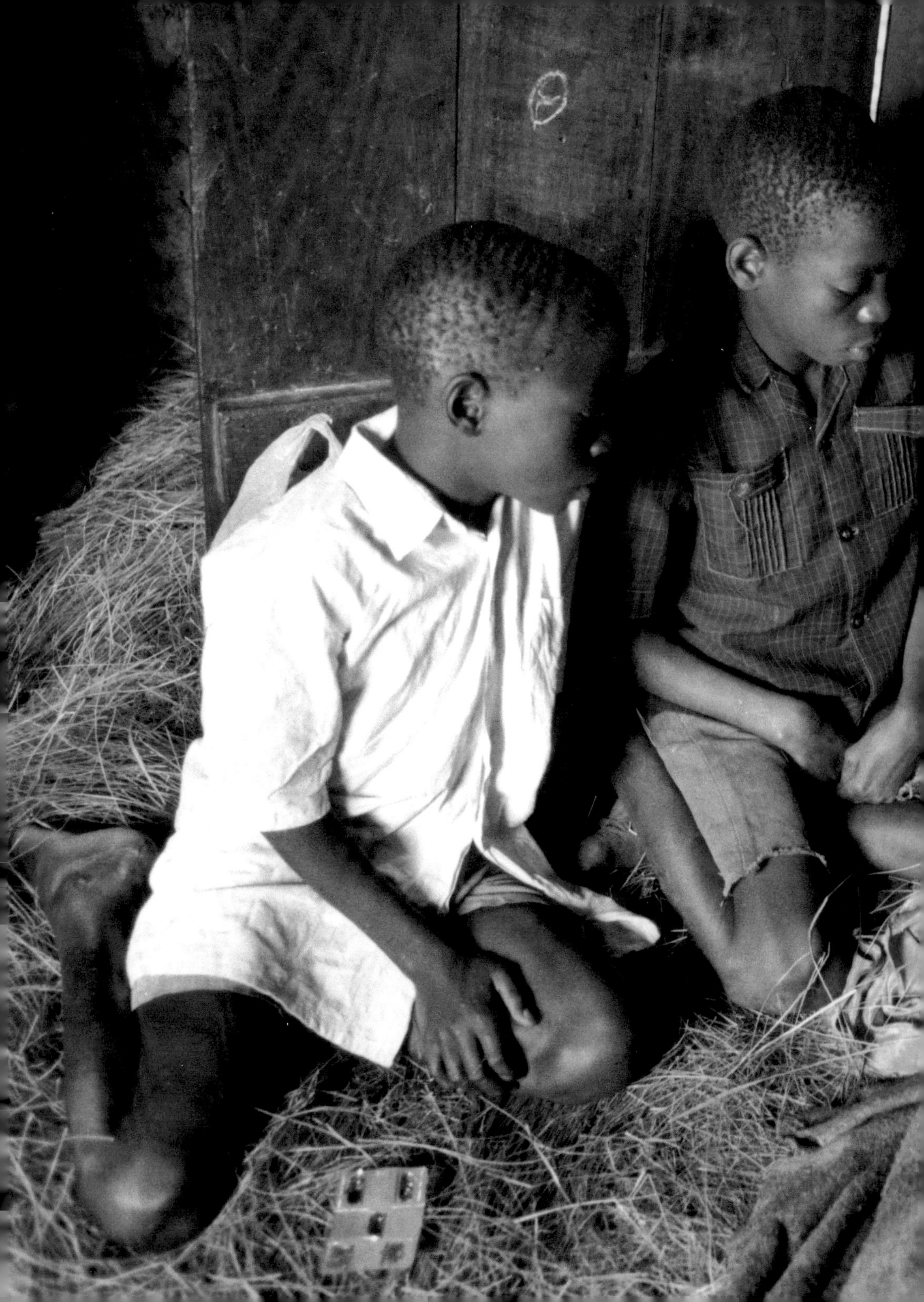

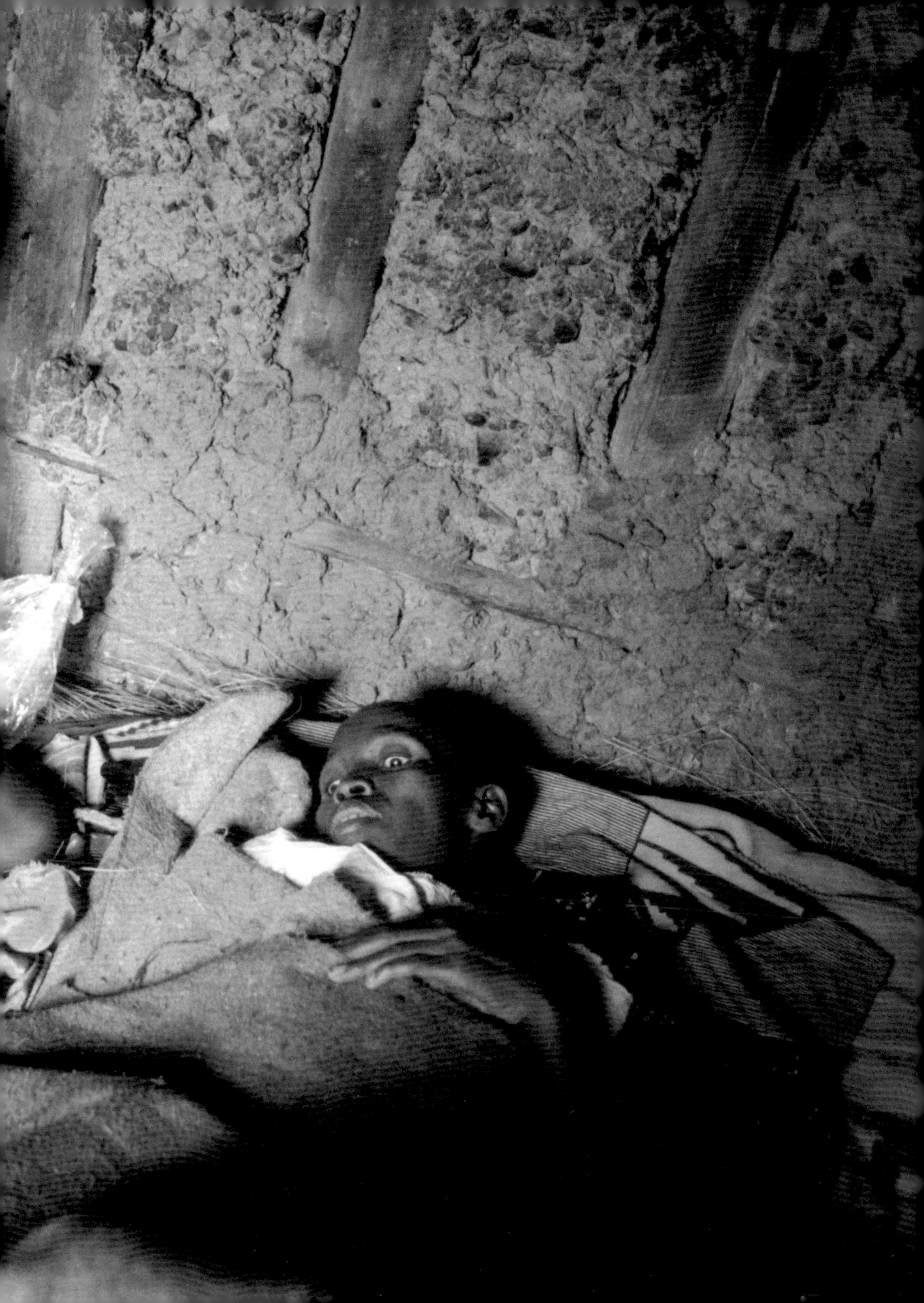

Huérfanos. Copperbelt, Zambia

Las familias numerosas están a punto de extinguirse como consecuencia del número de muertes a causa del sida.

Yo nací en 1934. Tuve nueve hijos. Todos mis hijos están muertos excepto uno. Sufro mucho. Cuido de ocho nietos huérfanos: Cecilia, Cholo, Dyanis, Joseph, Jennifer, Douglas, Lydia y Grace. Y de mi marido ciego. No soy feliz. Esta enfermedad destruye a tantas personas. Los padres de los niños que cuido murieron de sida. En ocasiones me enfado mucho, sufro intentando cuidar de tantos y sin ingresos. Estos niños no tienen ropa. No tengo una cama. No tengo sábanas, sólo sacos vacíos. El suelo no está pavimentado en nuestra cabaña; los niños duermen sobre el polvo. Únicamente puedo dar a cada uno una pequeña cantidad de comida. Si no quedan satisfechos comienzan a pelear y yo me pongo a llorar. La hija que me queda está enferma a menudo y temo que tenga el VIH. Sólo de pensarlo, lloro. Me hago mayor y ya no puedo cultivar alimentos para mí misma. Puedo morir en cualquier momento. ¿Quién cuidará a mis nietos? ¿Cómo sobrevivirán cuando yo falte?

ANNA MULENGA

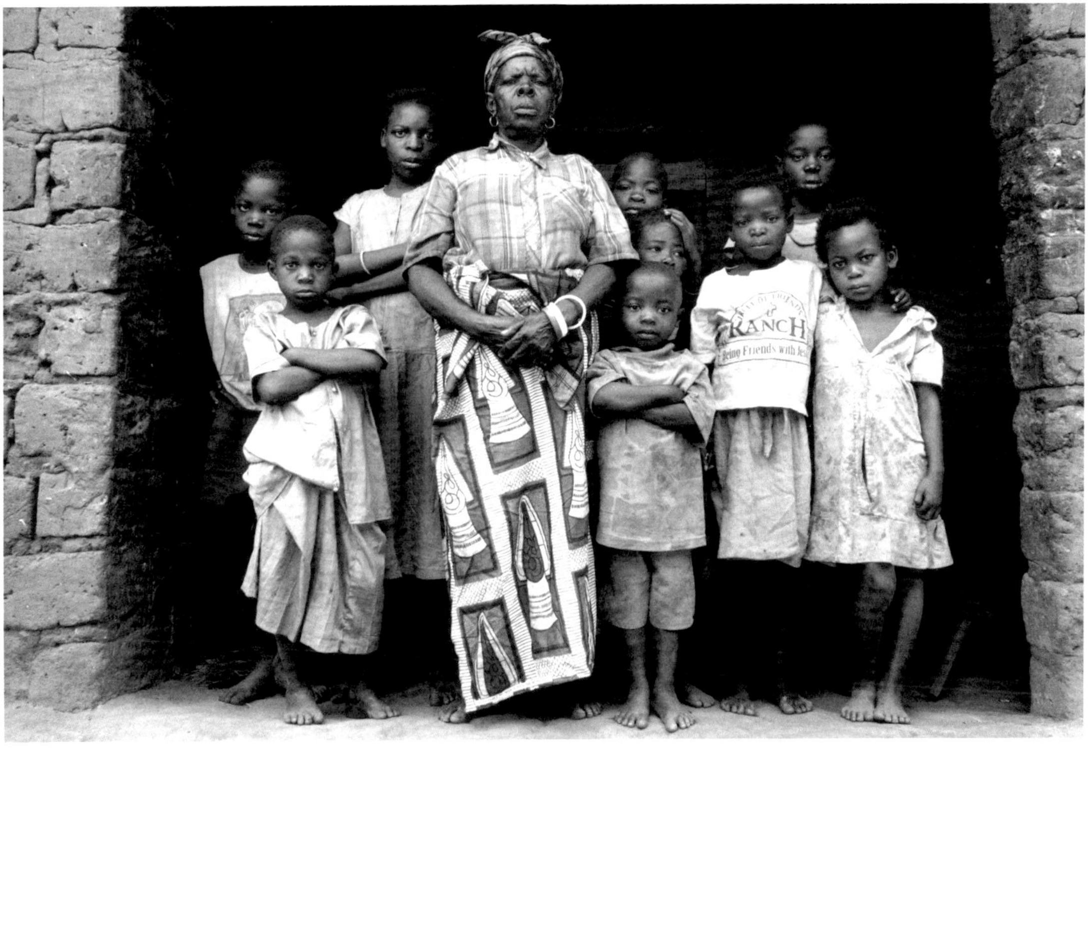
RANCH
Being Friends with

Me llamo Sola Chasala. Tengo diez años y vivo con mi hermana mayor Beatrice, ya que mi padre y mi madre están muertos. Mi madre murió en diciembre. Cada día lavo los platos, limpio la casa y voy al colegio.

Me gustaría vivir junto a todos mis hermanos y hermanas pero no hay suficiente comida para que todos vivamos en la misma casa. A veces los veo a todos durante el fin de semana, pero me gustaría verlos más a menudo.

Me gusta vivir aquí. Me gustan las casas y los árboles, y me gusta ver los aviones que nos sobrevuelan. Pienso que quizás el presidente se halla en su interior. Si fuera presidente reduciría el precio de la comida y del carbón para que todos pudieran comer.

SOLA CHASALA

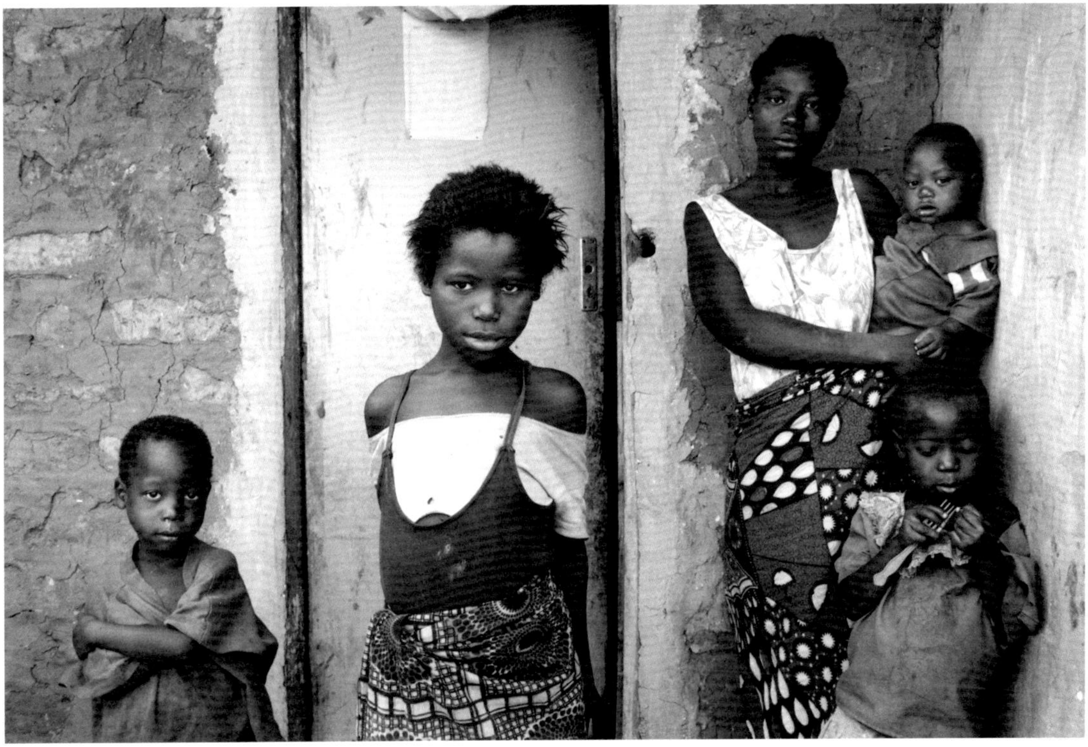

Mi nombre es Janet Mbulo y tengo 20 años. Desde que nuestro padre murió, el 17 de noviembre de 1997, todo ha sido muy difícil para nosotros. Era el único que ganaba un sueldo. Nuestra madre murió en 1998. Incluso antes, nuestros padres estaban enfermos; los hijos mayores tuvieron que hacerse cargo de los más pequeños. Hemos logrado mantenernos juntos en nuestra casa familiar en el recinto de Kwacha, pero es difícil sobrevivir. Solíamos hornear tortas para venderlas en el mercado, pero ya no podemos hacerlo porque nos han cortado la electricidad ya que no pagamos las facturas. Ganamos algo de dinero tejiendo felpudos para vender, y también nos ayuda una organización que nos da harina de maíz cada mes y paga el colegio de Joyson, Joshua y David. Monica, Musonda y Christopher se quedan conmigo en casa.

JANET MBULO

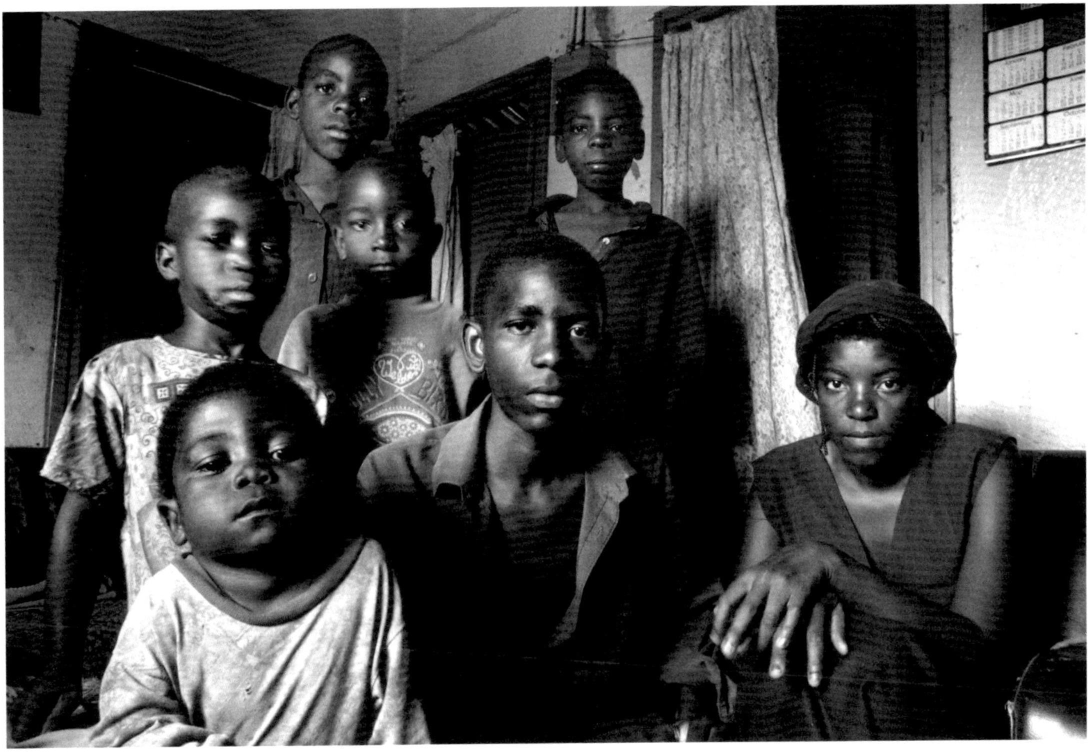

Este colegio es gratuito y yo soy maestra voluntaria. Los alumnos vienen aquí porque no pueden pagar las tarifas de los colegios estatales. Deben pagar las colegiaturas y los zapatos y uniformes, mientras que aquí no lo exigimos. Sólo pedimos una contribución mínima de 1.000 kwacha al mes. Algunos no pueden pagarla, pero les dejamos venir al colegio. Es por el bien de la comunidad, pues si no estuviera aquí, muchos niños no recibirían ninguna educación.

Hay 87 niños en mi clase, de los que 47 son huérfanos. Informamos a los niños sobre el sida para que intenten protegerse de esta enfermedad cuando crezcan.

PRISCA MWANSA, ESCUELA COMUNITARIA DE IPSUKILO

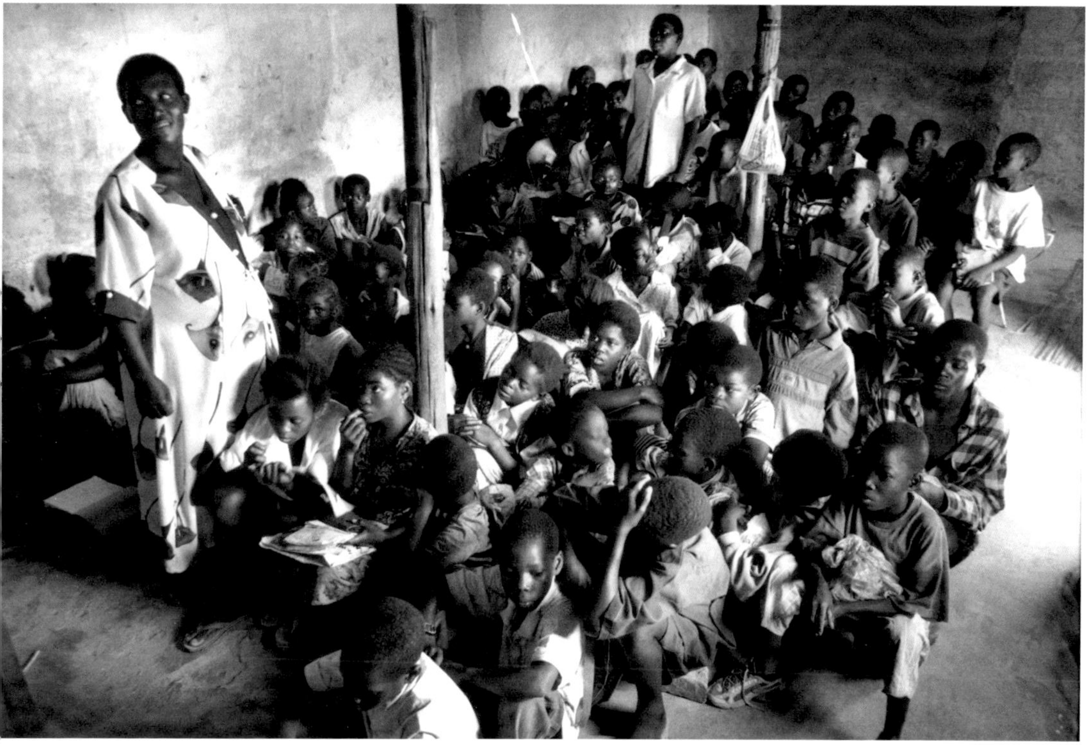

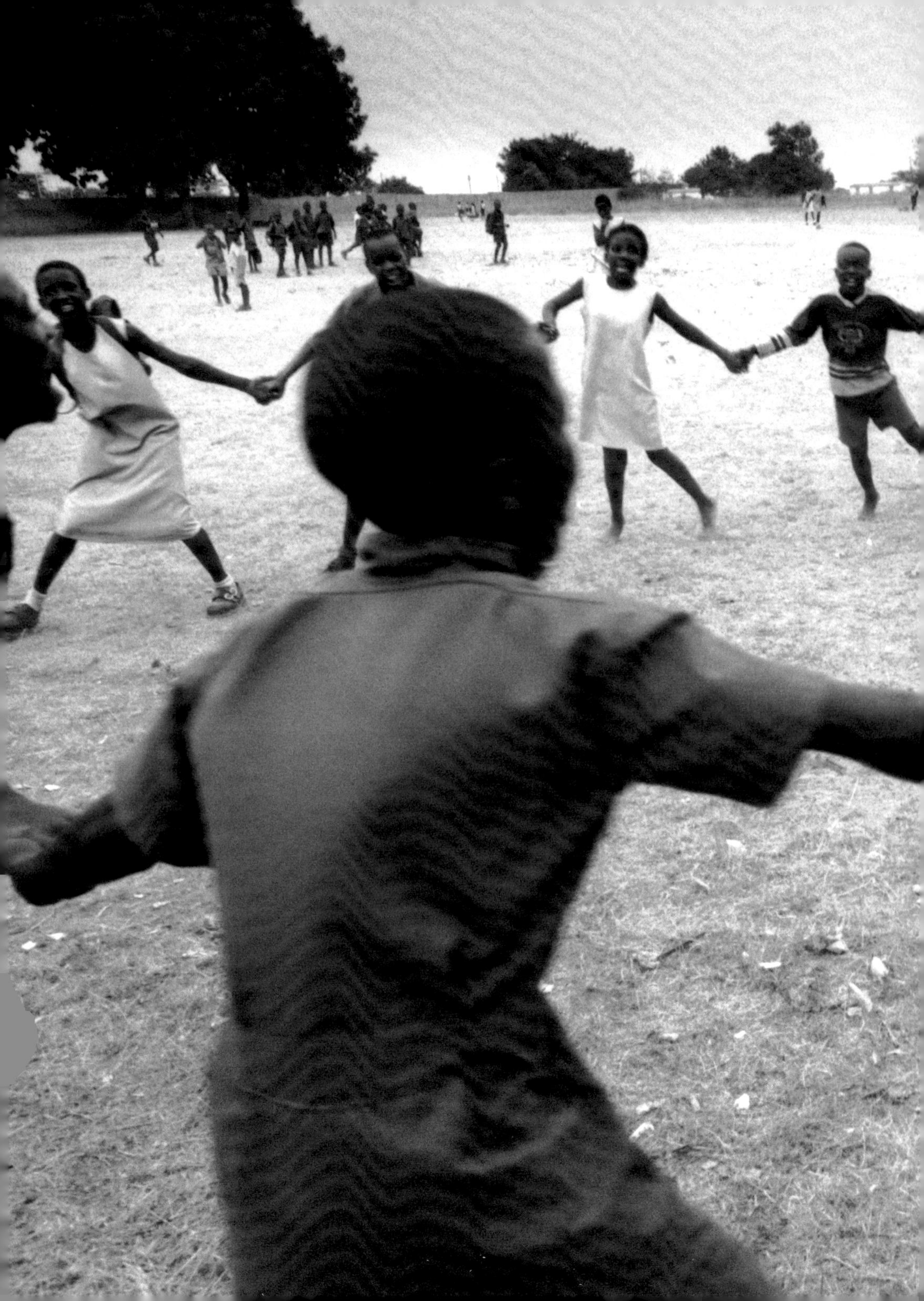

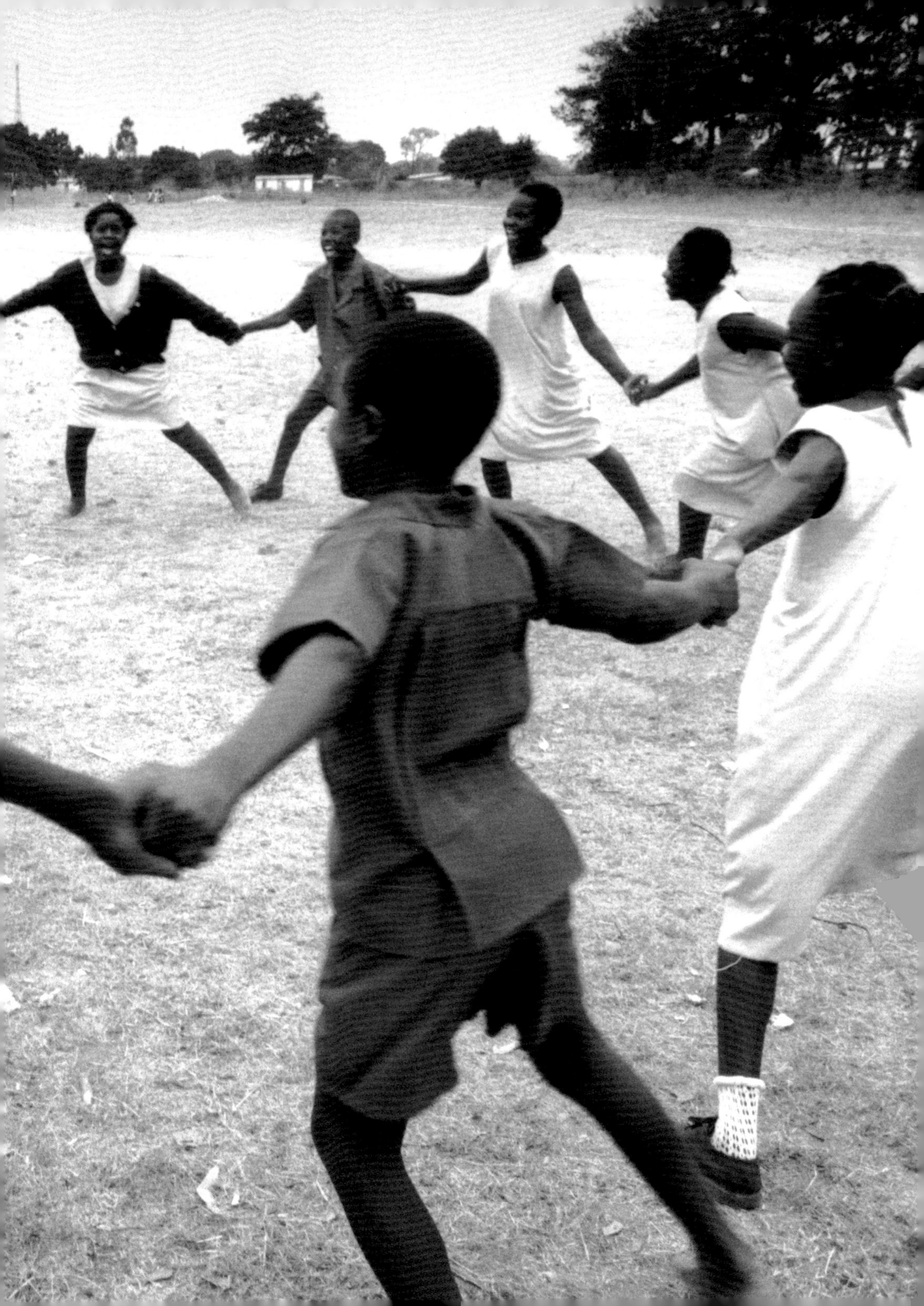

Familia Mwila. Chilonga, Zambia

Joanna, de 7 años, Joseph, de 9, y Aaron, de 2, son huérfanos desde 1995.

Cuando mis seis nietos vinieron a vivir conmigo, me sentí muy mal, pero acepté el hecho de que era la única persona que podía cuidarlos. No he recibido ayuda de otros familiares, sólo la del convento y la unidad antisida. Tengo 72 años y me resulta muy difícil cuidarlos adecuadamente, por lo que en ocasiones tengo que pedir ayuda a los vecinos. He empezado a cultivar judías, cacahuetes y maíz. Tengo cerca de una hectárea de tierra. Nadie me ayuda con la labranza, excepto los niños.

Lo más difícil es darles de comer. Les observo jugando a su juego favorito, que es imaginarse que cocinan y comen. No sé lo que les pasará cuando crezcan, pero espero que sólo sean cosas buenas.

ALBINA MWILA

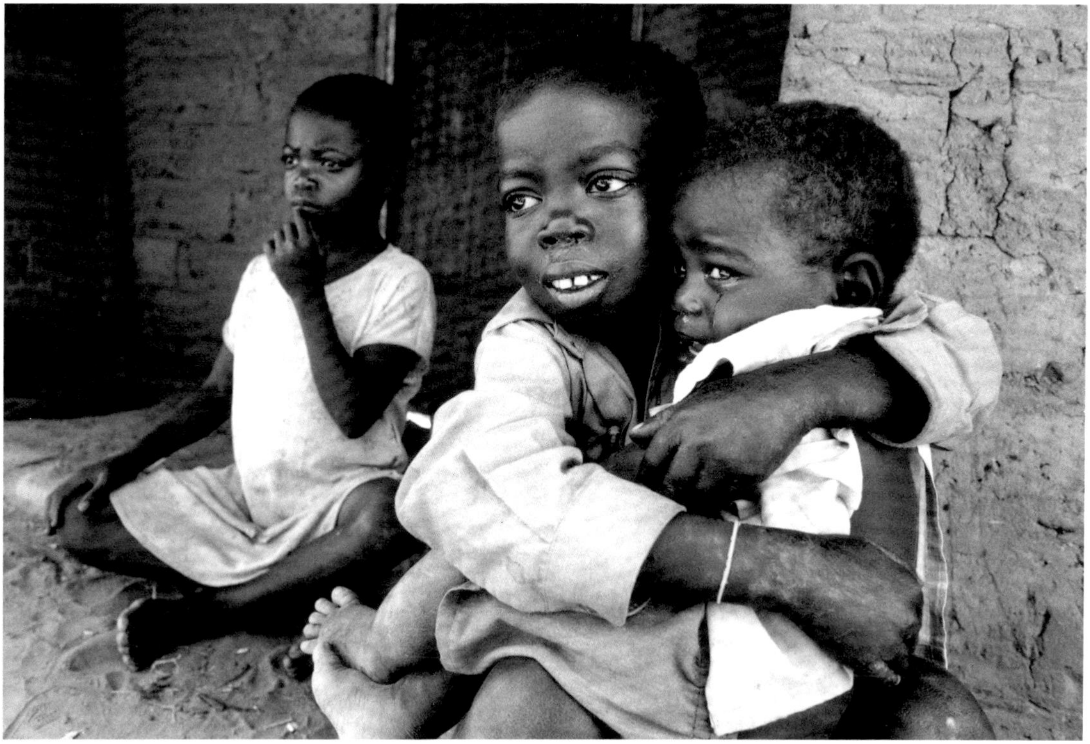

Familia Gatsi. Chinhoyi, Zimbabue

Melody, de 7 años, Russie, de 12, Joshua, de 11, Bianca, de 10, y Esnath, de 13, han perdido a sus dos padres. Cuidan de sí mismos mientras su hermana mayor Grace, de 17 años, trabaja como empleada doméstica en Harare.

A veces pasamos tres días sin comer. A veces no vamos al colegio porque los otros niños se burlan de nosotros porque nuestros uniformes suelen estar sucios, y a veces tenemos demasiada hambre para hacer los deberes. Tengo 12 años y estoy en sexto curso. Grace nos da algo de dinero, pero no es suficiente. Habitualmente trabajamos en casas particulares a cambio de ropa y comida.

JOSHUA GATSI

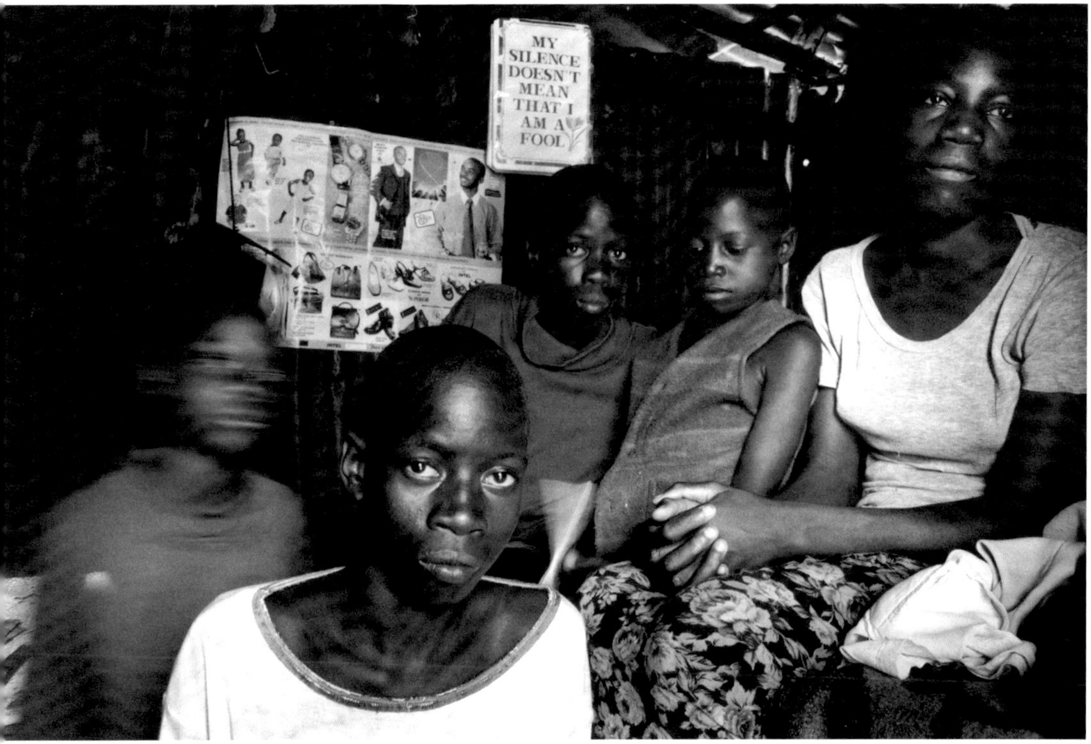
MY
SILENCE
DOESN'T
MEAN
THAT I
AM A
FOOL

Thembeka Sibiya. Municipio de Enseleni, Sudáfrica

Thembeka, de 11 años, cuida de su hermano menor Sandile y de sus hermanas Lihle y Quinn.

Cuando mis padres murieron nos quedamos solos en la habitación. Los vecinos nos dieron comida. Para ganar dinero llevaba mensajes a la gente. En las épocas en las que no teníamos comida mis hermanas pedían las sobras a nuestros vecinos. A veces la mujer del reverendo nos traía comida, y ahora nos hemos trasladado a su casa. Antes de eso vivimos solos durante dos años. Mi deseo es poder estudiar para convertirme en trabajador social y poder ayudar a otros niños de la misma manera que me han ayudado a mí.

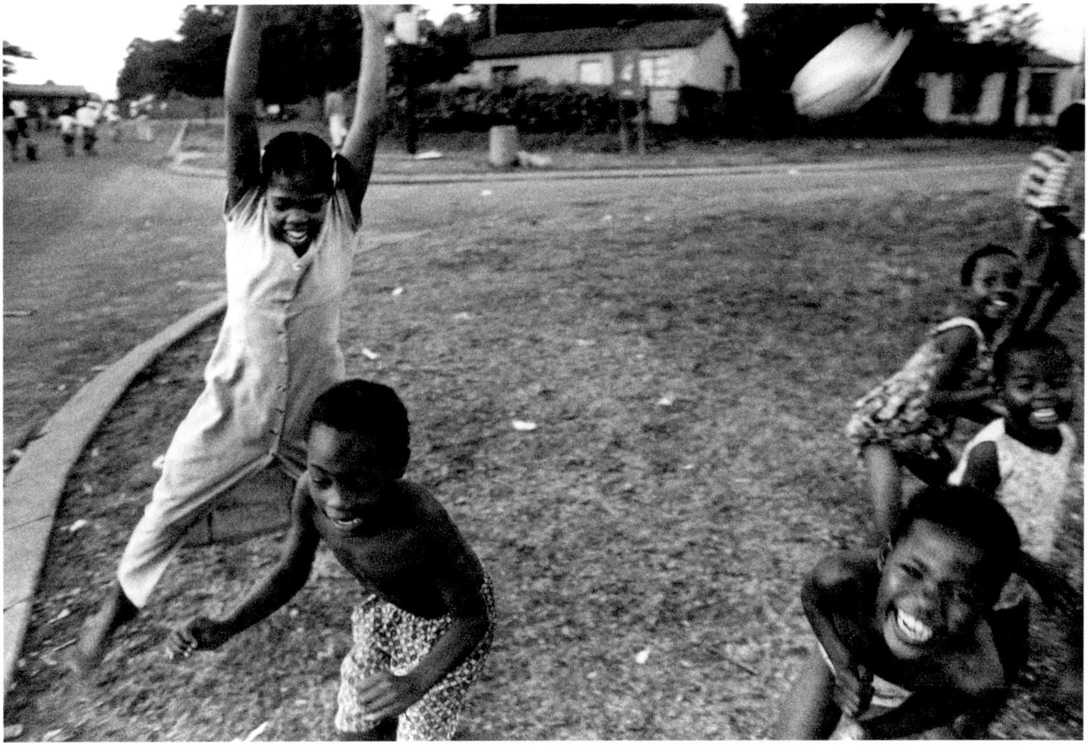

Peter Mguni. Municipio de Enseleni, Sudáfrica

Peter, de 8 años, está enfermo de sida y está a cargo de su tía abuela Emily Mguni.

Toda la familia de Peter está muerta. Es el único superviviente. Su abuelo es el hermano de mi padre. Cuando sus padres murieron se quedó solo. Yo cuidé a su madre hasta que murió porque nadie quería cuidarla. Tengo una familia de diez miembros a mi cargo y sólo percibo una pequeña pensión. Intento llevarlo a la clínica porque está cubierto de llagas, pero está muy lejos y es caro ir en taxi. A veces no puede caminar y tengo que llevarlo sobre mi espalda.

EMILY MGUNI

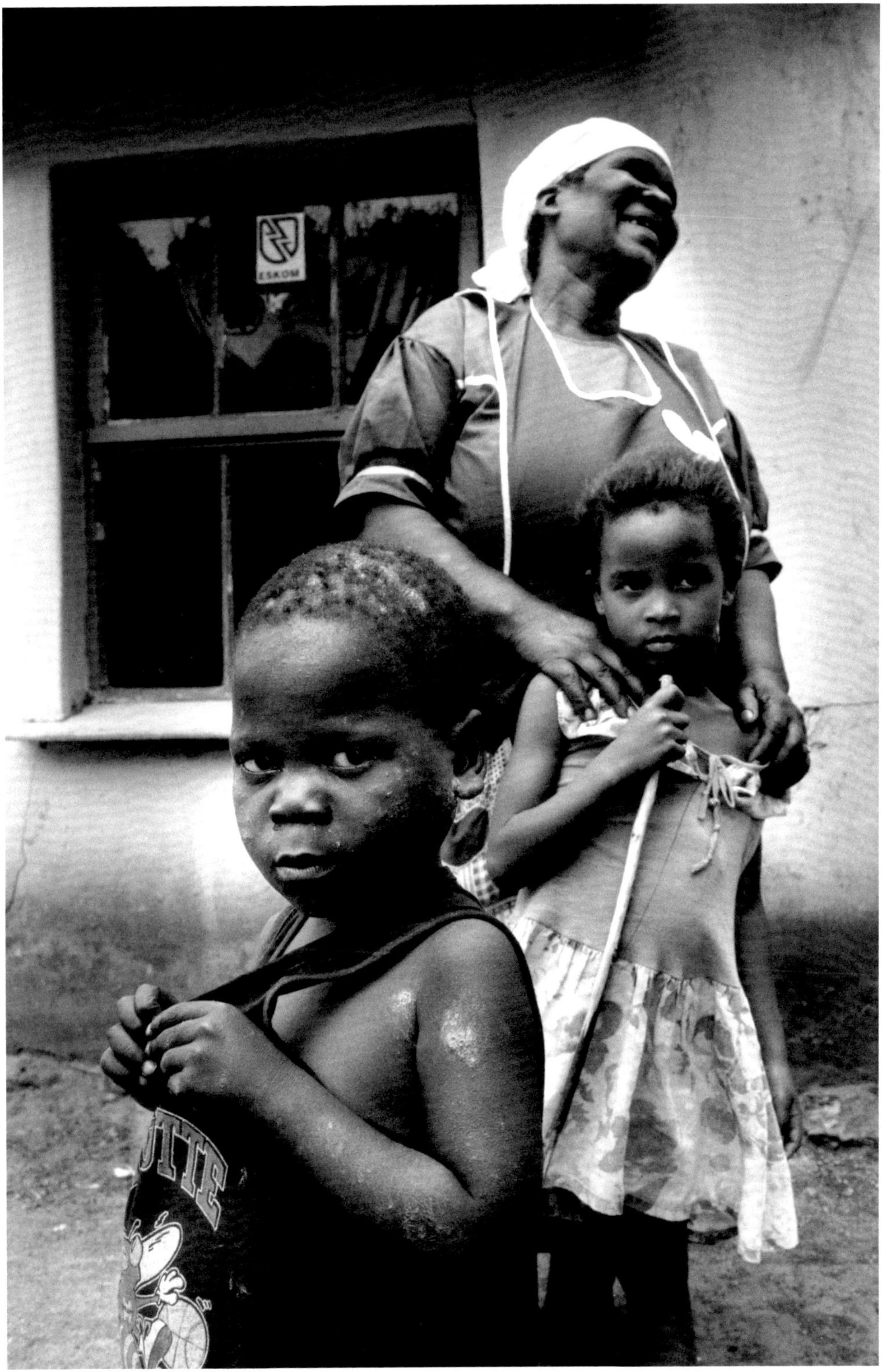
ESKOM

Placidia Karugendo. Bukoba, Tanzania

Placidia, de 13 años, sufre una infección cutánea relacionada con el sida.

Cuando llevo a cabo mi trabajo visitando y cuidando a los pacientes en sus casas, algunas personas realmente me conmueven. Placidia es una de ellas. Padece una infección fúngica en la piel que cubre la mayor parte de su cuerpo. Sin embargo su mirada es brillante y siempre tiene un montón de cosas que contarme. Tiene 13 años. Considero que es muy inteligente.

Creemos que se contagió con el VIH cuando recibió una transfusión de sangre de su padre para tratar un caso de anemia por malaria que sufrió cuando era una niña. Su padre murió el año pasado y su madre está muy enferma, por lo que son extremadamente pobres. Placidia ha estado enferma a intervalos durante el último año. Su rendimiento escolar en el último año ha sido bueno a pesar de su enfermedad y sus ausencias. Durante nuestra visita más reciente nos mostró sus notas con orgullo. Era la primera de su clase.

AMRI BYARUGABA, ASISTENTE MÉDICO

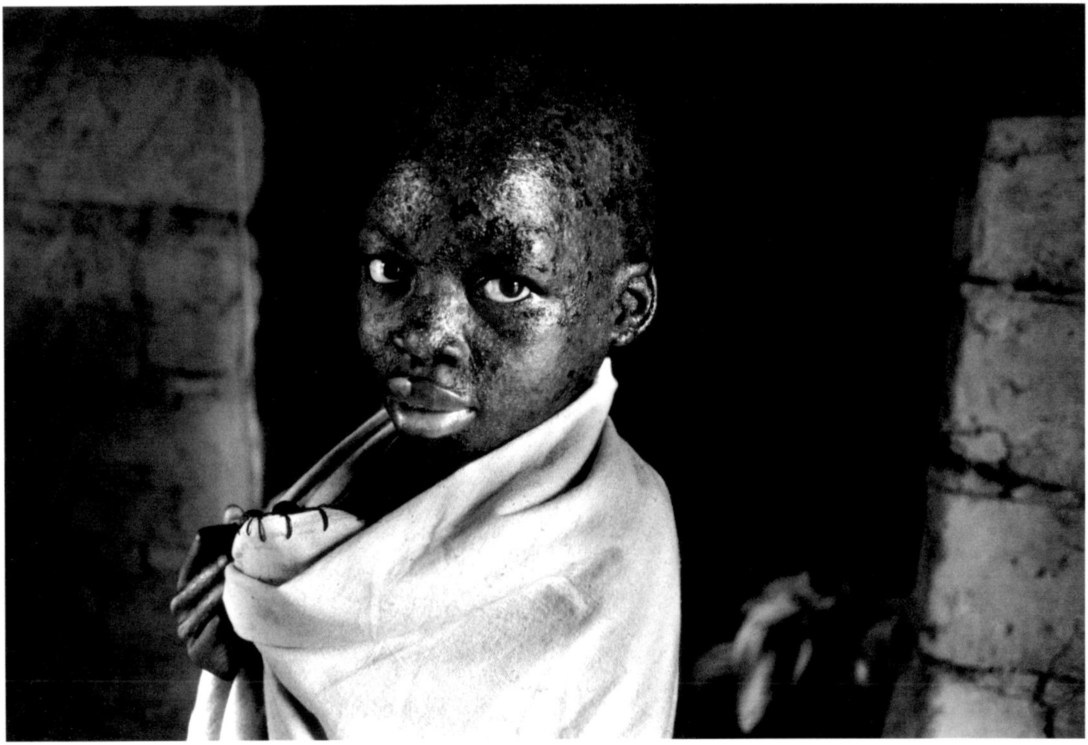

Unidad infantil. Ngwelezane, Sudáfrica

La mitad de las admisiones en la unidad infantil del hospital de Ngwelezane se deben a enfermedades relacionadas con el VIH.

Esta enfermedad ha incrementado enormemente nuestra carga de trabajo.

No sólo son los niños, sino también sus familiares, los que necesitan ayuda, pero no tenemos recursos para ayudar a los padres. Vemos a muchas madres adolescentes con un trauma psicológico por haber perdido a sus bebés y descubrir al mismo tiempo que son seropositivas. No están preparadas para ello. Vemos cómo el sida afecta a menudo a la relación familiar debido a que uno de los progenitores, habitualmente el padre, no acepta el hecho de que el hijo esté infectado. Suelen culpar a la pareja. Ello desencadena violencia y separaciones.

Tengo tres hijos, y como madre me causa un gran dolor ver a tantos bebés y niños agonizando. En ocasiones, como enfermeras, nos sentimos traumatizadas al ver tanto sufrimiento, pero no podemos permitir que esto nos impida desempeñar nuestro trabajo. Debemos continuar poniendo nuestras vidas al servicio de la comunidad y proporcionando los cuidados que se esperan de nosotros.

JUNE MNGADI, ENFERMERA JEFE, UNIDAD DE RESUCITACIÓN PEDIÁTRICA

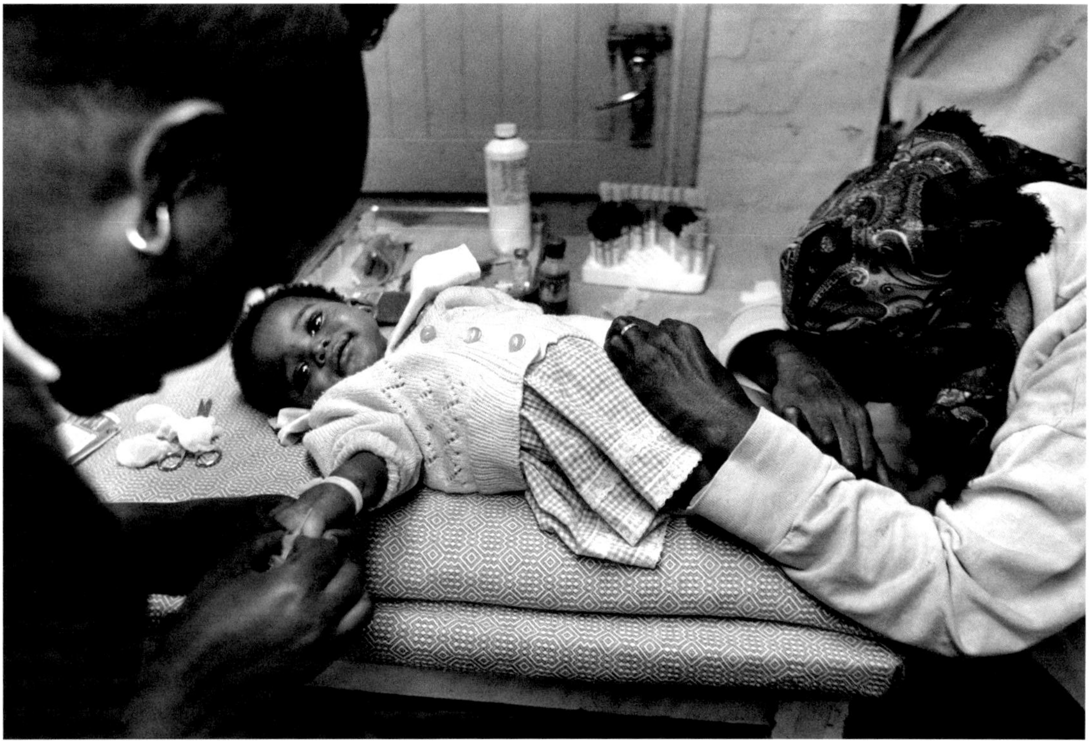

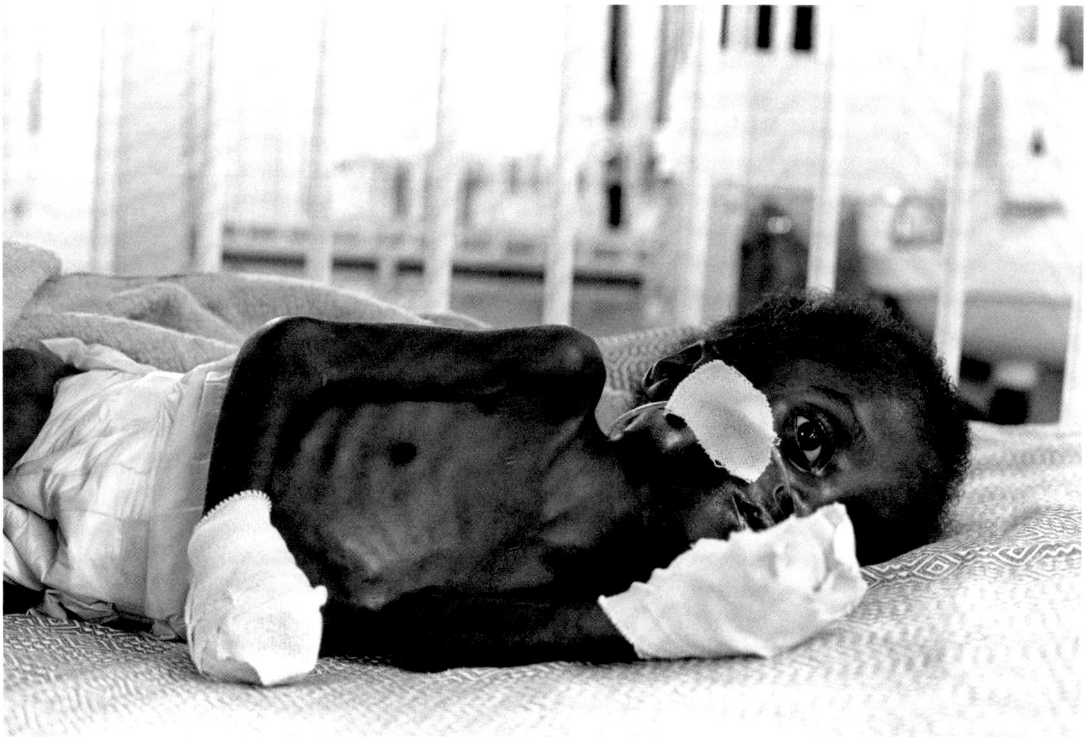

Nazareth House. Ciudad del Cabo, Sudáfrica

Nazareth House es una casa de acogida infantil que cuida de niños abandonados afectados con VIH o sida.

Esta fotografía muestra a Josaphat, que llegó a nosotros en los últimos estadios del sida después de haber sido abandonado en el hospital. Aquí disfruta del masaje que le aplica un aromaterapeuta voluntario. Desafortunadamente, éste debía llevar guantes, ya que el cuerpo del niño estaba cubierto por una infección fúngica. En general estamos convencidos de que hay que dar a los niños tanto contacto físico como sea posible sin emplear guantes. Nos concentramos en ayudar a los niños a vivir su infancia de la manera más normal, feliz y sana posible. Para la mayoría de nuestros niños, éste es el único hogar que conocerán, por lo que debemos asegurarnos de que es un hogar auténtico, lleno de amor y felicidad.

Cuando a un niño le llega el momento de morir, creemos firmemente que se les debe dejar hacerlo con dignidad en un entorno familiar y no en un hospital. Es importante que sea el cuidador con quien el niño tenga un vínculo más intenso quien esté con él para ayudarle en los últimos momentos.

HERMANA MARGARITA, MADRE SUPERIORA

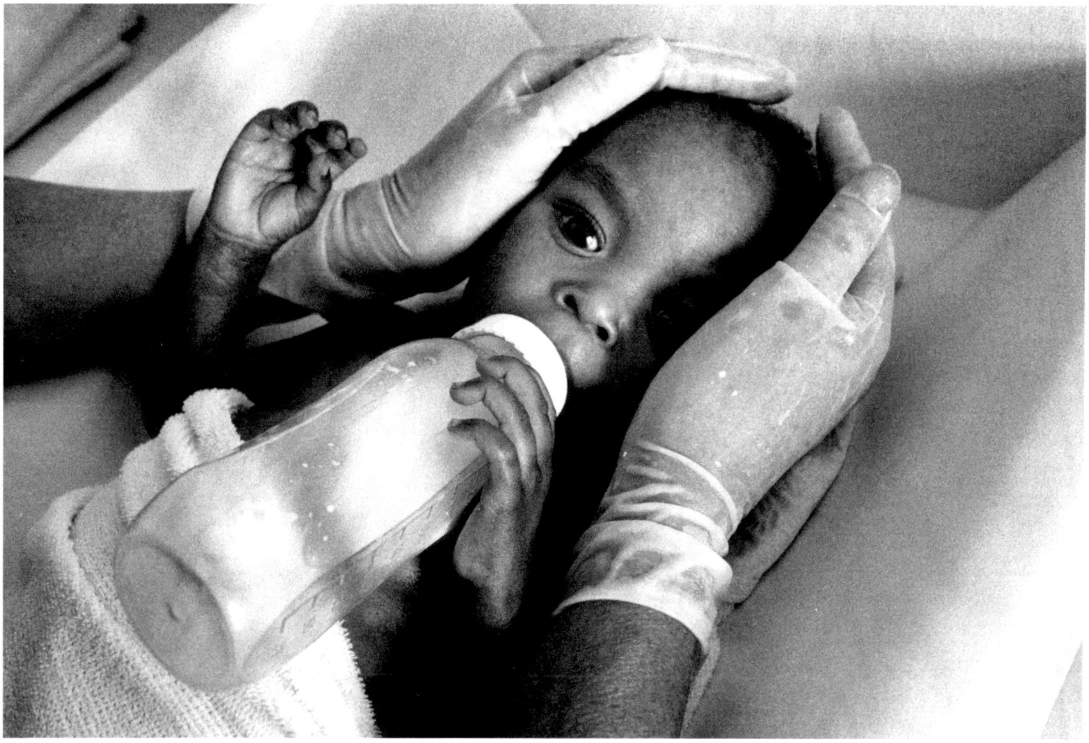

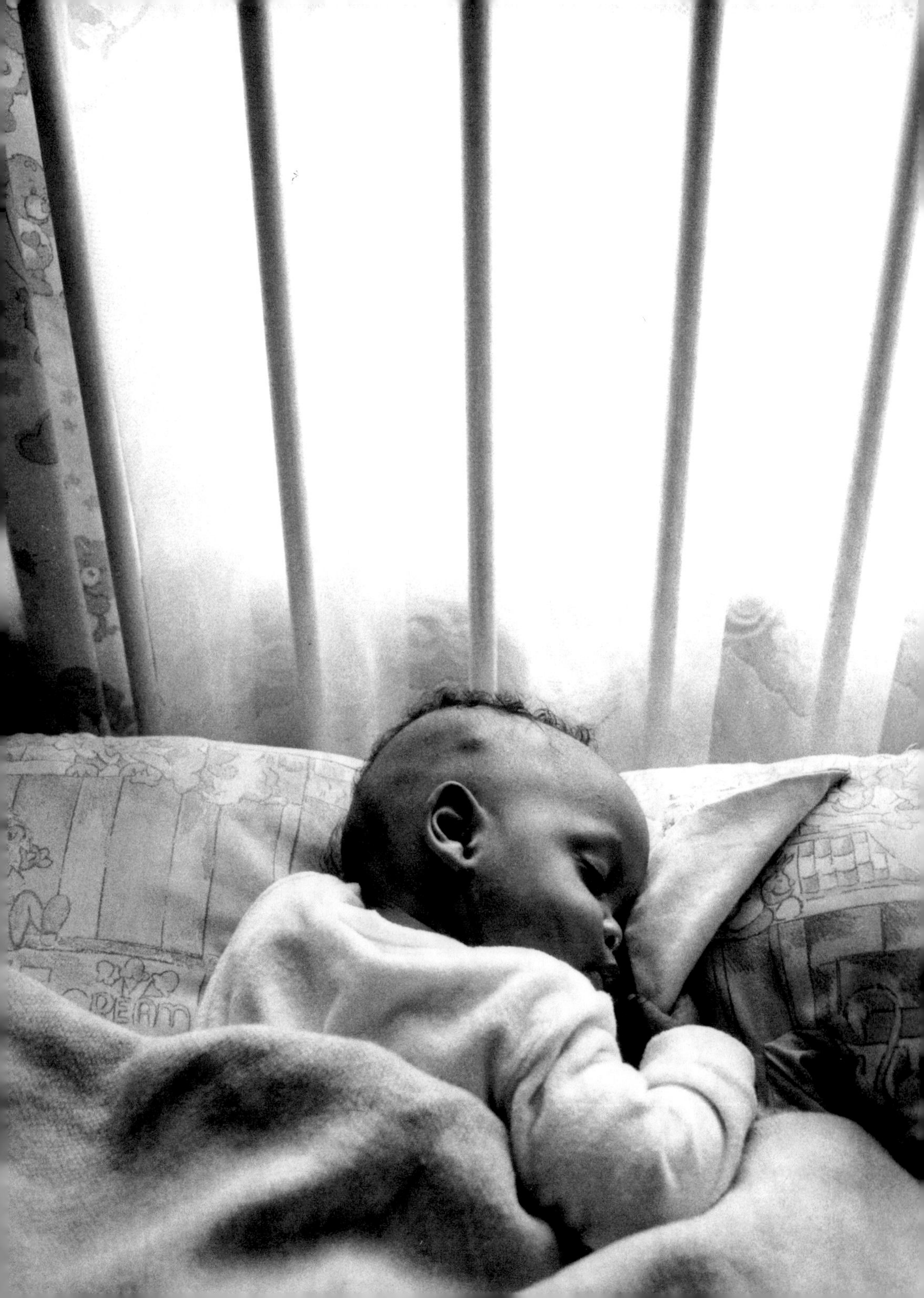

Planta de montaje de Ford. Silverton, Sudáfrica

El Sindicato Nacional de Trabajadores del Metal y el equipo directivo de la planta de Ford participan activamente en la educación y prevención del sida.

Hasta finales de 1998 esta compañía, como tantas otras, se encontraba aún en proceso de especulación sobre la realidad del sida. Nos pusimos en acción en noviembre de 1998, cuando se cerró toda la planta durante una tarde para impartir un cursillo de concienciación. Se dejaron de fabricar 150 automóviles ese día. A partir de abril de 1999 comenzamos a repartir preservativos gratuitos. Al principio nos limitábamos a la clínica, y distribuíamos 700 preservativos al mes; lo extendimos a la entrada de la fábrica y los lavabos. Ahora distribuimos 17.000 preservativos al mes.

JOHAN STRYDOM, JEFE DE RECURSOS HUMANOS

Estoy orgulloso de que en nuestra empresa el sindicato y la gerencia hayan llegado a un acuerdo y llevemos esta política tan acertada. He sido encargado de taller durante siete años. A menudo me dirijo a los trabajadores durante las reuniones generales del sindicato para hablar sobre el sida. Lo hacemos con los empleados de la planta, pero debemos llegar a la comunidad y extender nuestras palabras de todas las maneras. Nuestro plan consiste en intentar involucrar a todos.

JIMMY KHUMALO, SINDICATO DE TRABAJADORES DEL METAL (NUMSA)

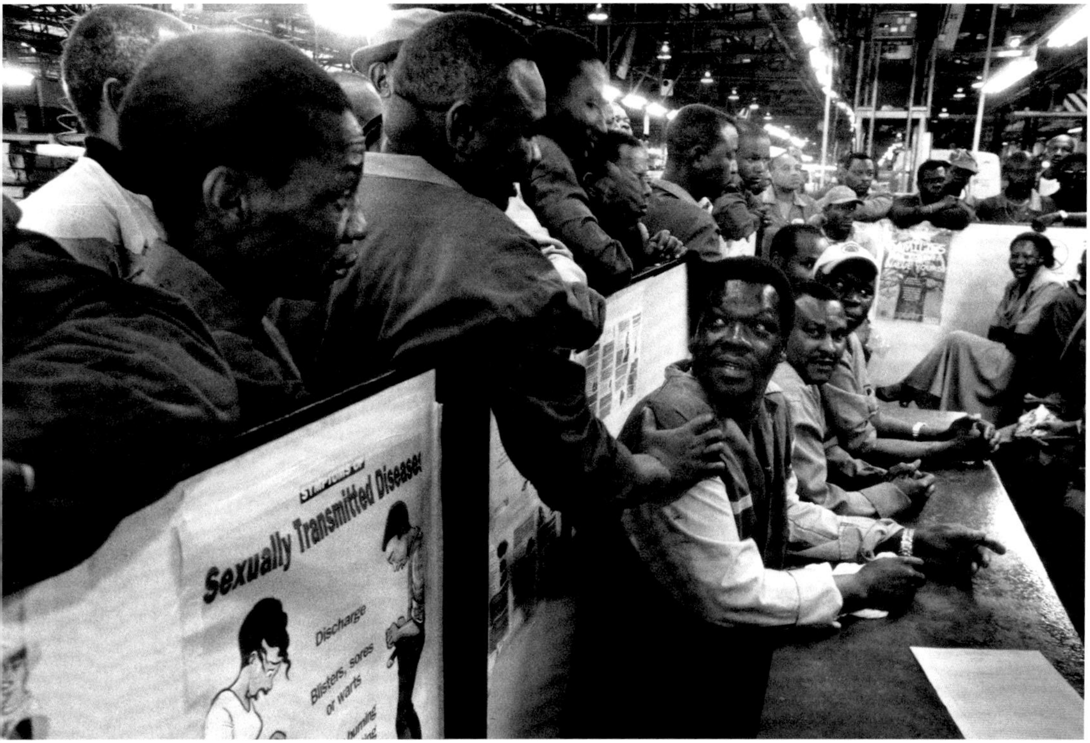
Sexually Transmitted
Discharge
Blisters, sores or warts

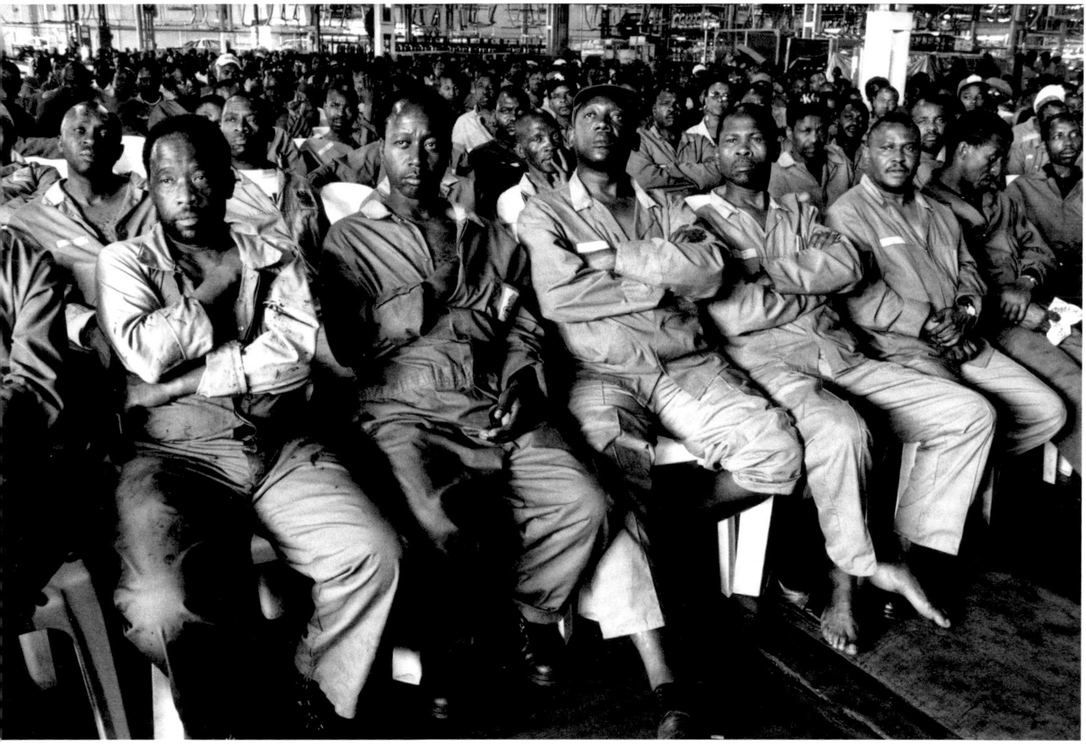

Mina Lonmin. Rustenberg, Sudáfrica

Lonmin es una mina de platino donde, de acuerdo con un sondeo realizado en el año 2000, el 24 % de los mineros son portadores del VIH.

Soy de un pequeño pueblo llamado Tabase, cerca de Umtata, en el Transkei. He trabajado en la mina Lonmin durante nueve años. Superviso las máquinas de perforación así como los desagües y la ventilación. Trabajo 12 horas bajo tierra cada día para ganar tanto dinero como pueda, pues tengo una gran familia bajo mi responsabilidad. Gano 2.200 rand al mes y envío por lo menos 1.000 rand a mi familia. Me alojo en un hostal donde 12 hombres comparten una habitación. En esta mina hay algunas habitaciones en las que se permite a la esposa de un hombre alojarse durante un mes al año, pero es difícil para mi mujer, ya que debe quedarse en casa para cuidar de la familia.

En la mina nos han dado explicaciones sobre los preservativos. Muchos de los hombres van a buscar mujeres. Nuestro sindicato nos ha explicado los peligros del sida. No quiero utilizar preservativo. Los hombres dicen que si mantienen relaciones sexuales, debe ser «piel con piel». Prefiero quedarme en el hostal y no salir. Quiero protegerme para poder mantener y satisfacer a mi familia.

NTOBEKO NGWENZE, SUPERVISOR DE EQUIPO

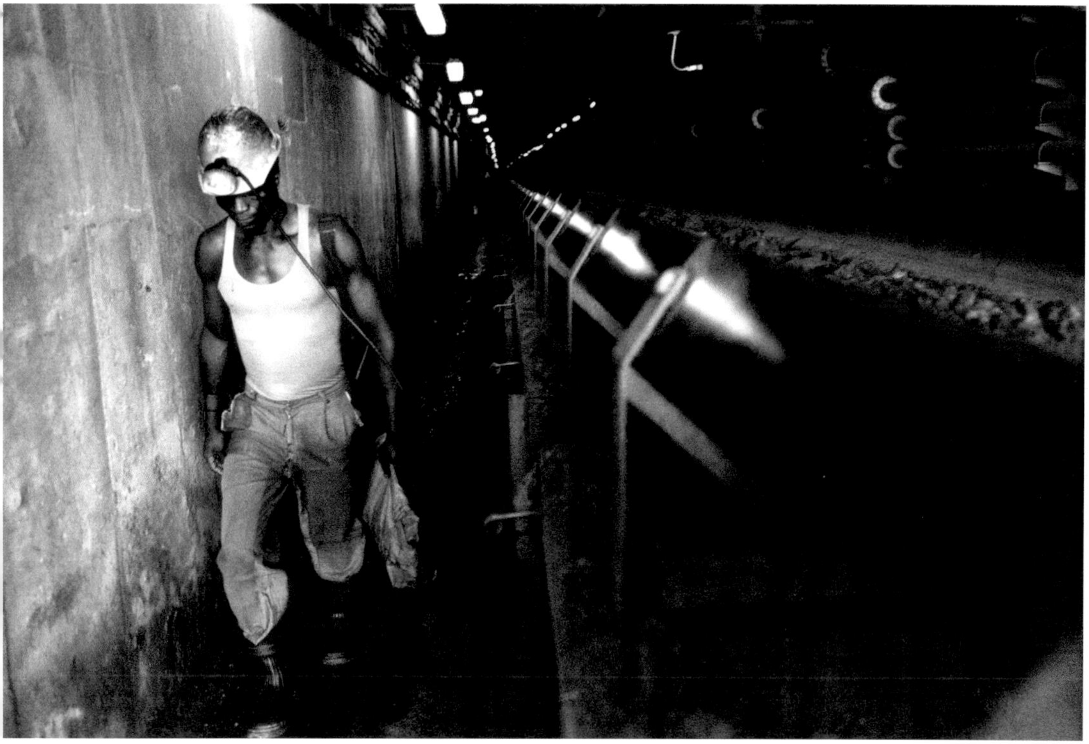

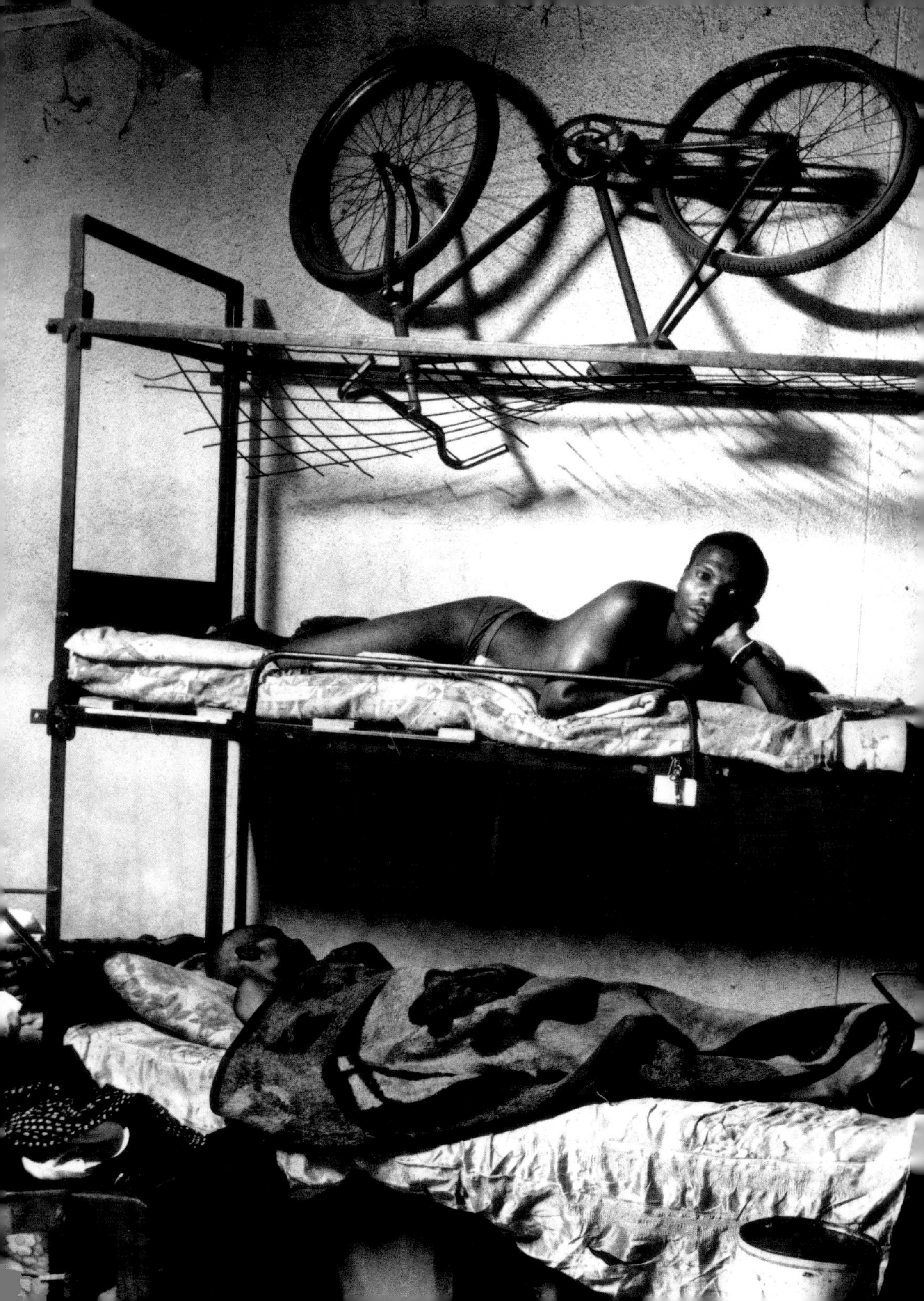

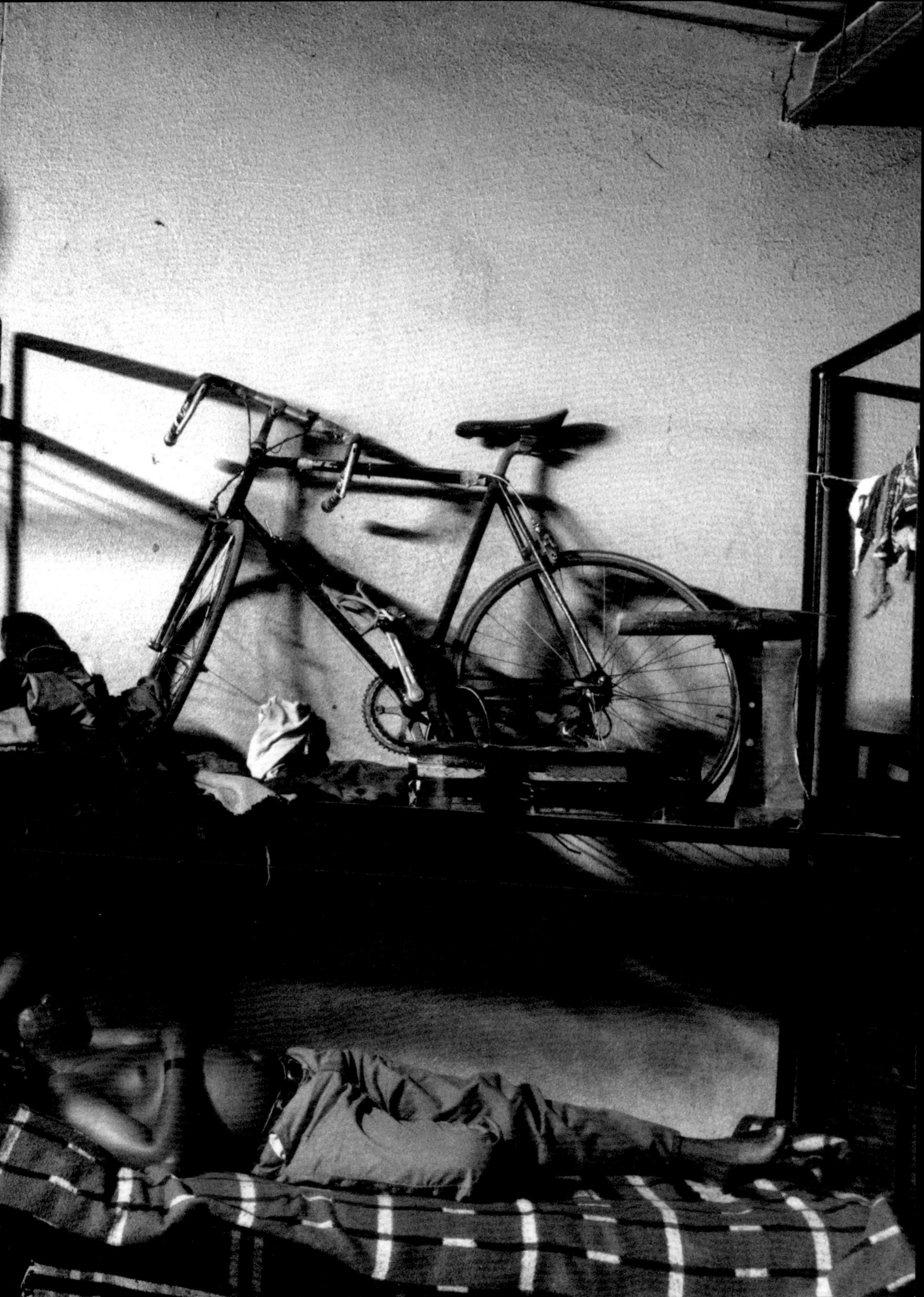

Tengo 31 años. He venido aquí a la mina a trabajar en el negocio del sexo para mantener a mi familia y a mis hijos en el Transkei. Nadie en mi familia tiene un empleo, por lo que soy la única que gana dinero. En septiembre del año pasado conocí a Busi y me habló sobre los preservativos y el VIH. Me invitó a unirme al grupo de compañeros educadores. En el proyecto, informo a la gente sobre las enfermedades de transmisión sexual y el VIH, y realizo demostraciones del empleo de los preservativos. Enseño tanto a los trabajadores de las minas que vienen a los prostíbulos de los alrededores como a las otras chicas que venden sexo.

Resulta difícil para muchas de las chicas convencer a los hombres para que utilicen preservativo. Algunos únicamente pagan la mitad; dicen: «No me acuesto contigo, lo hago con el plástico». Como ahora soy seropositiva, si un hombre quiere mantener relaciones sexuales sin preservativo le digo: «No». A veces los hombres no me creen cuando se lo digo. Dicen que parezco demasiado hermosa y sana. Me cuido mucho.

NOSIPO MPETSHWA, COMPAÑERA EDUCADORA

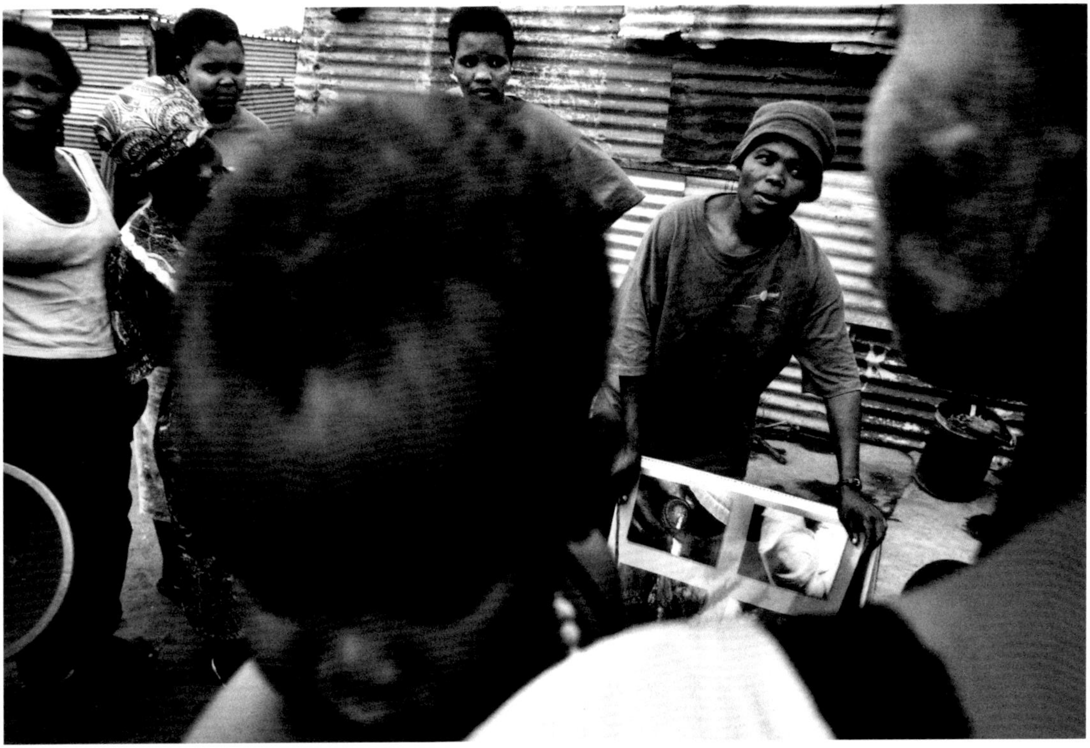

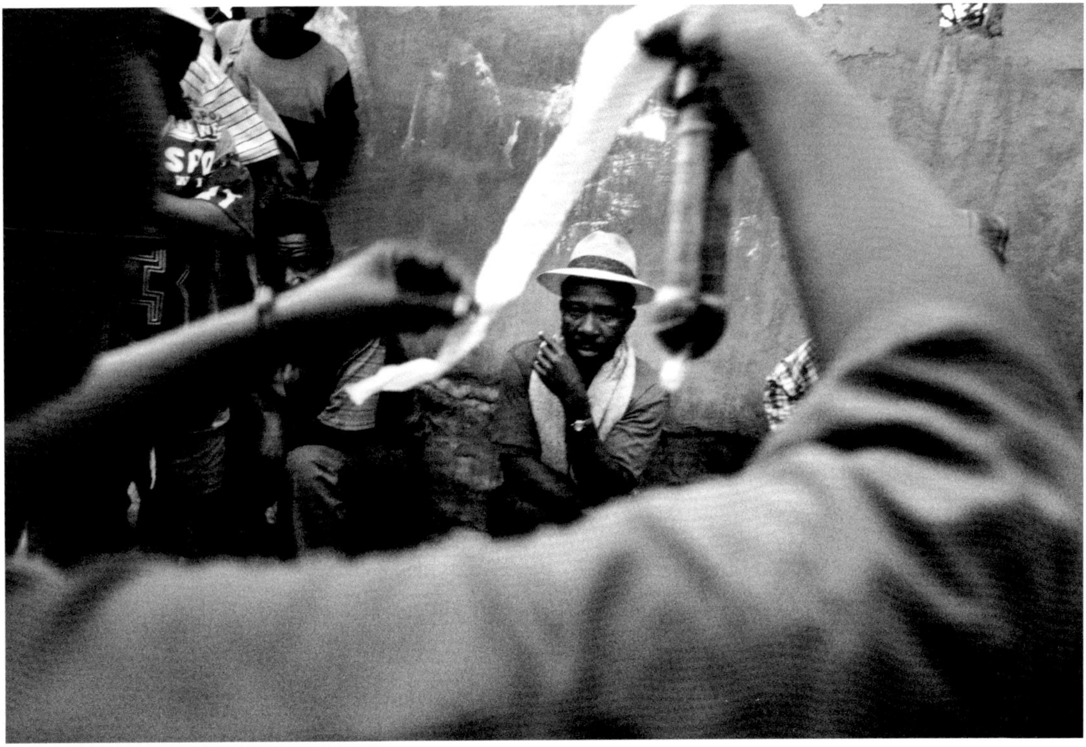

Rose Kamuriwo. Chinhoyi, Zimbabue

Rose vive abiertamente con el sida y participa en un grupo de teatro que educa a los trabajadores del negocio del sexo y a sus clientes sobre el VIH y el sida.

Janet y yo trabajamos en el negocio del sexo durante más de 15 años. Nunca recibimos educación y nunca conseguimos un trabajo decente. No teníamos otra manera de ganarnos la vida. Cuando Janet enfermó, su casero la echó y ahora vive en mi casa. La madre de mi prima Fungai murió de sida y entonces Fungai también vino a vivir con nosotros. Mi novio Clemence también vive con nosotros, y tiene sida. Janet no puede caminar. Mi hija de cuatro años, Eva, también está enferma. Fungai nos cuida a todos.

Disfruto participando en los programas de compañeros educadores, principalmente debido al hecho de que recibo ayuda emocional de mis iguales. También recibo una pequeña compensación a fin de mes, pero más que nada deseo educar a la gente, en especial a los jóvenes, en medidas de prevención. No quiero ver crecer el número de personas infectadas. Quiero dar esperanza a los infectados y ayudarles a vivir de forma positiva y optimista. En muchas ocasiones la gente me señala con el dedo diciendo que tengo sida, pero esto no me aparta de mi trabajo.

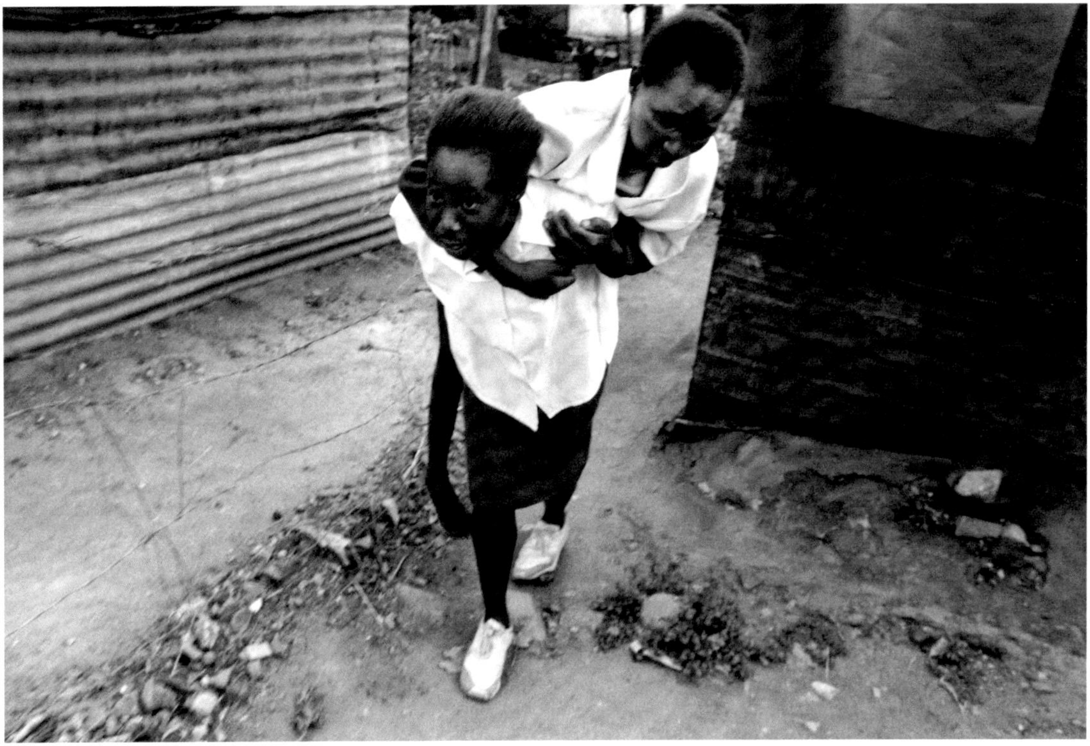

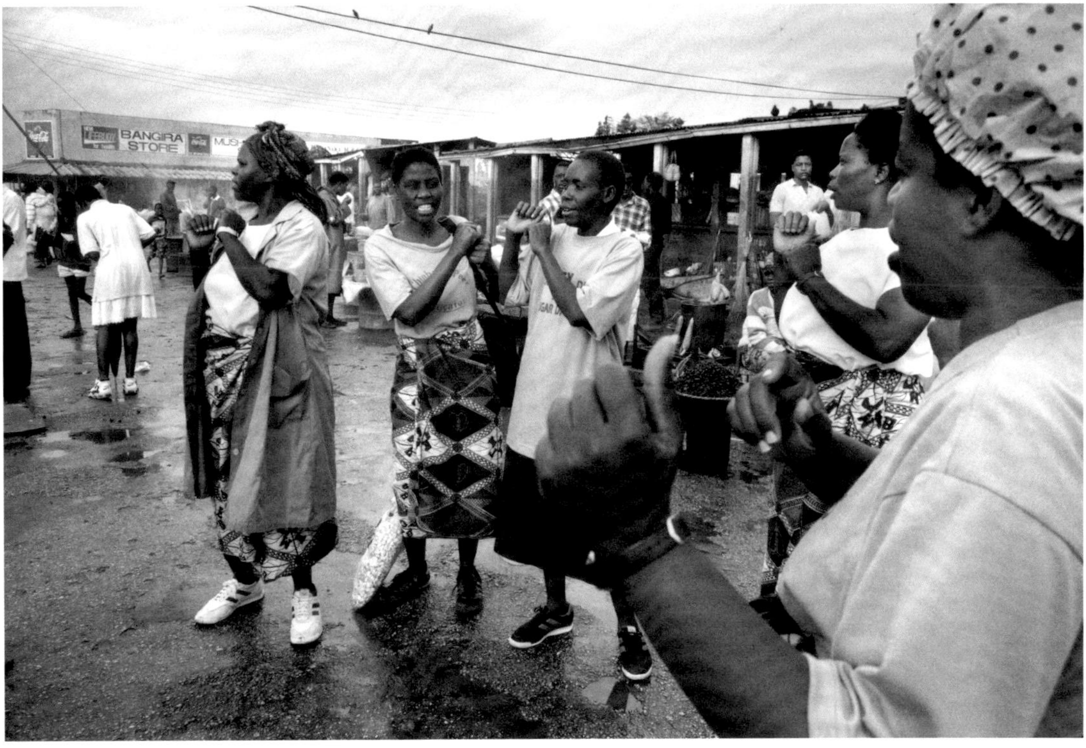
BANGIRA
STORE

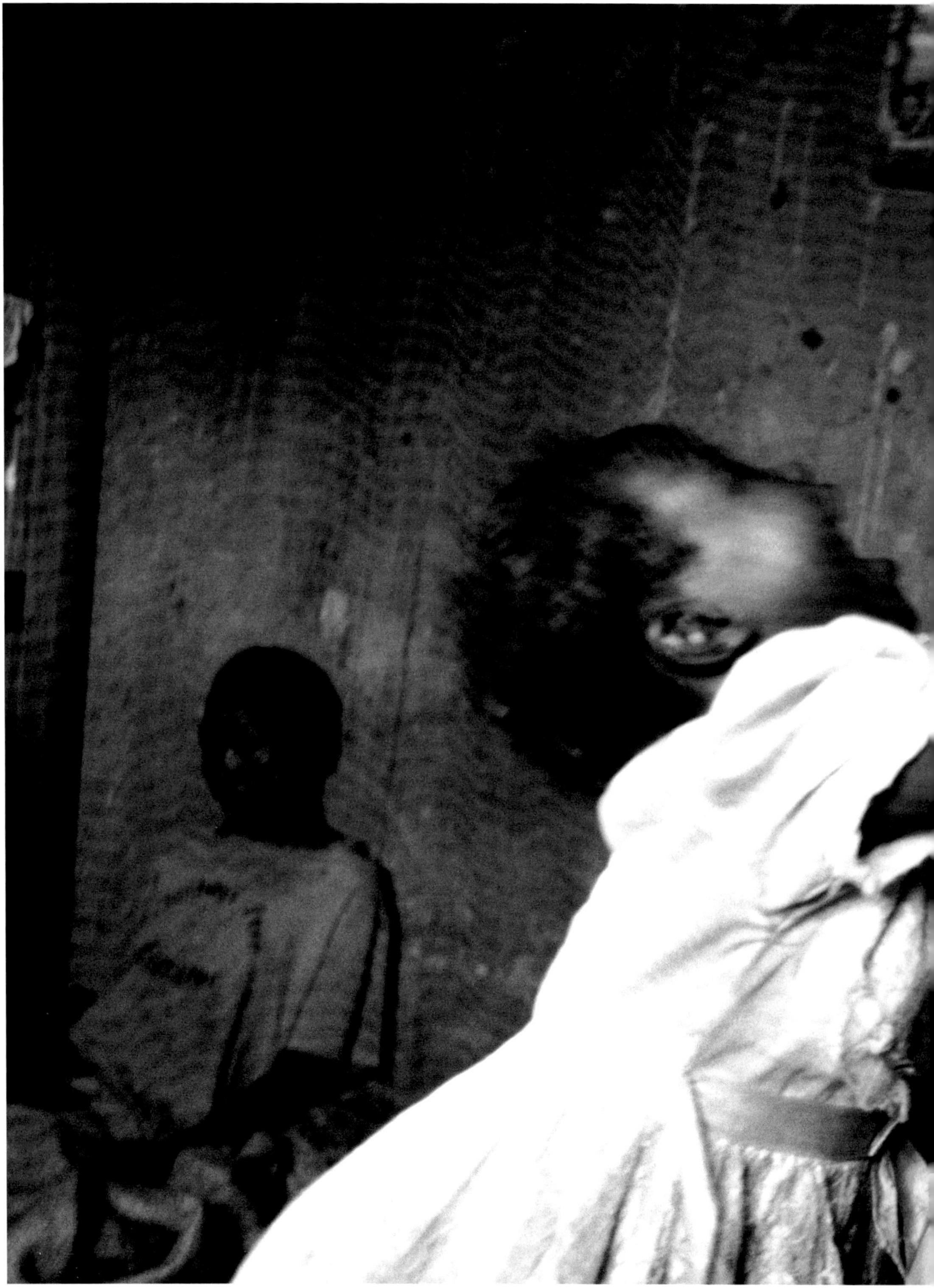

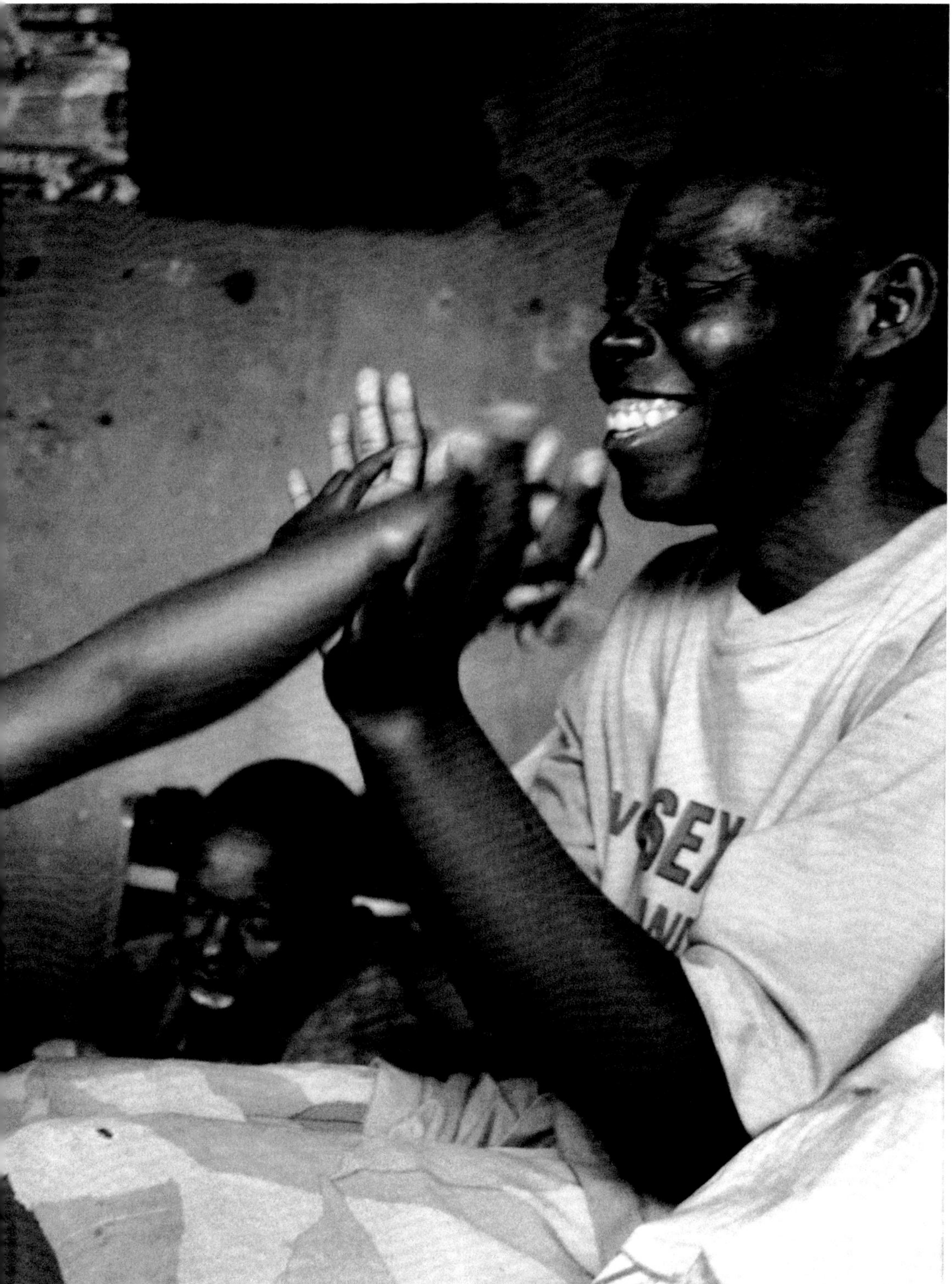

Educadores sobre el sida. Hlabisa, Sudáfrica

En el hospital de Hlabisa, un grupo de pacientes desveló su estatus de seropositivos a la comunidad y recibió formación como educadores sobre el sida. Promocionan la educación sexual en su propia comunidad y en las de los alrededores.

Fui al hospital y me dijeron que tenía sida. Al principio quería suicidarme. Una enfermera habló conmigo y me preguntó si quería ver a otras personas con VIH, para que no me sintiera sola en esta enfermedad. Seis de nosotras nos conocimos en el hospital de Hlabisa. Nos dimos cuenta de que nos sentíamos más fuertes por el simple hecho de conocernos. Recibí formación como educadora sobre el sida junto con los otros miembros del grupo. Gané confianza en mí misma para poder enfrentarme a otras personas e informarles sobre el virus. Enseño a usar preservativos en los centros de salud locales.

Ahora mi salud ha mejorado y recupero peso. A veces incluso olvido que tengo el VIH. Mi marido Jaconia también está luchando contra la enfermedad conmigo, y nuestra relación se ha fortalecido.

JABU SHEZI

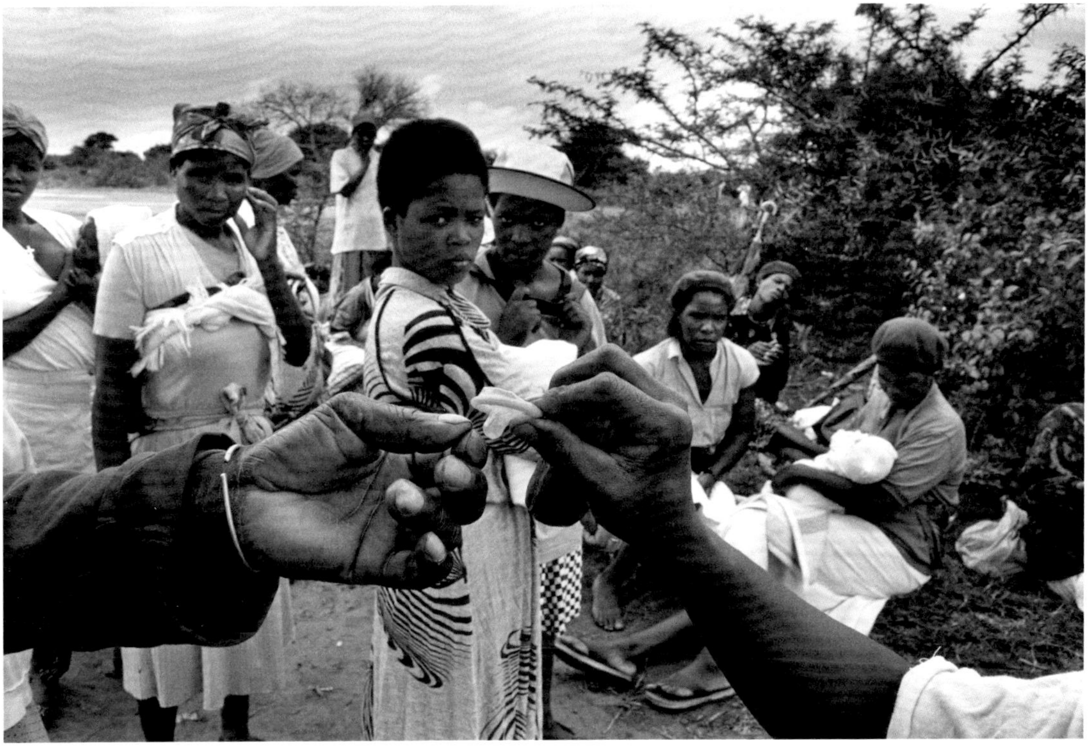

Cuando mi esposa Jabu me dijo que tenía esta enfermedad, pensé que era el fin. Entonces recapacité y me di cuenta de que no era así. He aprendido a vivir con la enfermedad y ahora quiero a mi mujer cada vez más. Es cierto que tengo algunas novias, pero ahora utilizo preservativo. Se tienen las mismas sensaciones. Es lo mismo que piel con piel.

No sé cuándo llegará el momento en que el sida me mate. Me siento presionado y por eso estoy construyendo esta nueva casa para mis hijos. Quiero terminarla pronto ya que estoy perdiendo peso y me estoy debilitando.

Aprendí yo solo a tocar la guitarra zulú en mi juventud y he escrito canciones sobre el sida para advertir a la gente sobre este peligro. Mi hijo de 15 años toca el bajo conmigo. Espero que mis canciones le hagan compañía cuando yo me haya ido.

JACONIA SHEZI

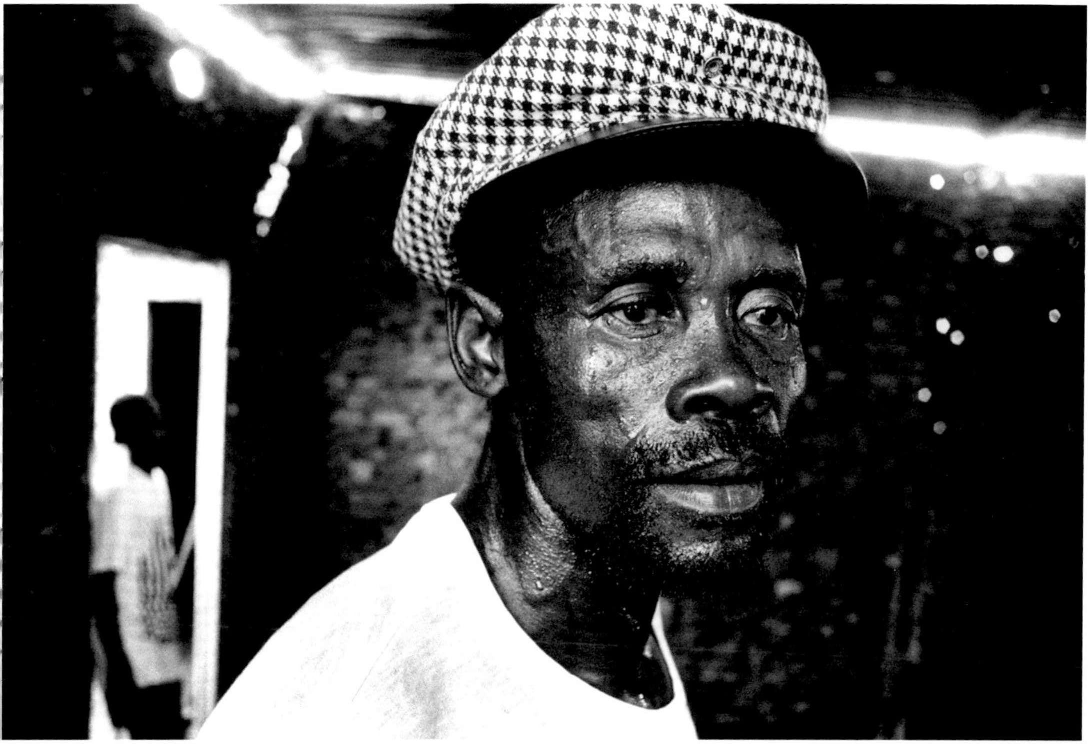

Debí de contagiarme cuando vivía en KwaMashu, en Durban. Vendía ropa y tuve muchas novias, así que no sé exactamente quién me contagió. Desearía que alguien me hubiese explicado algo sobre el sida y los preservativos. Soy un hombre joven y me debería quedar toda la vida por delante.

Mi familia es pobre. Cuando vinimos a esta zona tuvimos que construir estas cabañas para vivir en ellas. El suelo está muy sucio por aquí, pero es importante para mí el estar bien arreglado cuando voy al pueblo para informar a la gente sobre el sida, o al acudir a la iglesia. Cuando le digo a la gente que tengo esta enfermedad a menudo no me creen porque tengo buen aspecto y parezco fuerte. En una ocasión, cuando estaba informando en un colegio, los alumnos me dijeron que volviera cuando estuviera enfermo y que sólo entonces se convencerían. Sólo pude encogerme de hombros. Ya les he advertido. ¿Qué más puedo hacer? Me preocupa que una gran parte de los jóvenes de Sudáfrica se contagien de esta enfermedad.

GIDA MTHEMBU

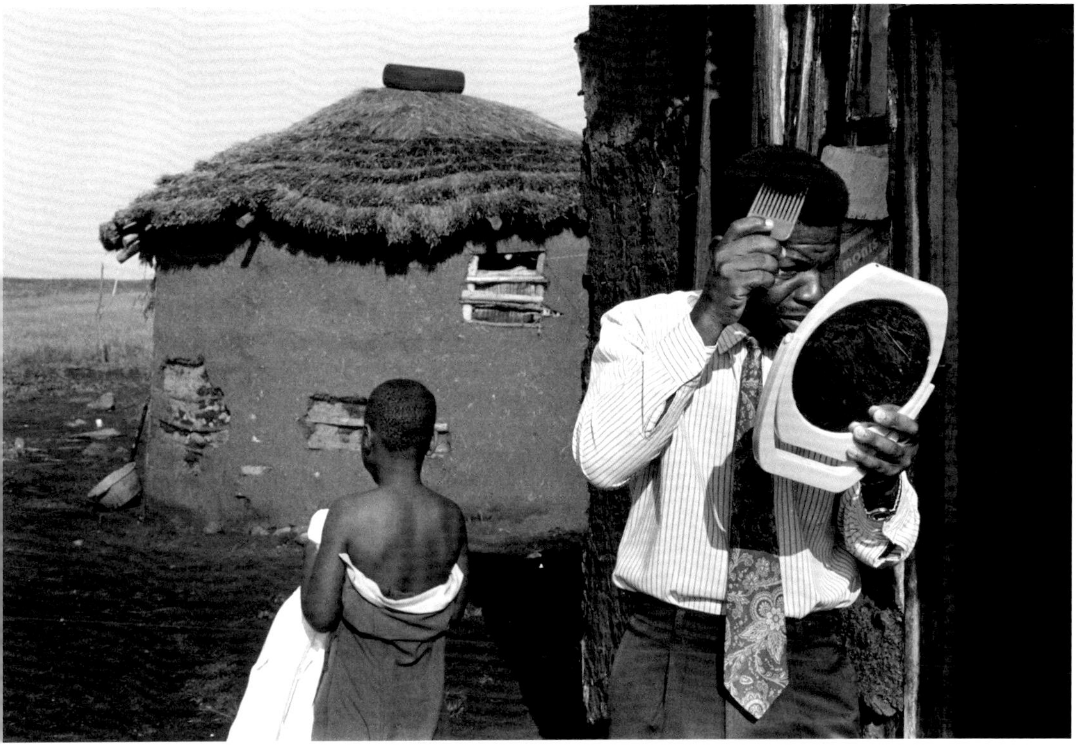

Reina Ntuli. Durban, Sudáfrica

Activa entre los sanadores tradicionales en lo que respecta a la educación sobre el sida, la Reina Ntuli atiende a sus clientes en su habitación.

Debo comenzar a partir del momento en que las personas aceptan el sida como tal y no como una maldición. Les digo que deben practicar el sexo seguro y seguir una dieta adecuada. Les puedo dar tratamientos con hierbas para potenciar su sistema inmunológico. Les puedo asesorar y explicarles lo que es el sida. Les digo que es importante seguir el tratamiento contra la tuberculosis, les enseño el uso de los preservativos y se los proporciono gratuitamente.

Yo aprendí lo que es el sida en un taller de concienciación sobre esta enfermedad impartido por el Departamento de Salud en 1996. Algunos sanadores tradicionales que no desean asistir a estos talleres dicen que pueden curar el sida, pero lo dicen porque no entienden lo que es. Creen que curando sus síntomas eliminan la enfermedad, pero están equivocados. Es perjudicial para la reputación de todos los sanadores el que realicen tales afirmaciones. Sé que en muchos casos puedo ayudar con los síntomas del sida, pero no puedo luchar contra el virus en sí. Potenciamos el sistema inmunológico, pero no curamos el sida.

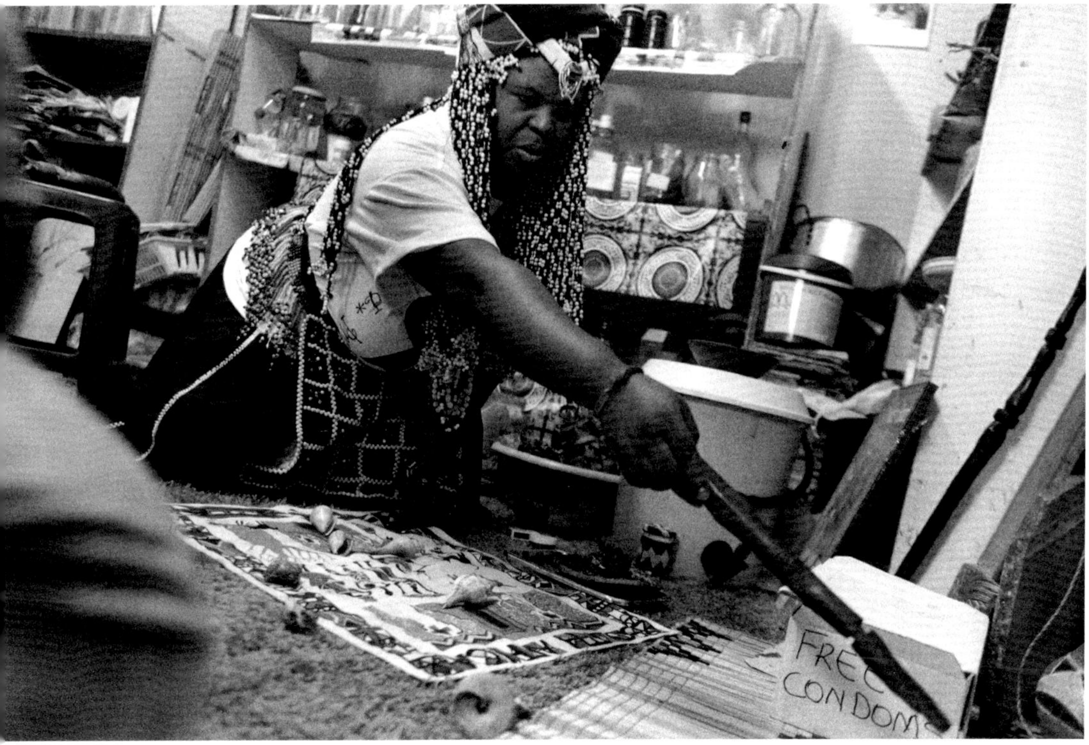

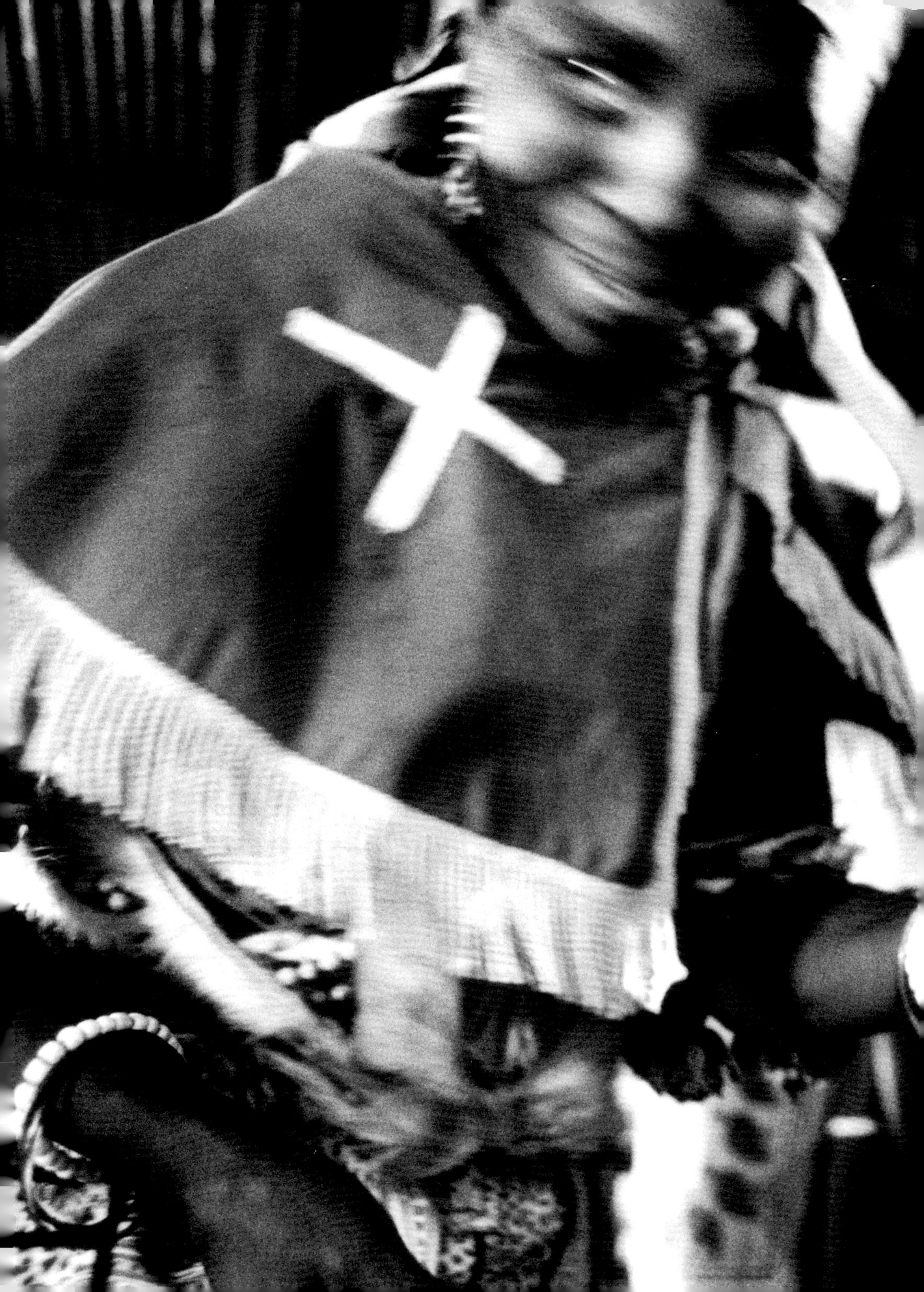

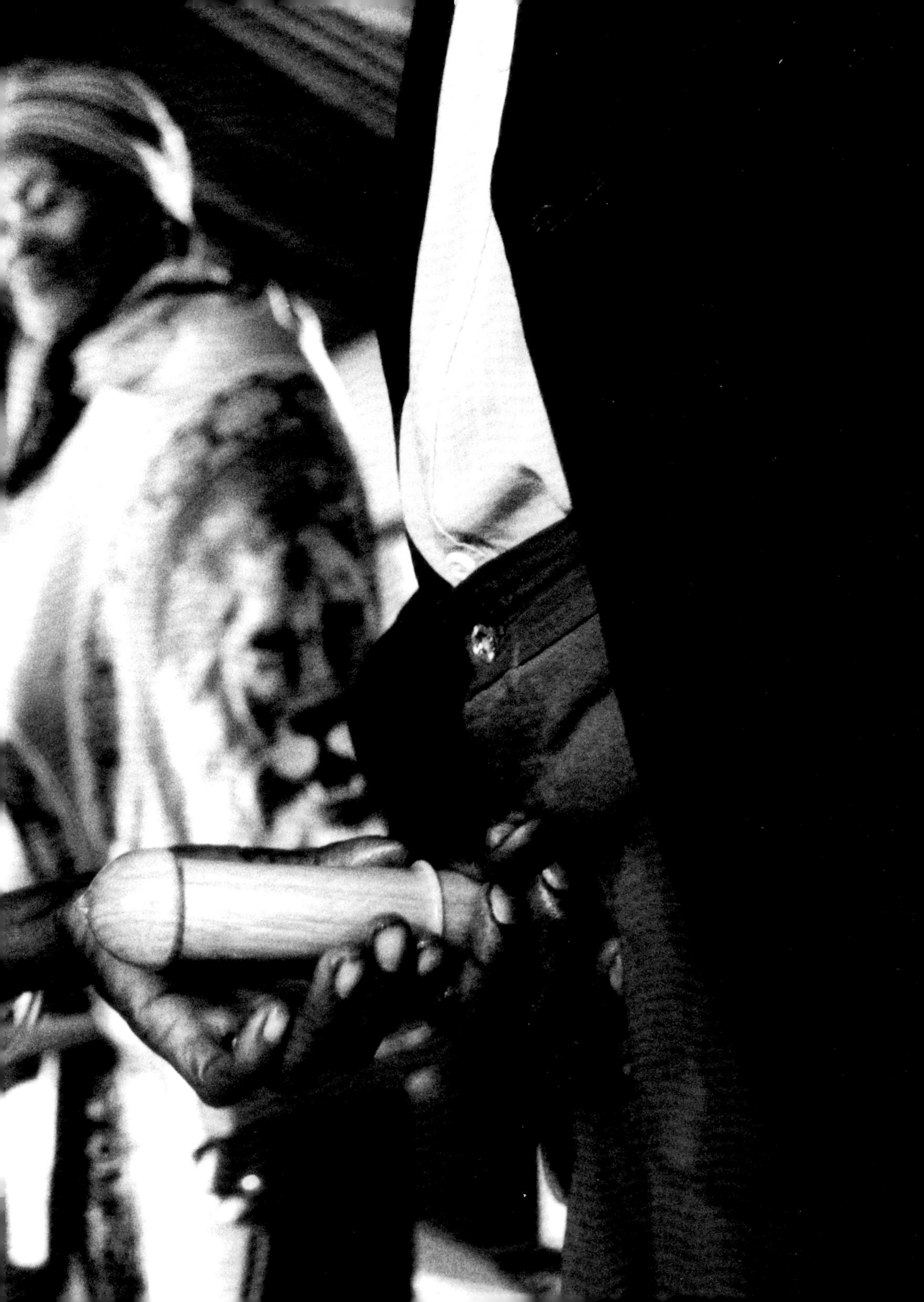

Compañeros educadores. Distrito de Bukoba, Tanzania

El equipo de atención domiciliaria de Bukoba ofrece programas de educación igualitaria, y proporciona asistencia a los enfermos.

Seleccionamos personas, a menudo alùmnos de las escuelas o miembros de grupos juveniles que están interesados y deseosos de enseñar a los demás. Les informamos sobre la transmisión del VIH y el sida, así como sobre las enfermedades de transmisión sexual, y discutimos acerca de las conductas responsables, incluyendo el empleo del preservativo y el retraso de la iniciación sexual. Queremos convertir conductas de alto riesgo en otras de bajo riesgo. También les enseñamos a convivir con personas que tienen VIH y a cuidar pacientes con sida, ya que la mayoría de las personas de esta región tienen algún familiar enfermo.

SPECIOSA RWAMAAG, ENFERMERA, EQUIPO DE ASISTENCIA DOMICILIARIA DE BUKOBA

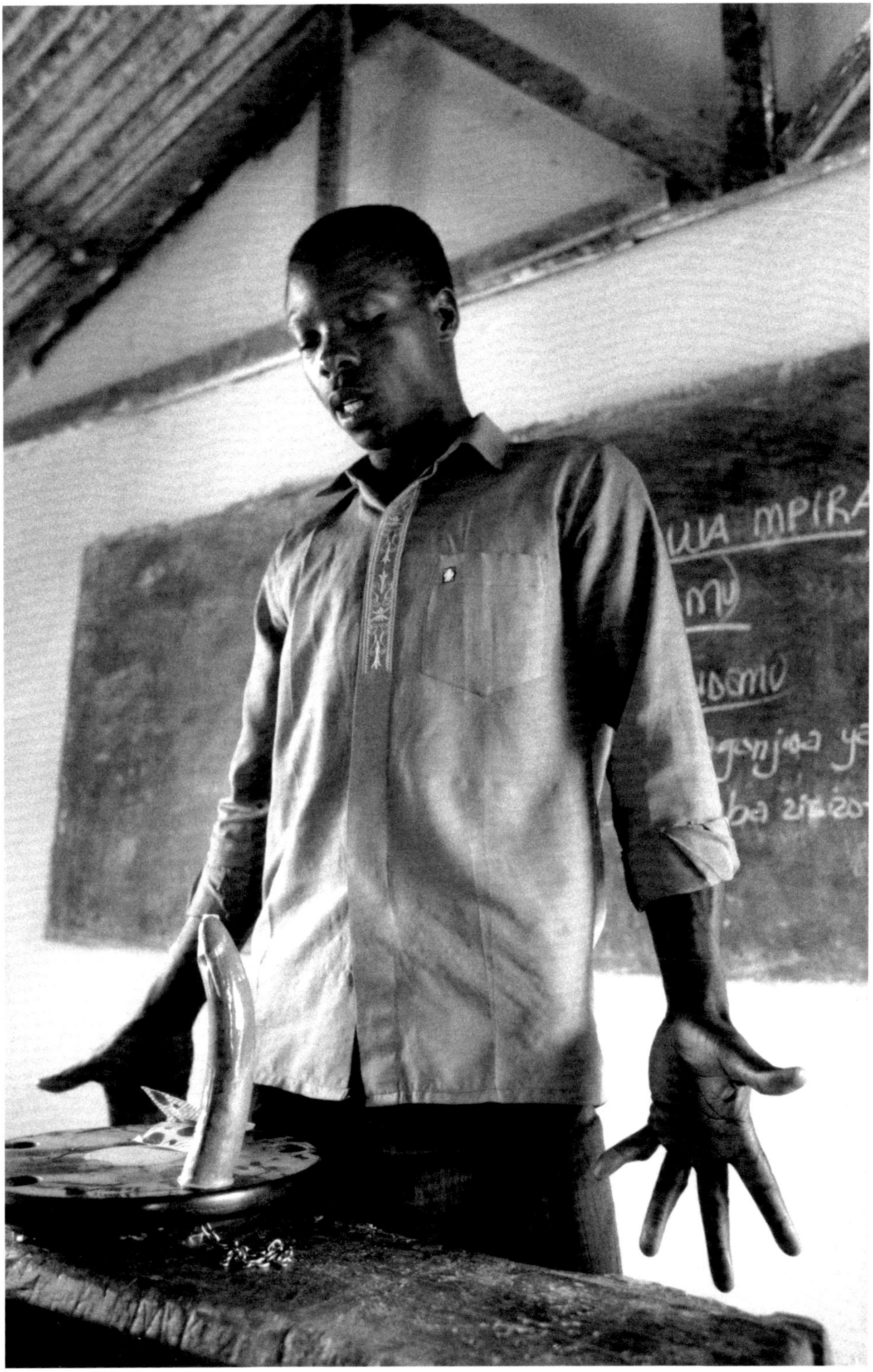

Teatro Angeningo. Pangani, Tanzania

Fundado por el hospital de Pangani en 1997, el grupo de teatro Angeningo utiliza la dramatización para promocionar el empleo del preservativo y prevenir el sida y otras enfermedades de transmisión sexual.

Representamos nuestras obras de forma que inviten a la audiencia a participar. Por ejemplo, hay una escena en la que la chica de la historia quiere mantener relaciones sexuales con su novio por primera vez porque le quiere y no quiere perderlo, pero él se niega a utilizar un preservativo. Entonces se gira hacia la audiencia y les pregunta qué debería hacer y seguir de ese modo su consejo. ¡La mitad de los espectadores le dice que lo deje, y la otra mitad que debe hacer lo que él le pide. La gente se ríe mucho y se organiza una gran algarabía! En la obra se acuesta con él, se contagia de VIH y sida, y muere.

La dramatización es una buena manera de educar a la comunidad con respecto al sida. Intentamos visitar todos los pueblos del distrito de Pangani al menos una vez cada tres meses. Ahora la gente tiene un mayor conocimiento. Hemos detectado una mayor utilización del preservativo, pero aún no puede decirse que estemos a salvo. Ocupamos el duodécimo lugar del país en lo que a niveles de infección se refiere.

DOCTOR ZUBARI MIHUNGO, DIRECTOR DEL HOSPITAL DE PANGANI

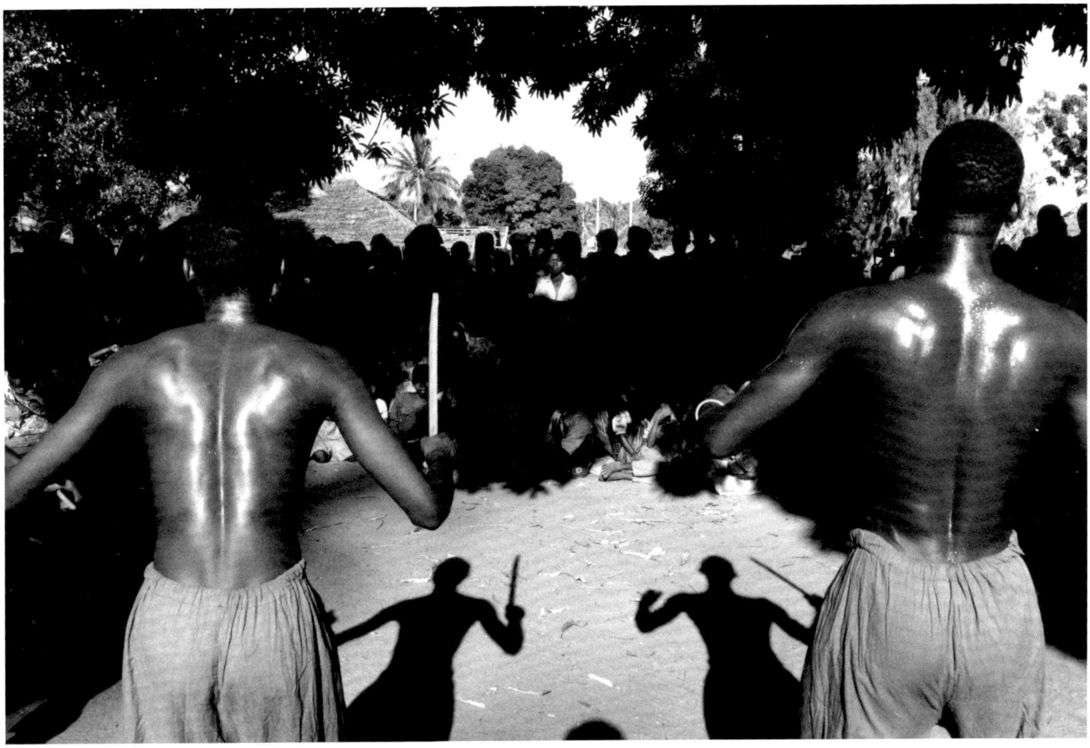

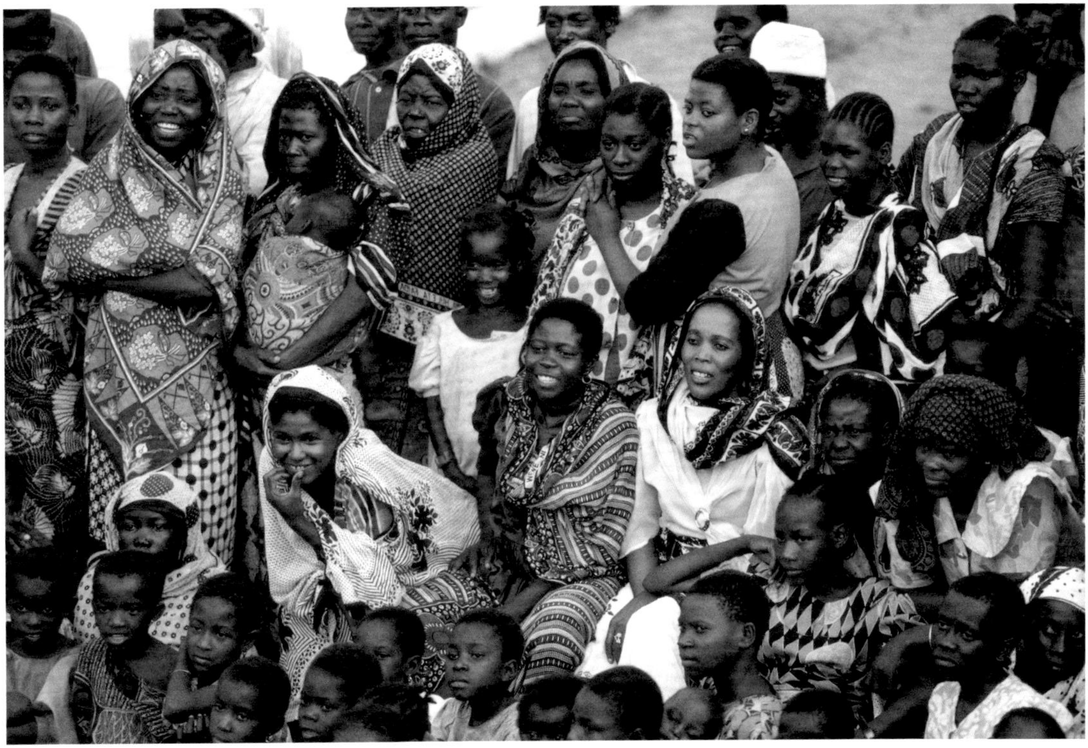

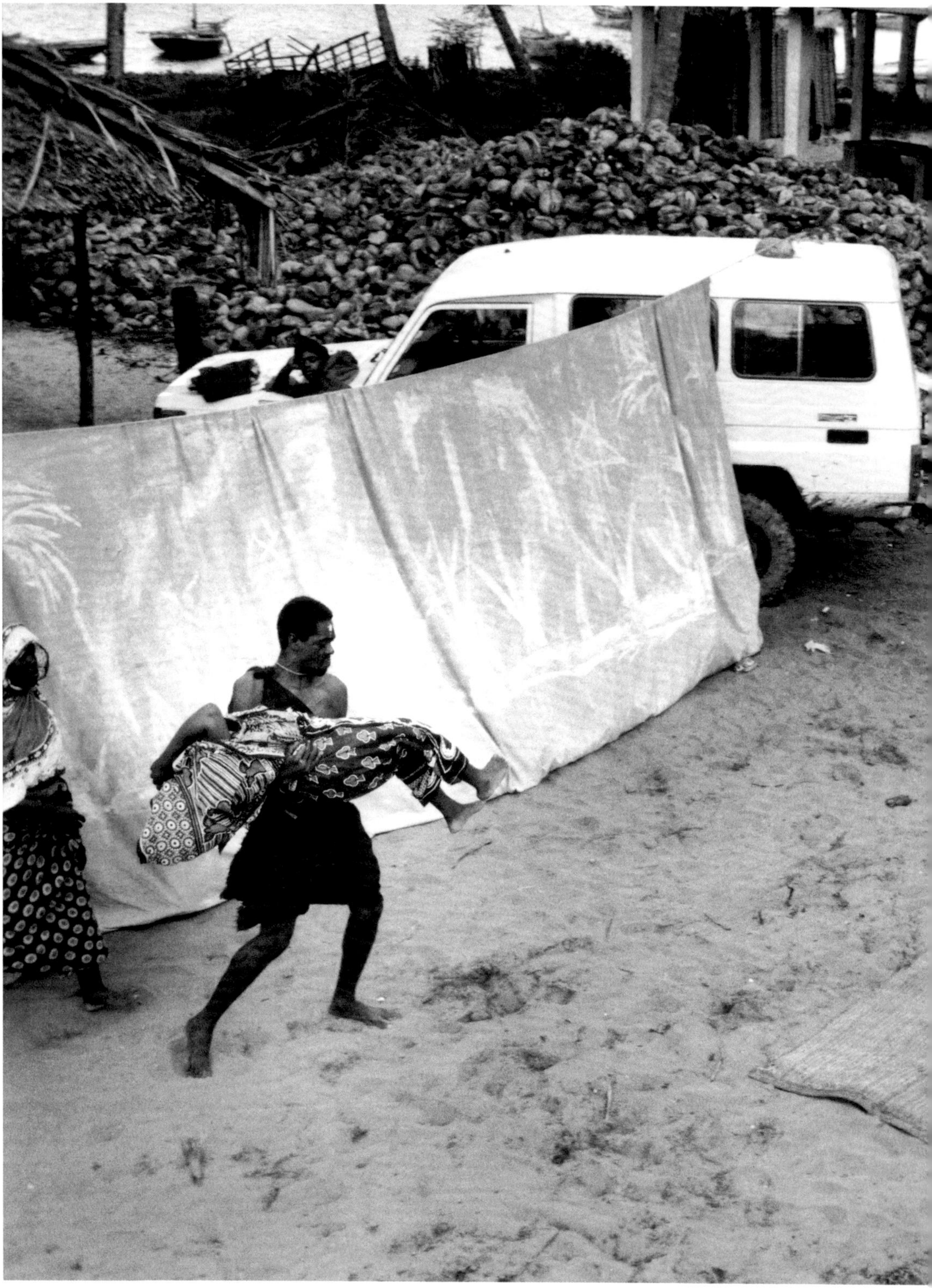

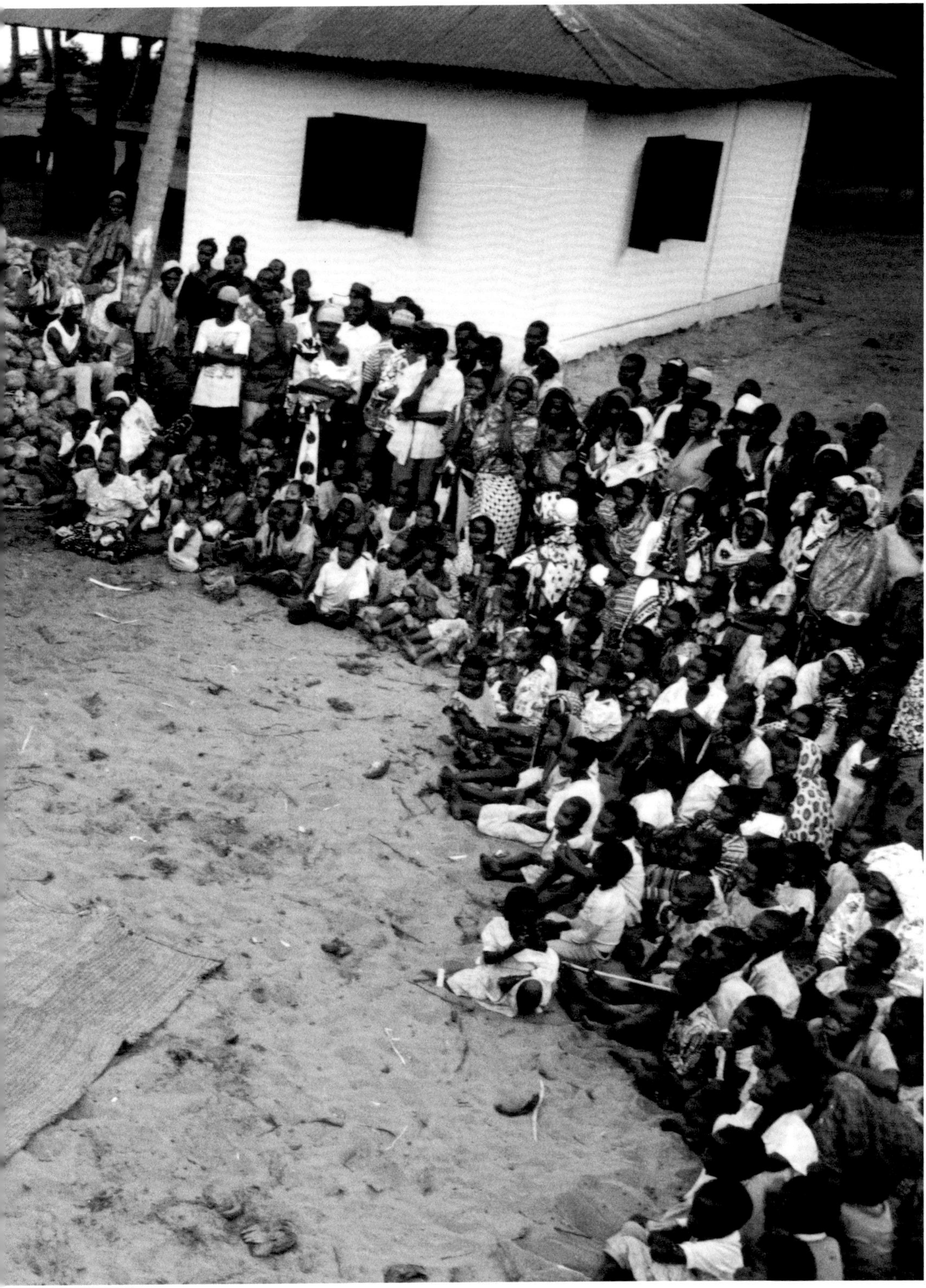

Florence Kumunhyu. Iganga, Uganda

Florence perdió a su marido en 1990. Padecía la enfermedad entonces conocida como «delgadez».

Fue un proceso muy largo tras la muerte de mi marido en 1990, pero recuperé la confianza en mí misma y decidí fundar un grupo aquí en el distrito de Buwolomera en Iganga, la Asociación para el Desarrollo de Buwolomera. Su objetivo es informar a la gente sobre la enfermedad. Actualmente tenemos veinte miembros femeninos y tres masculinos. La mayoría de nosotras convive con el VIH y he animado a otras mujeres a hacerse la prueba y, luego, a unirse a nosotras.

Los hombres son más reacios a unirse a la asociación. Sin embargo, algunos de ellos se quedan a dormir por aquí porque piensan que si se unen a nuestro grupo la gente sabrá que están infectados y nadie querrá acostarse con ellos. Sin embargo, creo que si al menos podemos enseñar a las mujeres, ellas volverán a casa y enseñarán a sus maridos lo que han aprendido sobre el sida.

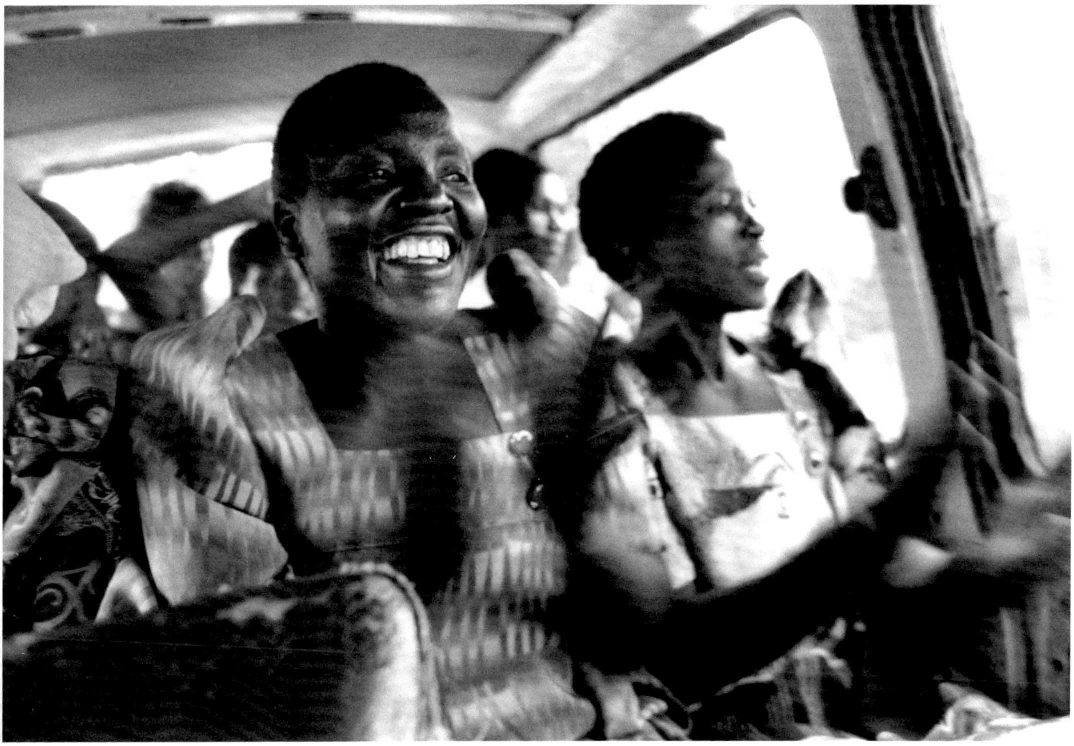

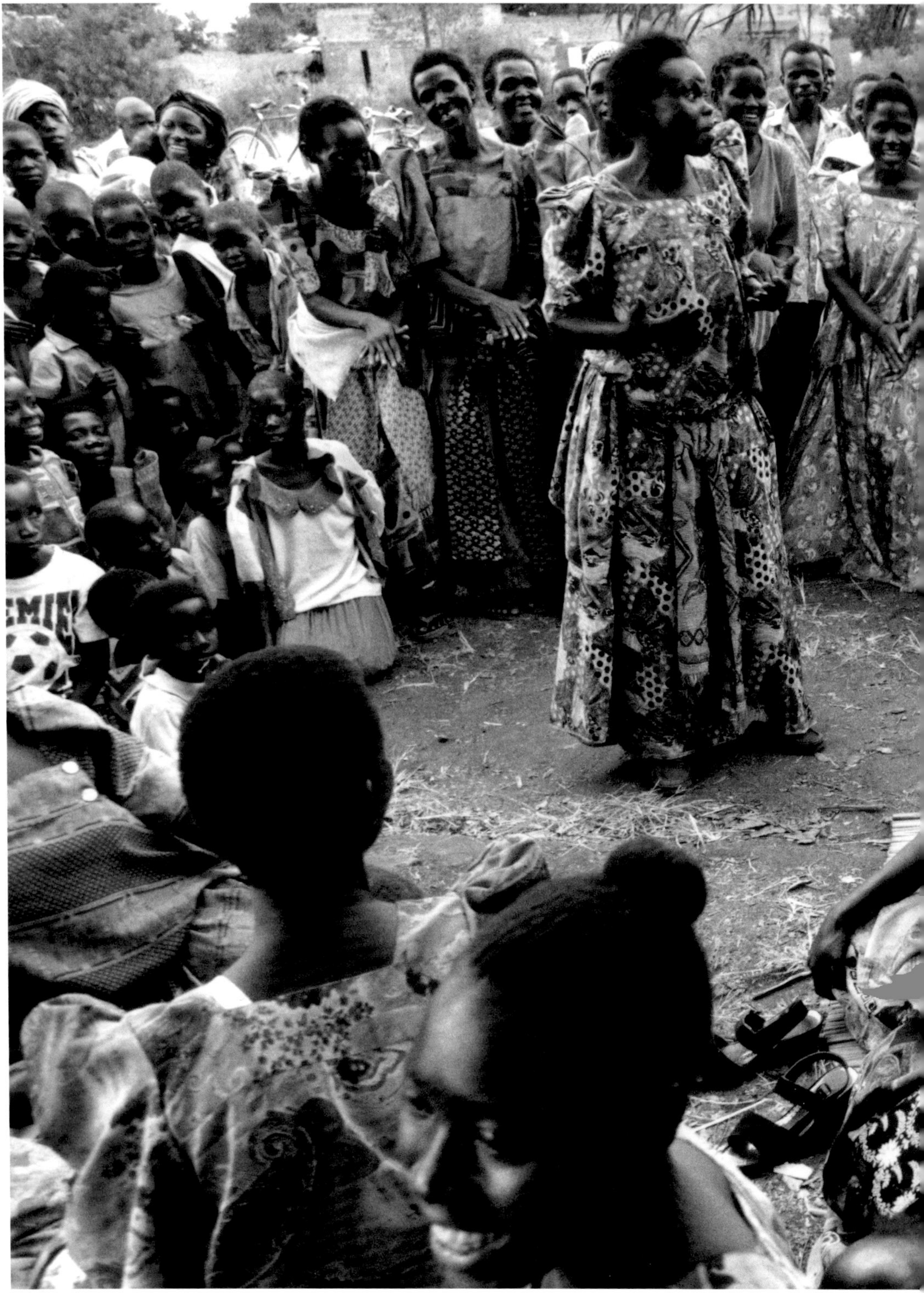

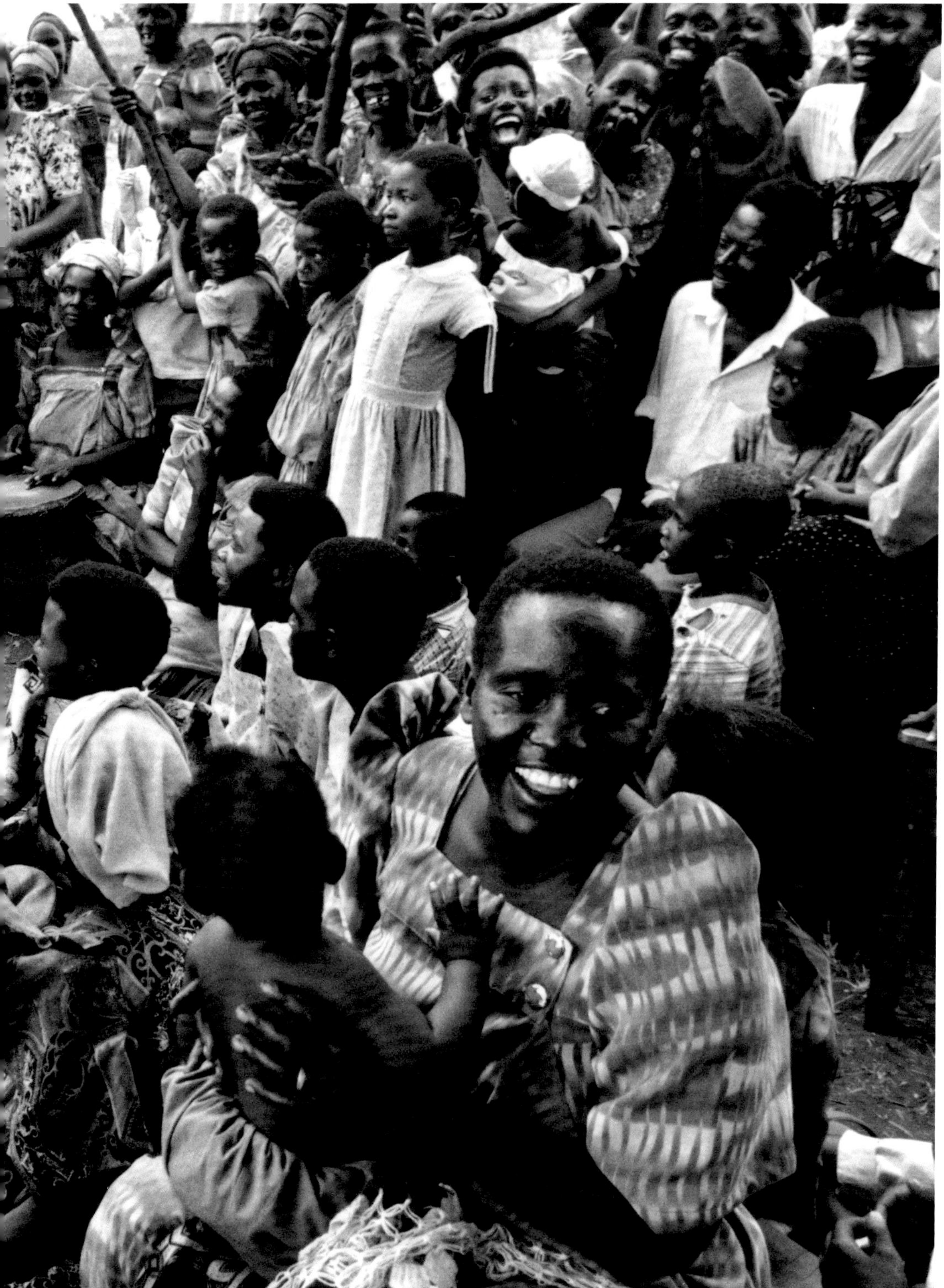

Cuando murió mi marido no estaba en paz conmigo misma porque quería saber si estaba a salvo o si también tenía sida. Algunas personas de la iglesia me animaron y decidí hacerme la prueba del VIH. Me asusté mucho cuando me dijeron que tenía el virus. Estaba muy triste y lloré mucho porque pensé que me había llegado la hora. Mis hijos aún eran muy pequeños e inmediatamente comencé a preguntarme quién los cuidaría.

Me decidí a explicarle a la gente de mi alrededor que tenía esta enfermedad. Era una viuda joven con ocho hijos a mi cuidado. Aunque sabía que mucha gente se volvería en mi contra, sentí que no tenía nada que perder. Tenía que encontrar gente que quisiera ayudar a mis hijos. Muchas personas se rieron cuando les dije que tenía la enfermedad. Cuando paseaba por el pueblo decían a mis espaldas: «Esa mujer tiene sida». Tuve muchas dificultades con mis cuñados. Algunos de ellos querían casarse conmigo y cuando les dije que tenía VIH dijeron que estaba mintiendo para evitarlos. Entonces se enfadaron y uno de ellos trató de alejarme de las tierras de mi difunto marido.

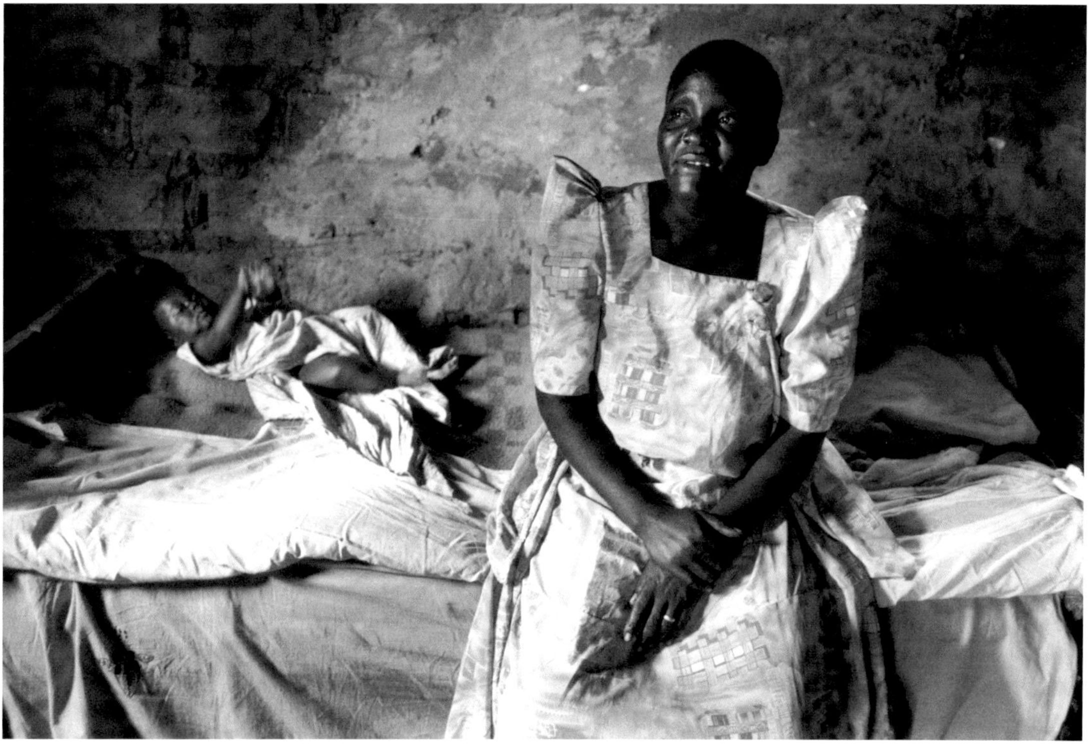

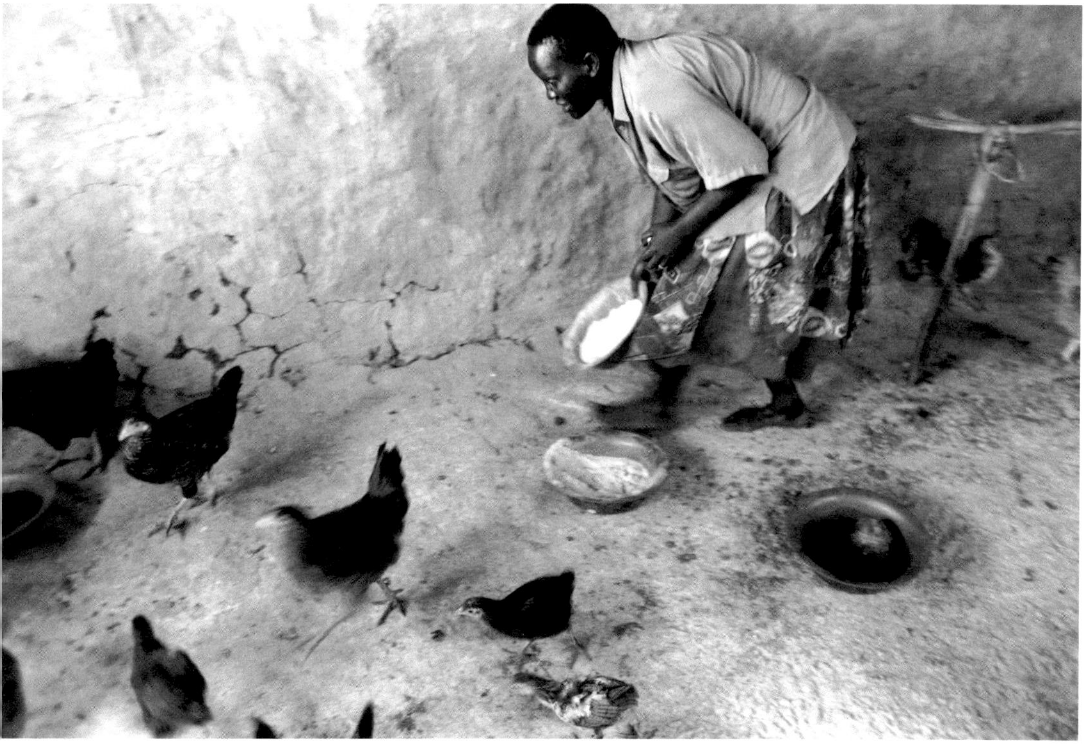

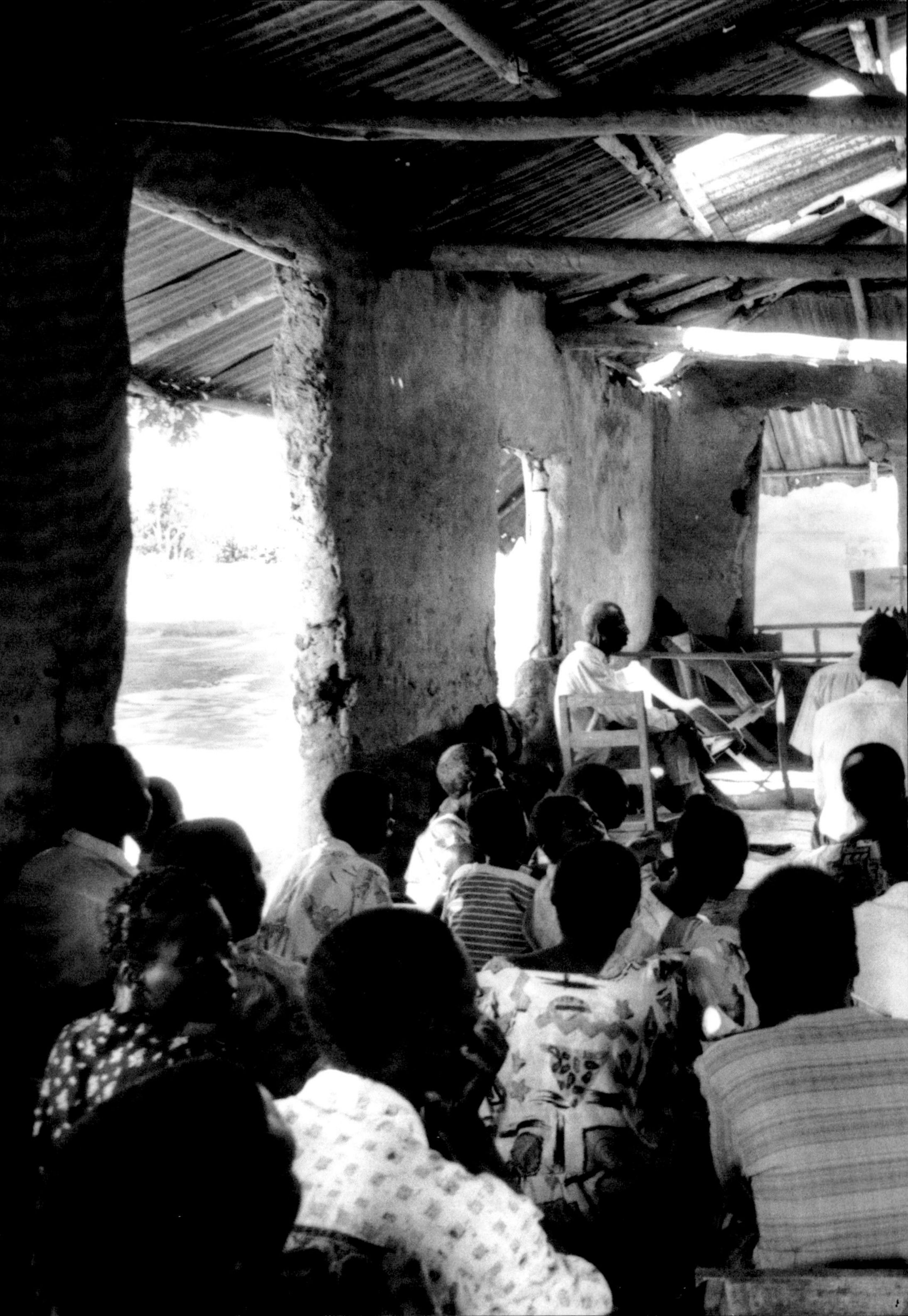

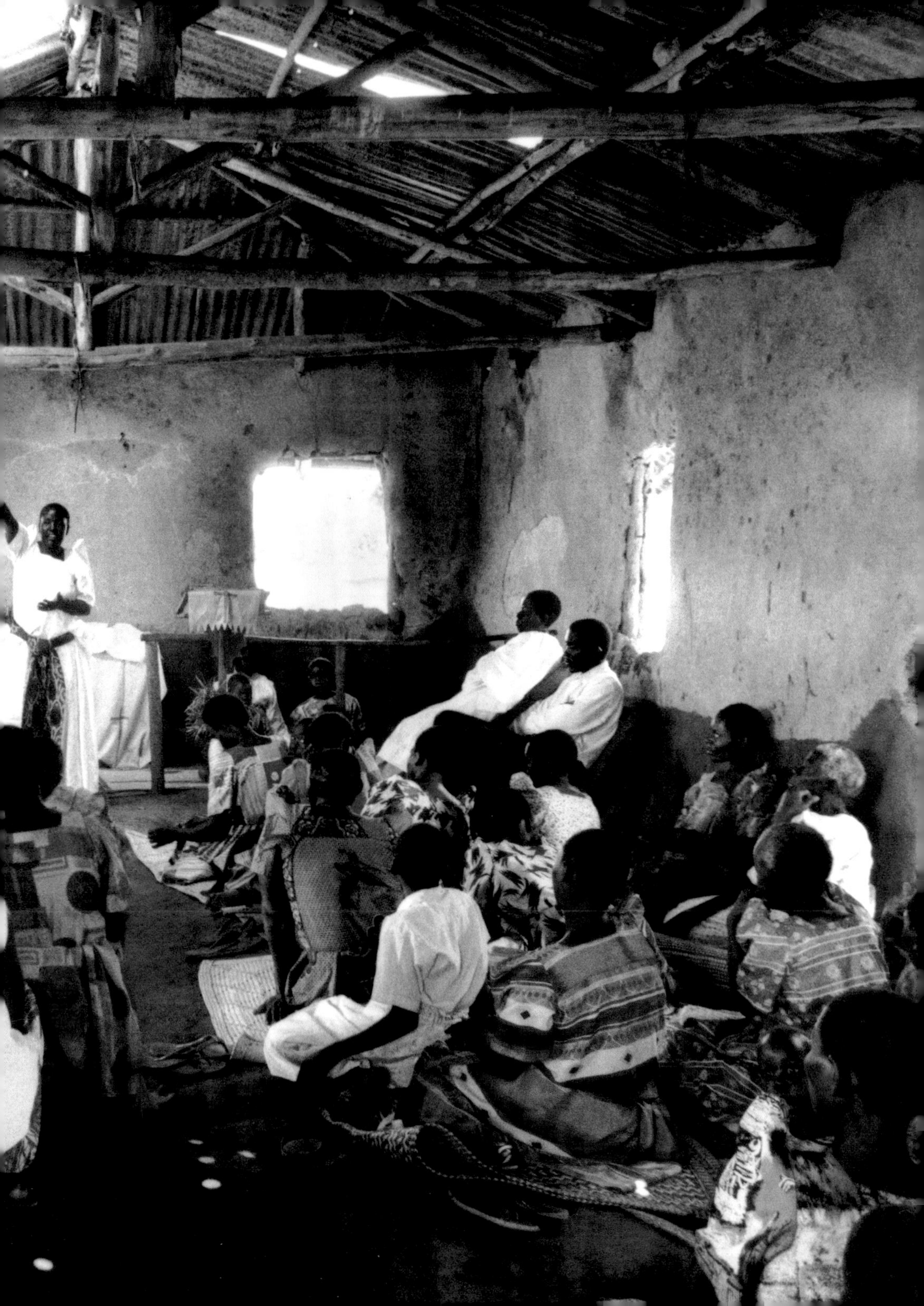

Esta enfermedad nos aprisiona con fuerza.
Queridos colegas, sentaos y escuchadnos.
La enfermedad está decidida a acabar con
nosotros.

¿Qué debemos hacer las chicas solteras?
Haceos una prueba, haceos una prueba junto
con vuestros pretendientes.

¿Qué debemos hacer los chicos solteros?
Haceos una prueba, haceos una prueba junto
con vuestras prometidas.

La enfermedad ha venido a acabar con nosotros,
tomad precauciones.

¿Quiénes sois los afectados por este desastre?
Somos de Iganga.
La enfermedad ha llegado a Kampala, a Jinja,
a Iganga y a todos los lugares.
Haceos una prueba, haceos una prueba.
Y haced que vuestras parejas se hagan una prueba,
porque la enfermedad es real.

CANCIÓN DE FLORENCE

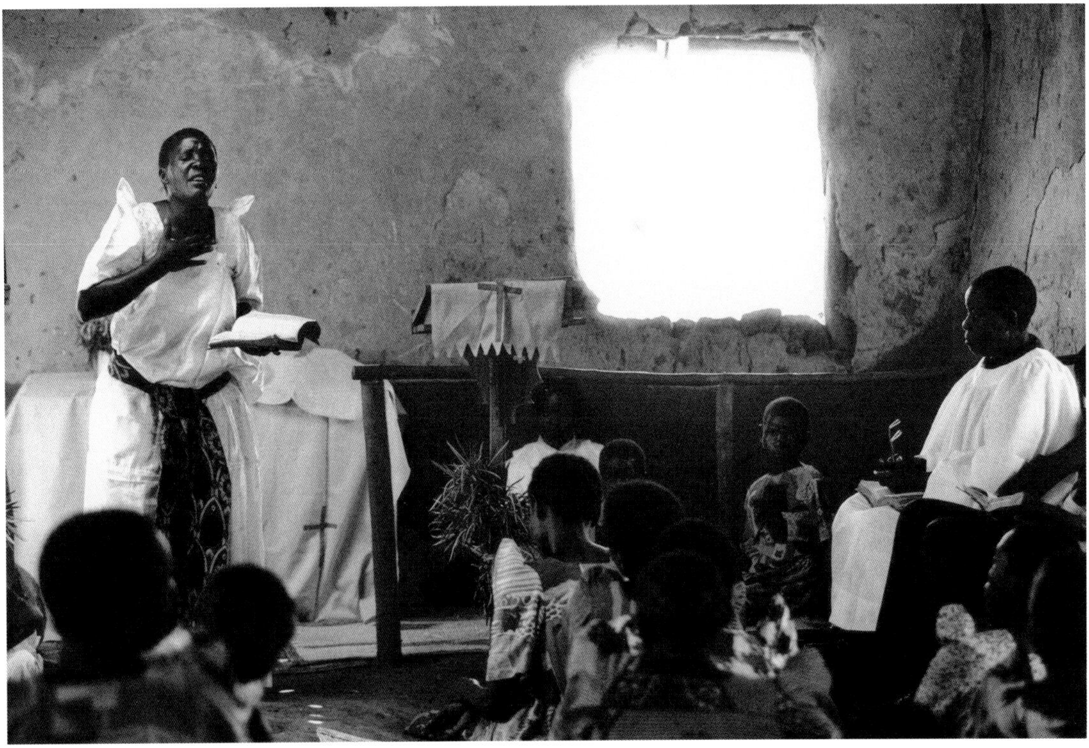

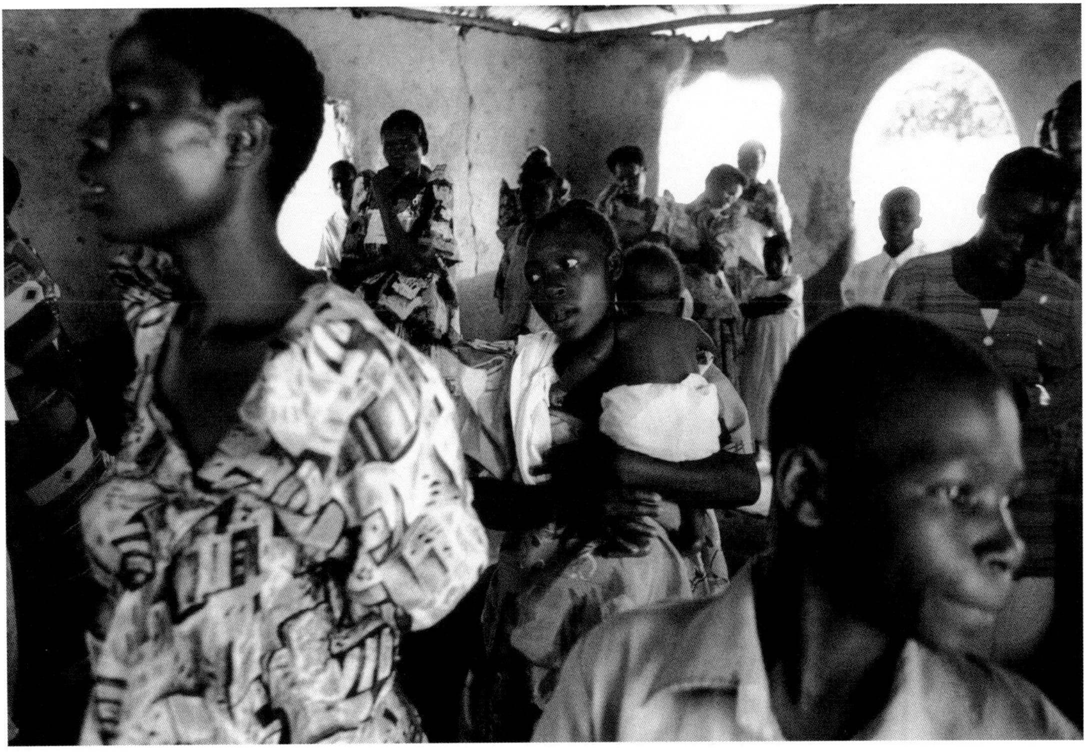

Susan Atuhura. Kampala, Uganda

Susan, de 16 años, inició el Club Hablar Claro (Straight Talk Club) en la Escuela Superior City View de Kampala. Su nombre es el mismo que el de una publicación para gente joven sobre temas de salud, sexo y reproducción, que se publica en Kampala.

Me di cuenta de que no sabía todo lo que debía sobre el VIH y el sida, y entonces pregunté a mis amigos, pero ninguno de ellos estaba seguro. Decidí formar un club en el que pudiéramos aprender unos de otros. Mis amigos se apuntaron, así como otros estudiantes de mi colegio. Ahora hace un año que lo formamos y somos 30 miembros, los cuales nos reunimos una vez por semana.

Como adolescentes, necesitamos que alguien nos guíe y nos informe sobre cosas que nos pueden salvar la vida, porque muchos de nuestros padres son demasiado tímidos para hablarnos sobre sexo y sexualidad, y no nos dan la información que necesitamos. En el club nos enseñamos unos a otros. Sobre todo hablamos. He aprendido mucho de mis amigos, y esto aumenta mi confianza, con lo que puedo expresar mi propia opinión.

Desde que se fundó el club, no ha habido más embarazos entre los estudiantes de la escuela. Creo que nos está ayudando mucho.

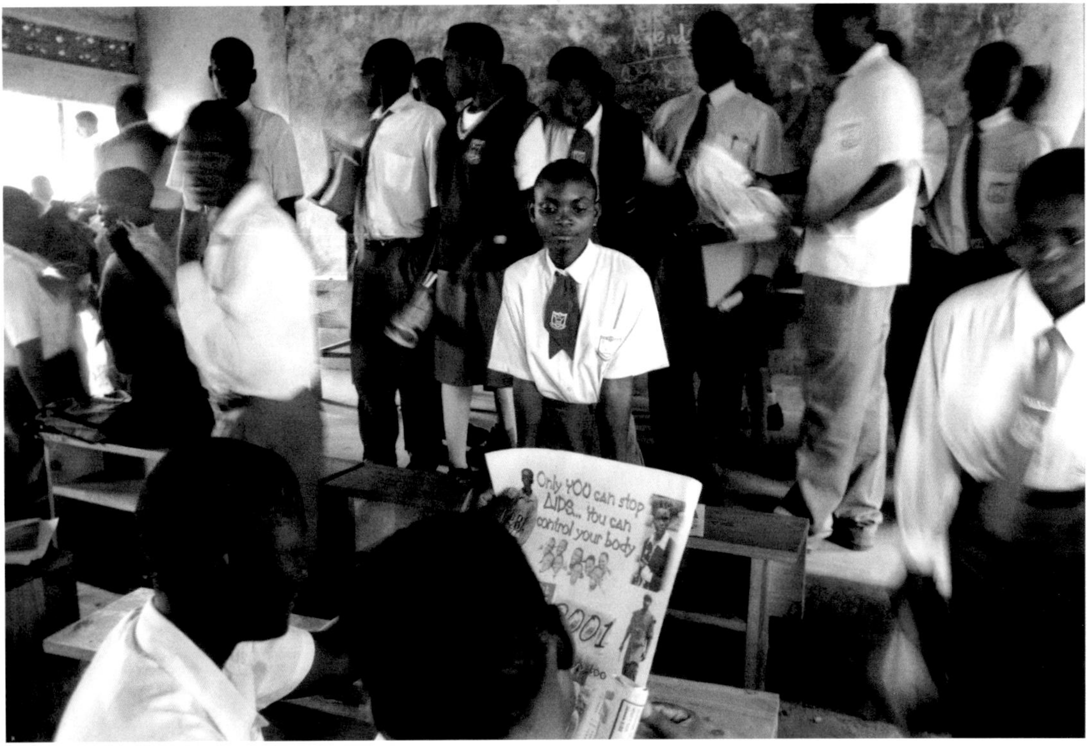
Only YOU can stop
AIDS... You can
control your body

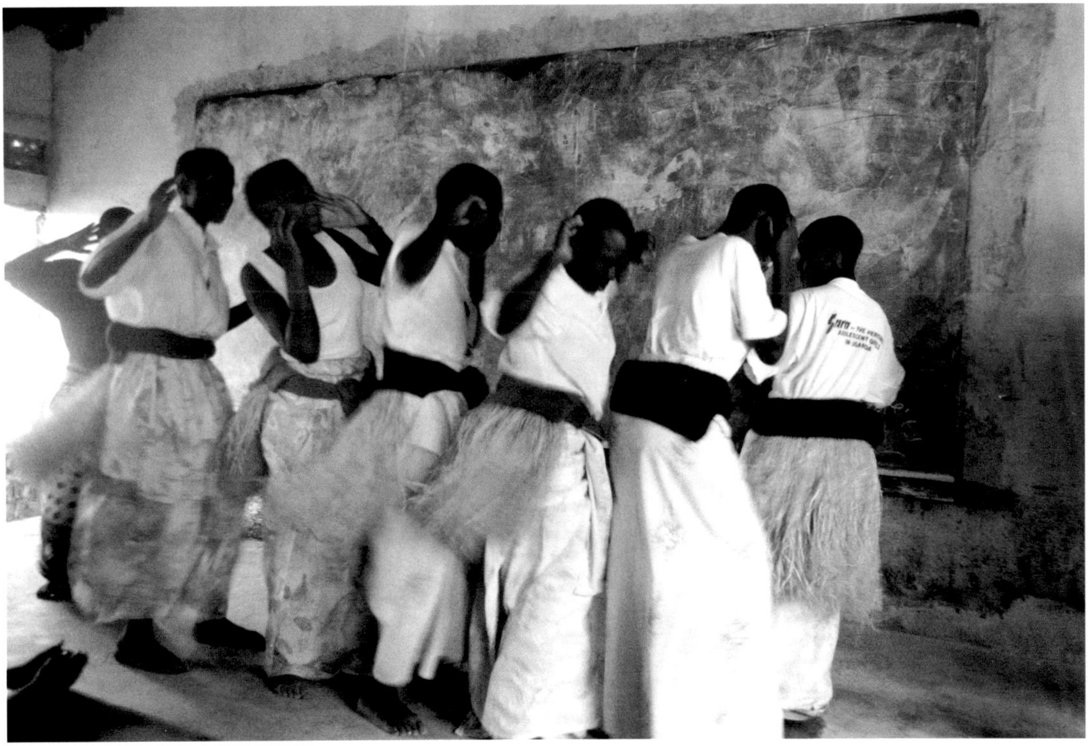

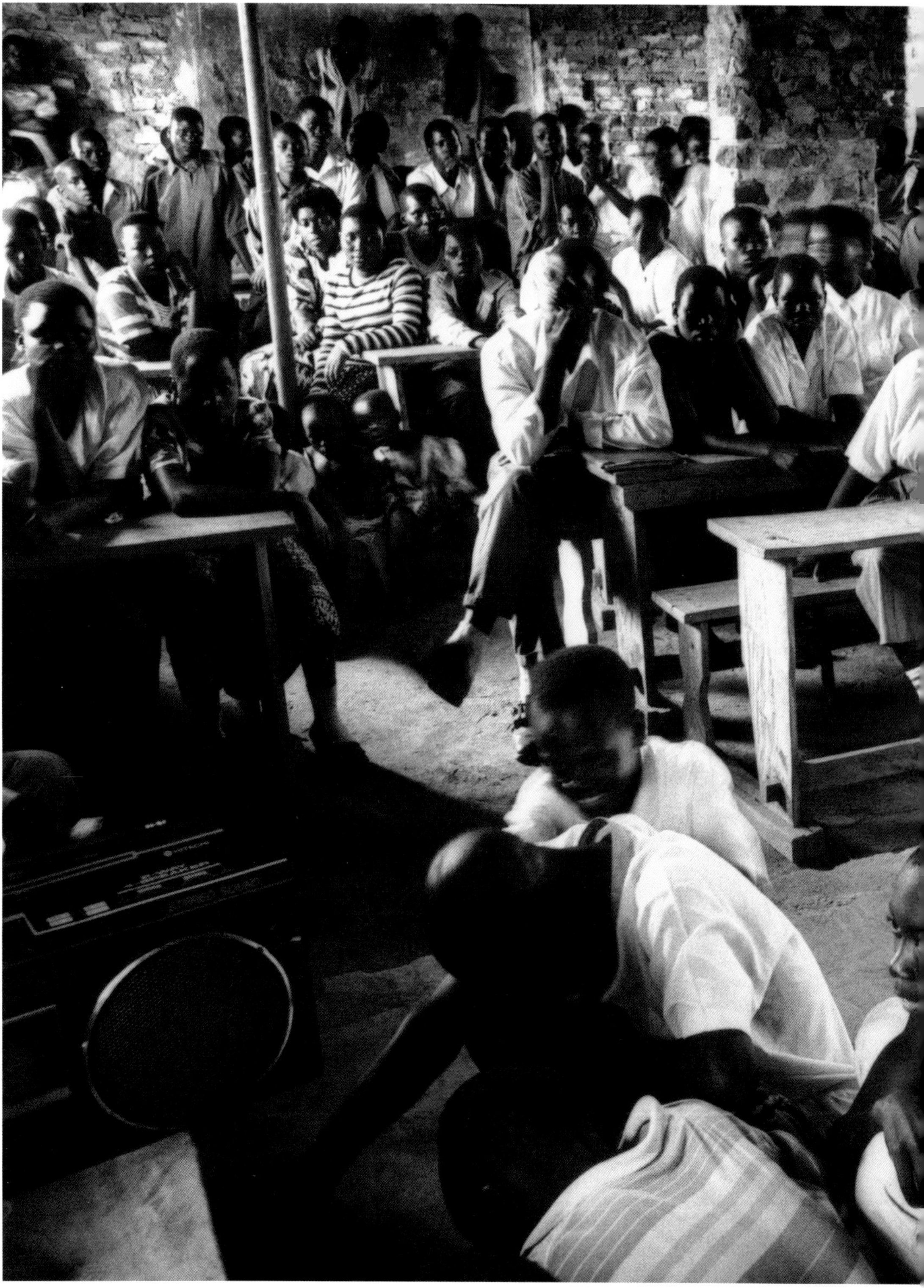

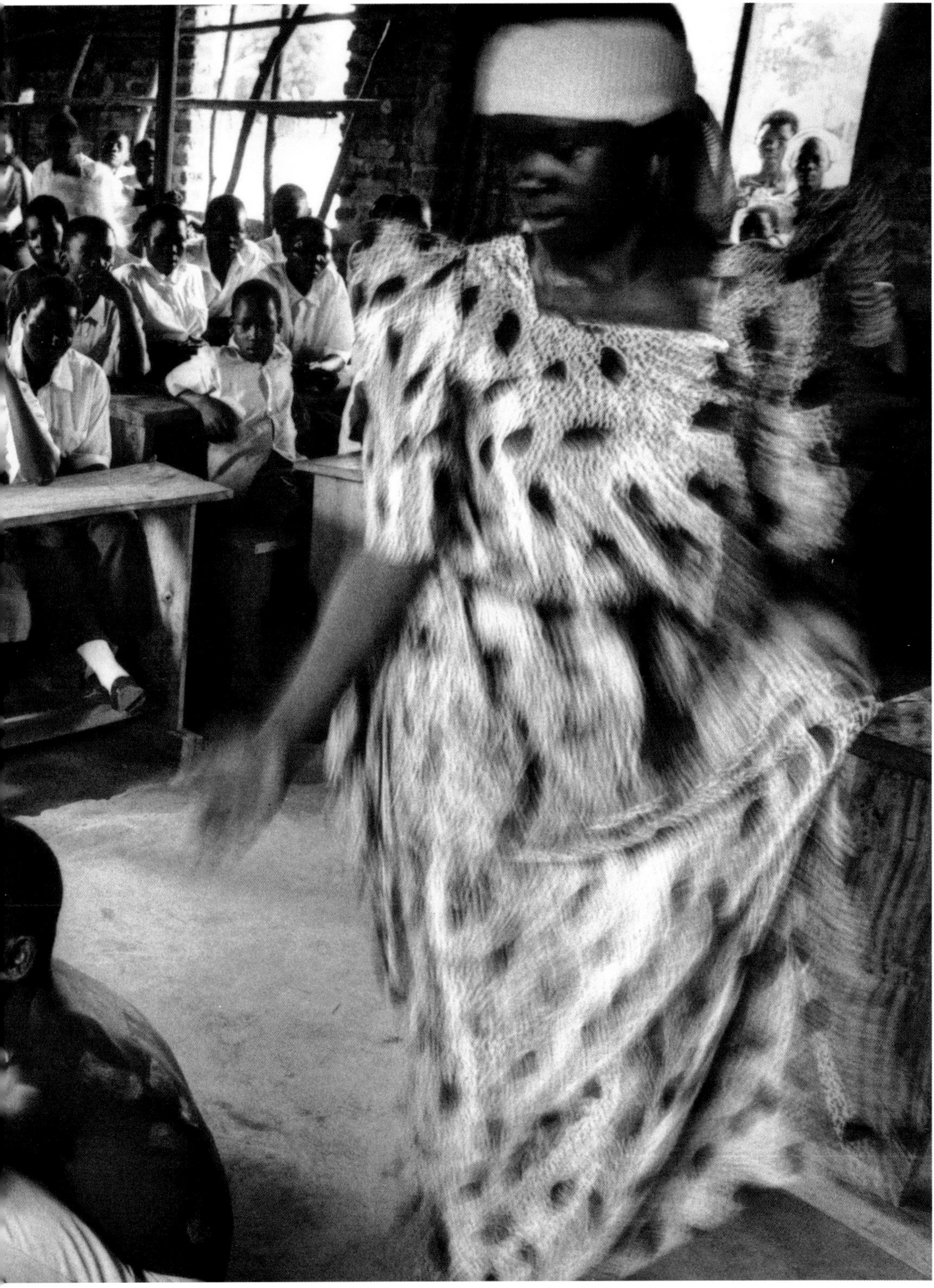

Club antisida. Lusaka, Zambia

En las escuelas de Zambia se han fundado 1.800 clubes antisida, entre los que se encuentra el de la Escuela de Secundaria de Munali.

Un club antisida suele ser una actividad extraescolar voluntaria que se desarrolla una o dos veces por semana. Animamos a los colegios a establecer clubes y los apoyamos proporcionándoles algunos materiales. Los miembros del club a menudo asumen la responsabilidad de intentar educar al resto de la escuela y a sus compañeros sobre la mejor manera de evitar contagiarse de sida. Los estudiantes que elijen formar parte del club suelen estar muy motivados, y muchos tienen una experiencia de primera mano en casa.

Nuestros esfuerzos deben considerarse en el contexto de las actividades sexuales propias de nuestros jóvenes. Una encuesta reciente mostró que el 12 % de las chicas de entre 15 y 16 años en Lusaka dió positivo en la prueba del VIH. Es de vital importancia educar a nuestros jóvenes sobre cómo se extiende el sida y acerca de la importancia de la abstención, pero también hay que decirles que si deciden practicar sexo deben utilizar preservativo. Las personas ideales para transmitir este mensaje a los jóvenes son los propios jóvenes.

PRISCA CHITOMFA, PROYECTO ANTISIDA DEL CONSEJO FAMILIAR DE SALUD

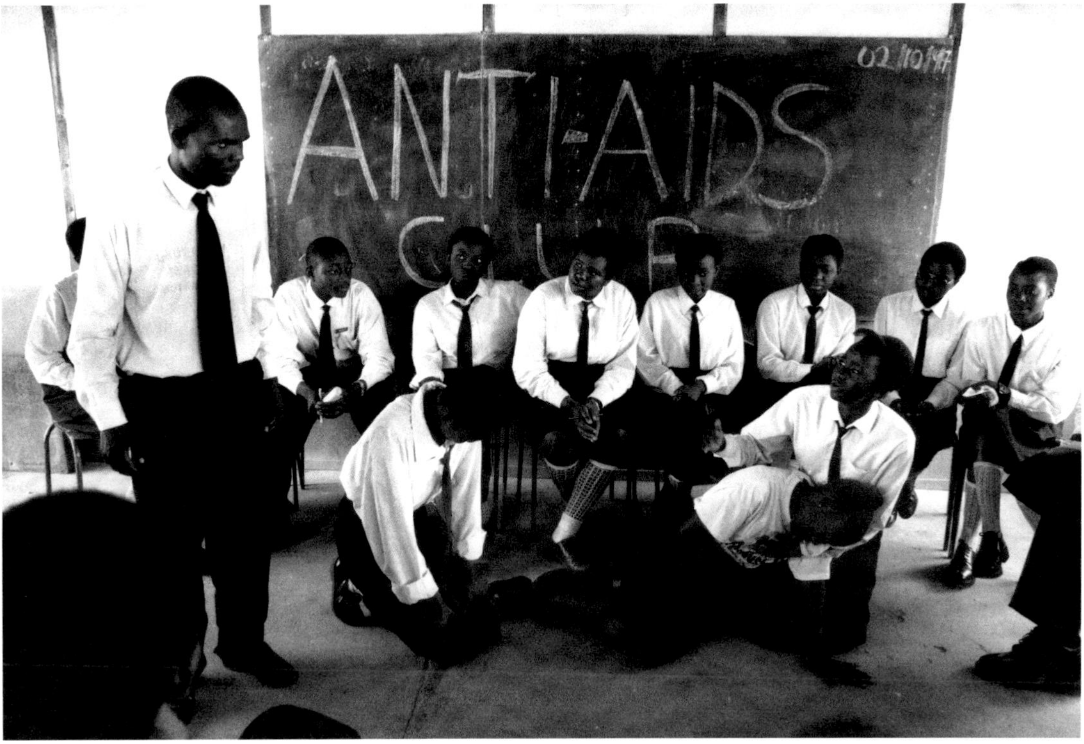
ANTI-AIDS

Nomhlahla Zungu. Ngwelezane, Sudáfrica

Nomhlahla es un miembro del grupo de compañeros educadores del hospital de Ngwelezane.

Primero padecí una tuberculosis en 1994, pero no supe que era seropositivo hasta 1998. Cuando el asesor sobre sida del hospital de Ngwelezane me dio la información, me preguntó cómo reaccionaría si me dijera que daba positivo en la prueba. Le dije que estaba preparado para afrontarlo.

Decidí que quería vivir con optimismo. Al principio mi madre no me creyó. Lo hizo cuando empecé a hablar en público. Entonces se preocupó porque pensaba que podían matarme. Pero la gente no se muestra hostil hacia mí. Quizá muchos de ellos saben o les preocupa ser seropositivo, por lo que respetan mi decisión de revelar mi situación, lo que he hecho para ayudar a educar a mi comunidad.

El grupo de compañeros educadores está muy unido. Organizamos conciertos y talleres y charlas en la comunidad, así como salidas a la playa. Nos ayudamos unos a otros.

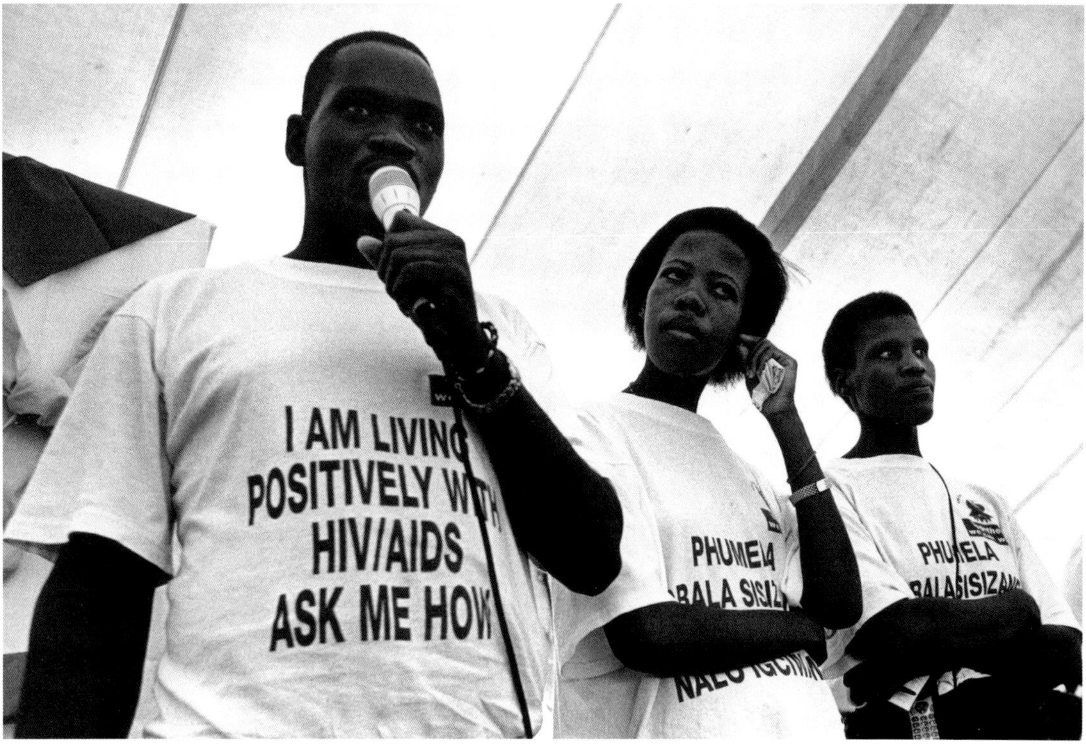
I AM LIVING
POSITIVELY W
HIV/AIDS
ASK ME HOW

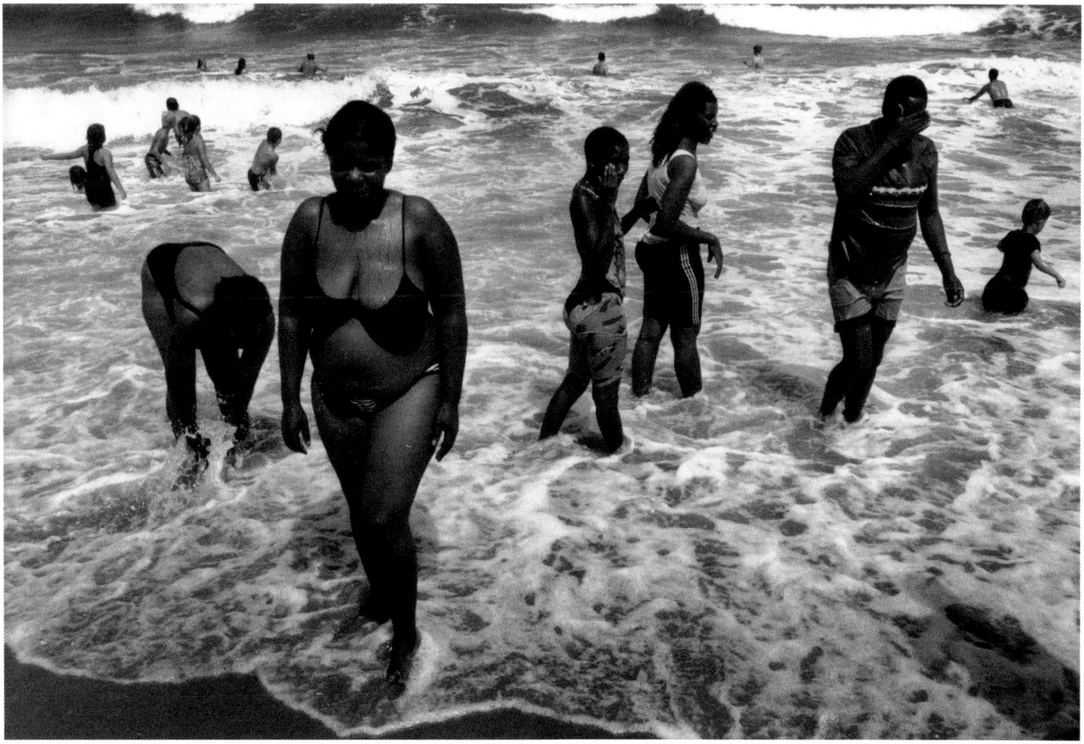

Fazeko Kuzwayo. Durban, Sudáfrica

Fazeko es miembro fundador de Voces Positivas, un grupo de teatro coral formado por seropositivos.

Cuando yo tenía tres años mi madre y yo seguimos a mi padre en el exilio, ya que era el representante especial del ANC en Lesotho. Cuando era adolescente, era activista del ANC, así que cuando descubrí que tenía el VIH, consideré natural convertirme en una activista contra el sida.

Recientemente actué como maestra de ceremonias en un servicio en el Ayuntamiento de Durban. Mi mensaje fue muy simple: el silencio ensordece y debe romperse. Fue una experiencia muy emotiva para mí, ya que mi madre era una de las conferenciantes a las que tenía que presentar. Ella hablaba como persona afectada por el sida. El título de su discurso era «¿La quiero lo suficiente?». Ella se preguntaba si aportaba suficientes risas y alegría a mi vida. Fue muy difícil presentarla porque yo lloraba desconsoladamente.

Si pudiera pedir un deseo éste sería que mi padre viviera. Si estuviera aquí le gritaría denunciando las políticas del ANC sobre el sida.

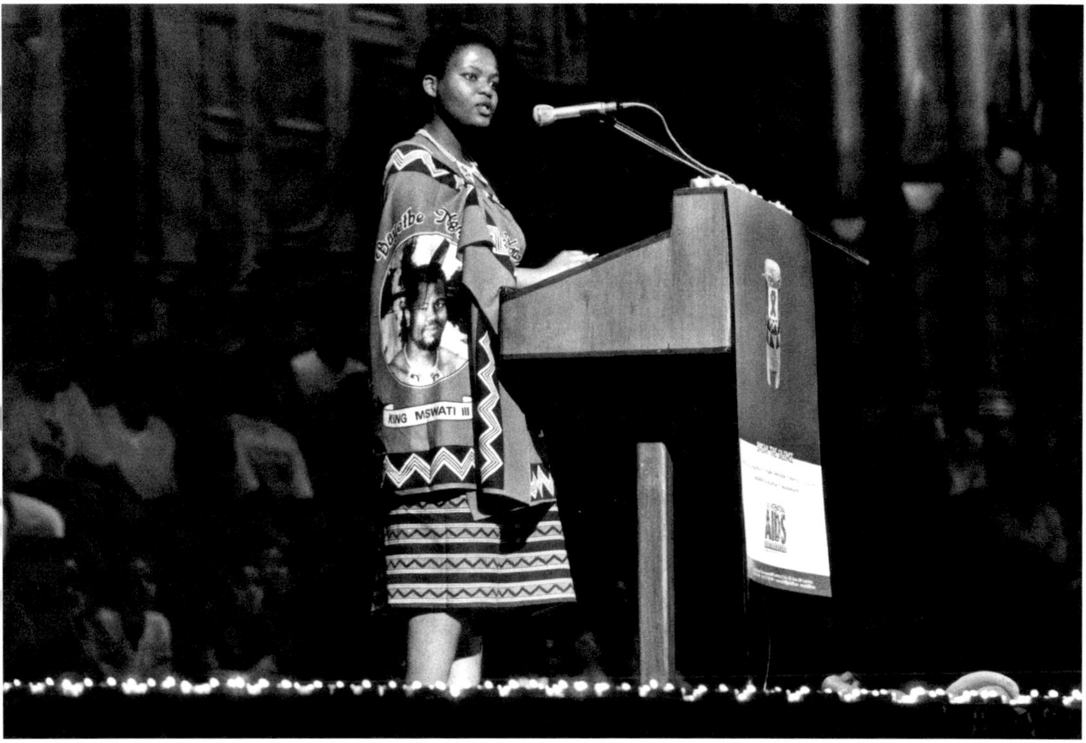
KING MSWATI III
AIDS

Reverendo Gideon Byamugisha. Kampala, Uganda

El reverendo Gideon es un sacerdote de la Iglesia de Uganda y encabeza el programa de VIH y sida en la diócesis de Namirembe. Fue el primer sacerdote en África que declaró ser portador del VIH.

Cuando le dije al obispo que era portador del VIH, en lugar de despedirme como yo esperaba se arrodilló y rezó por mí diciéndome que tenía una misión especial en la Iglesia.

Necesitamos integrar el VIH y el sida en el día a día de la Iglesia. Los líderes religiosos no sólo deberían condenar el sexo fuera de la ley, sino también el sexo no seguro. Esto es lo que deseo aconsejar. Algunos líderes religiosos sienten vergüenza de hablar sobre el sida, pero si queremos tener éxito debemos ser conscientes de que hay mucha actividad sexual en nuestra comunidad. Incluso el sexo ilegal debe ser seguro. Predico una cultura donde el sexo seguro sea fácil, aceptable y rutinario.

No siempre es fácil ser sincero. En una ocasión fui a Ruanda y no me permitieron dirigirme a un festival cristiano por ser portador del VIH. Se han burlado de mi hija en el colegio. La mayoría de las veces no me preocupa. Compro preservativos en las tiendas; la gente me ve y dice: «¡Aquí está el reverendo comprando preservativos!». Yo les digo: «Soy una persona portadora del VIH. Estoy casado y debo comprarlos».

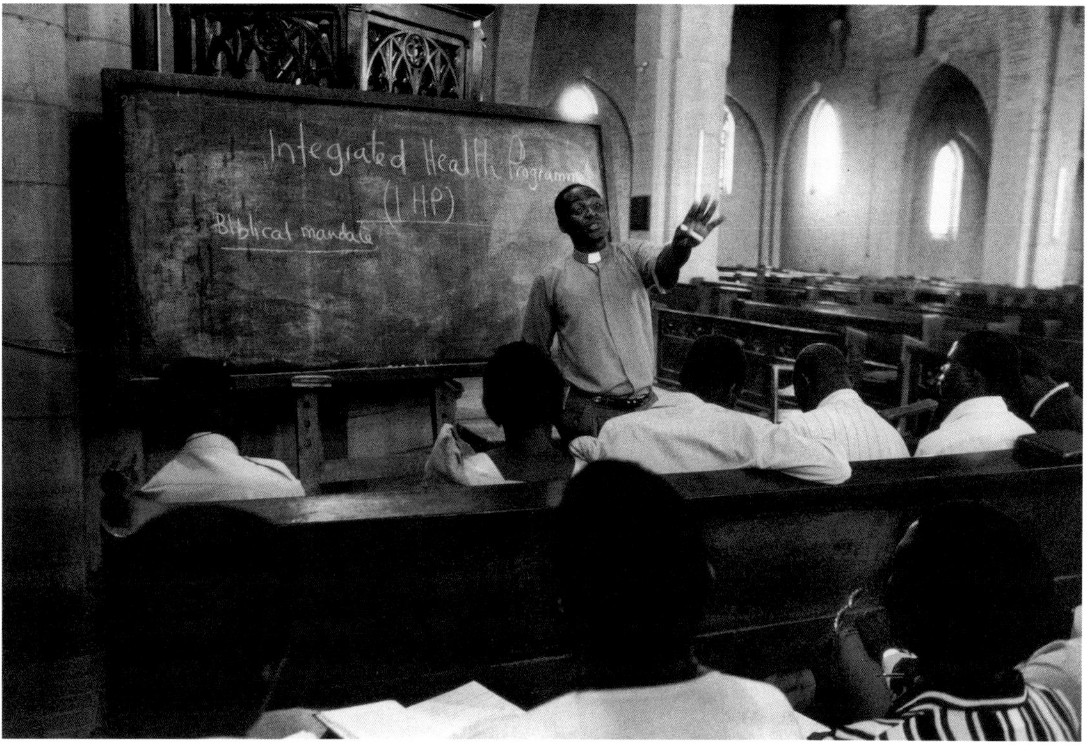
Integrated Health
(IHP)
Biblical mandate

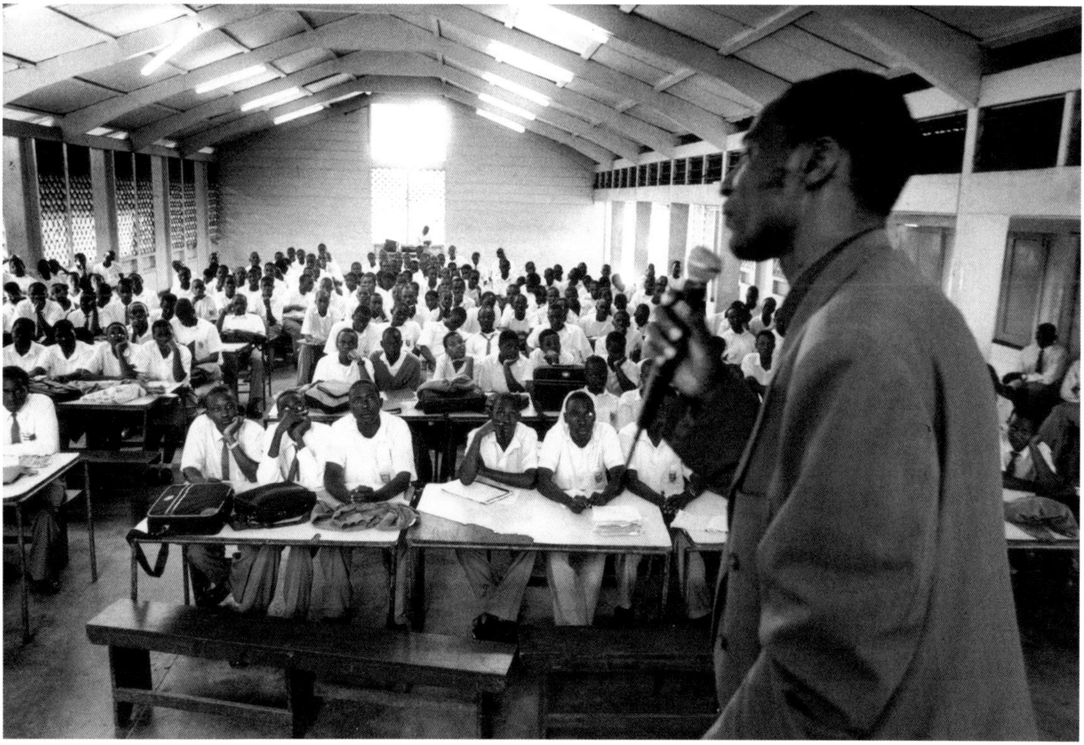

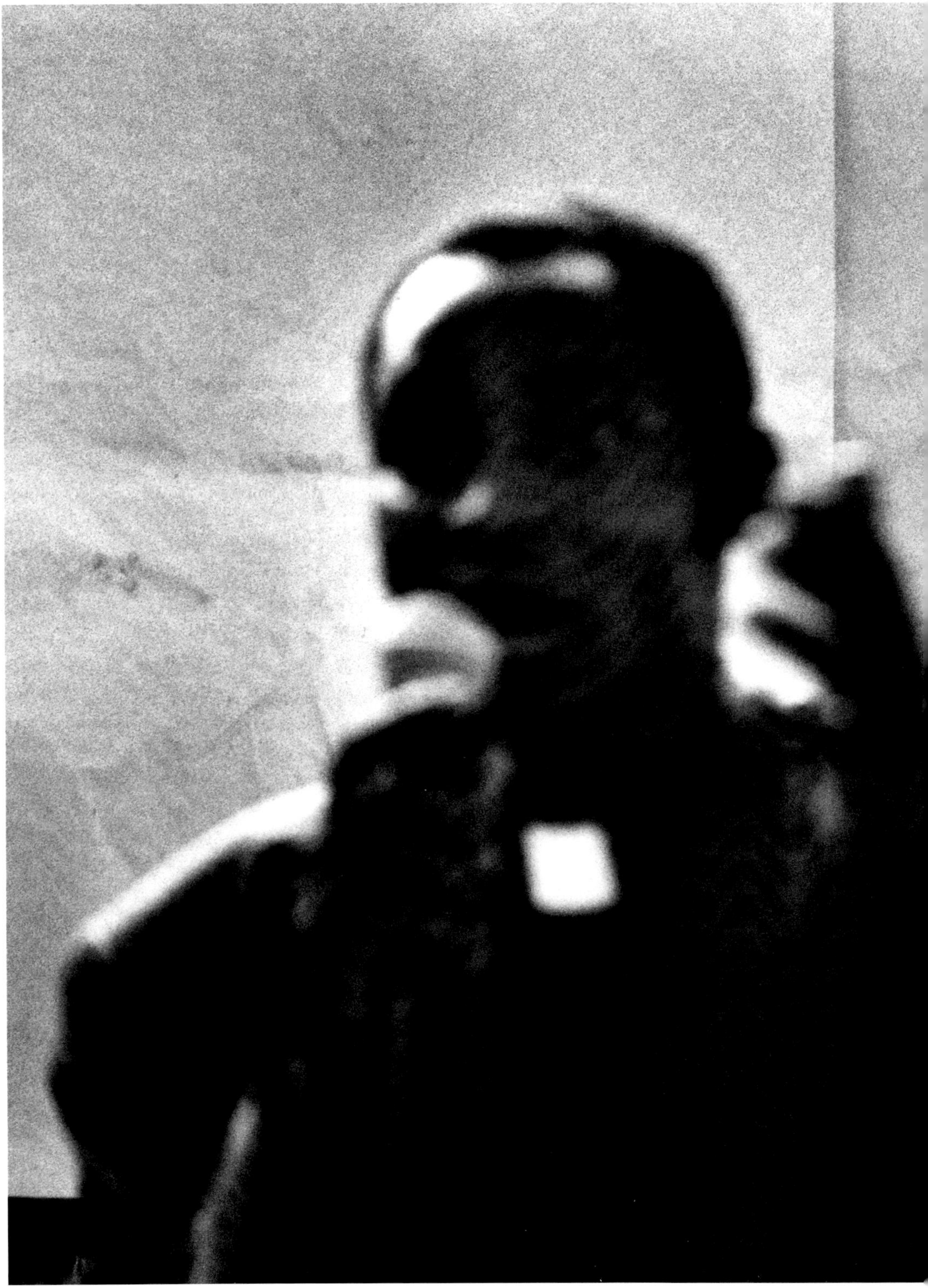

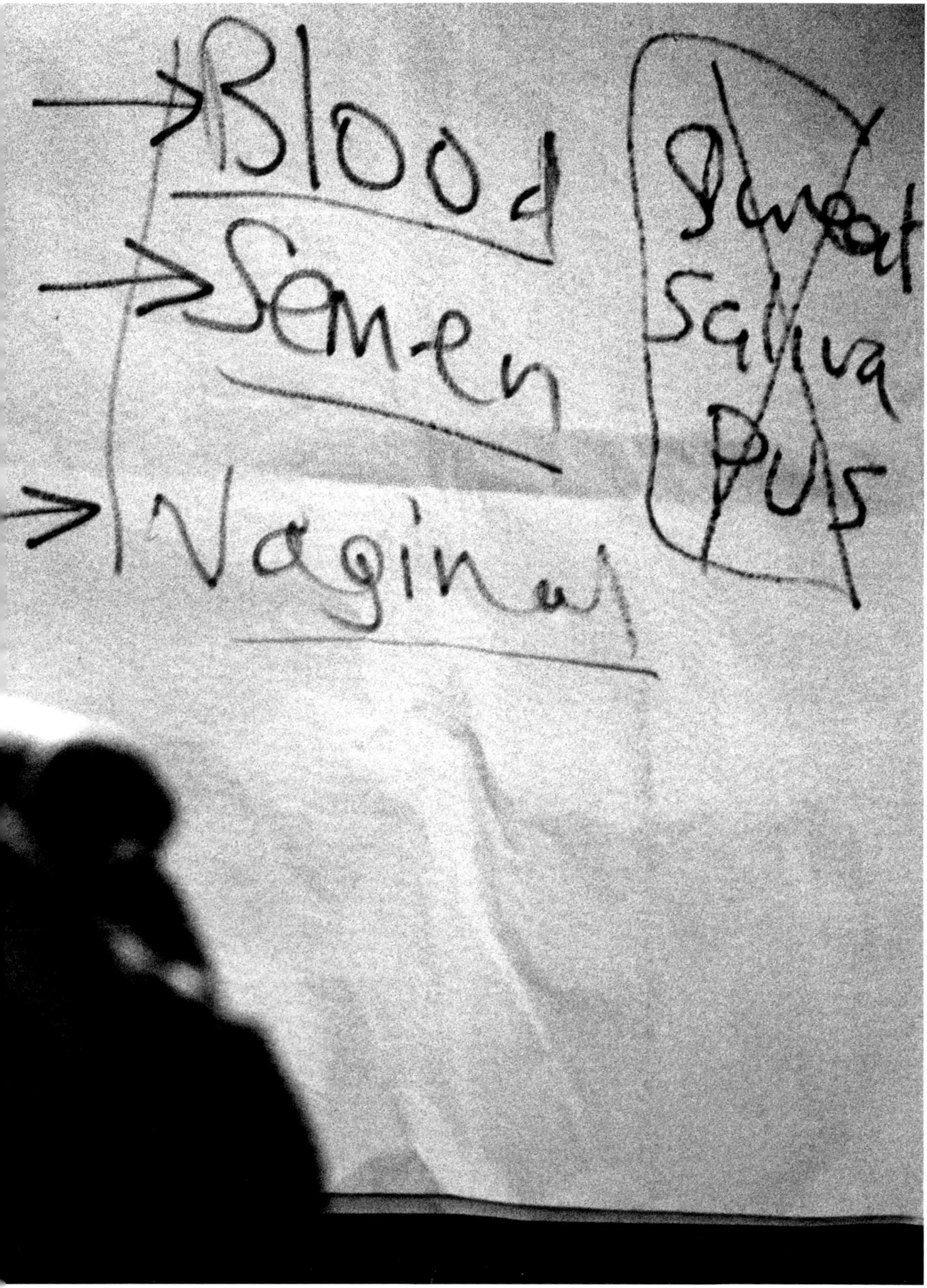
Blood
Semen
Vaginal
Sweat
Saliva
Pus

Festival juvenil contra el sida. Chinhoyi, Zimbabue

El festival juvenil de la provincia de Mashonaland Occidental, organizado por el grupo comunitario Batsarai (de la palabra shona, que significa ayuda), tenía como objetivo promover la concienciación y la prevención del VIH y el sida entre los jóvenes.

En Zimbabue la tasa de infección del VIH es del 25 %; diariamente se infectan por lo menos mil jóvenes más. Es muy importante implicar a los jóvenes de una manera seria: son nuestra esperanza para resolver el problema del VIH y el sida. Nos damos cuenta de que la mayoría de los jóvenes aún no están infectados, por lo que nuestra intención es informarles antes de que se contagien.

El festival de la juventud de Chinhoyi tuvo un gran éxito. Queríamos motivar a los grupos que ya están en marcha y atraer a más jóvenes para que formen parte del programa. El festival comenzó con un desfile encabezado por 35 *majorettes* que tocaban tambores y vestían camisetas espectaculares. Llevaban mensajes como «El sexo no garantiza el matrimonio», «Evita el sexo, las drogas y las malas amistades» y «Ríndete al sexo y te ignorarán». Queremos concienciar a los jóvenes para que se abstengan del sexo y pospongan su actividad sexual. Fue impresionante. Se representaron obras de teatro, poemas, música y bailes. Asistieron 3.000 jóvenes y formamos 15 nuevos grupos.

DANIEL GAPARE, DIRECTOR DEL GRUPO BATSARAI

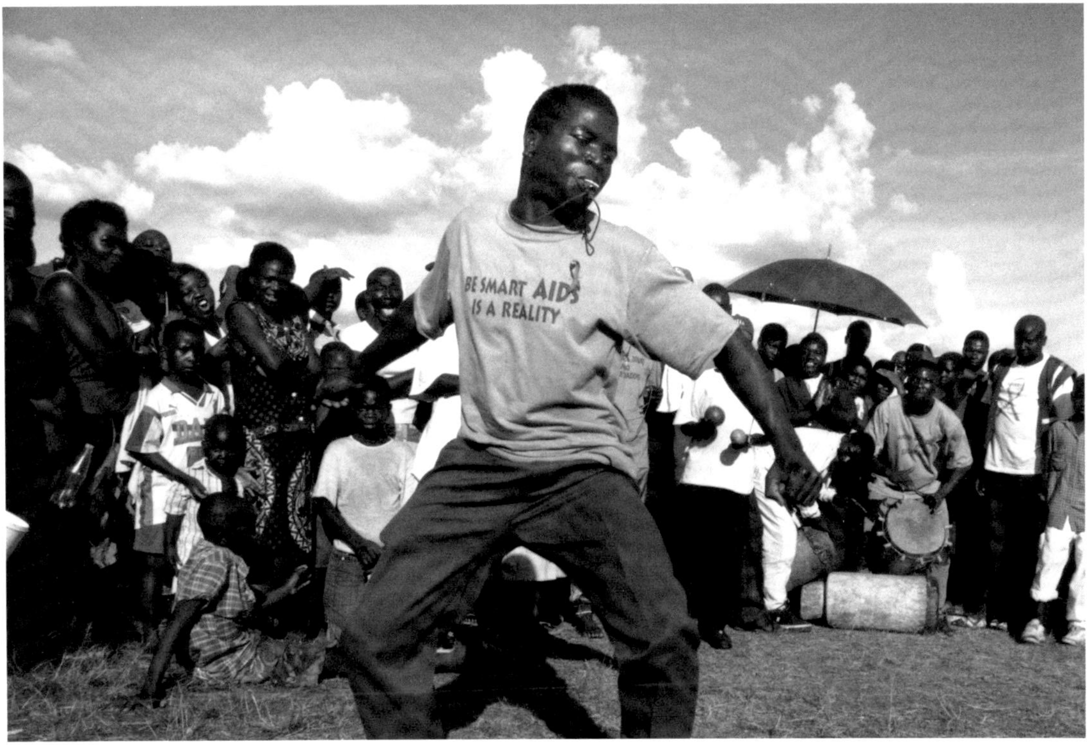
BE SMART AIDS
IS A REALITY

Campaña de Tratamiento en Acción. Durban, Sudáfrica

La marcha de la Campaña de Tratamiento en Acción durante la Conferencia Internacional sobre el Sida del año 2000.

La importancia histórica de esta marcha reside en que nos ayudó a cambiar la percepción internacional. En África la enfermedad se asocia con personas débiles y moribundas; la marcha demostró que muchos de nosotros estamos sanos y luchamos por continuar así. En Sudáfrica el trabajo ha comenzado a dar frutos. Los tratamientos antirretrovirales en el sector privado han disminuido de 3.500 a 900 rand al mes. Aún no es suficiente, pero muestra el poder de la acción colectiva.

Como activista he elegido no tomar antirretrovirales hasta que estén al alcance de todos. La gran mayoría de las personas infectadas con el VIH son pobres. Para mí es una cuestión de conciencia. No puedo vivir en un mundo en el que el gobierno, las empresas farmacéuticas y la comunidad internacional no hacen nada mientras ocurre este holocausto entre la gente pobre. Pero no es una actitud pesimista. Se basa en la creencia de que, si mostramos resolución en nuestra lucha, lograremos el acceso a los antirretrovirales para todos los enfermos y un sistema sanitario de calidad.

ZACHIE ACHMAT, PRESIDENTE DE LA CAMPAÑA DE TRATAMIENTO EN ACCIÓN

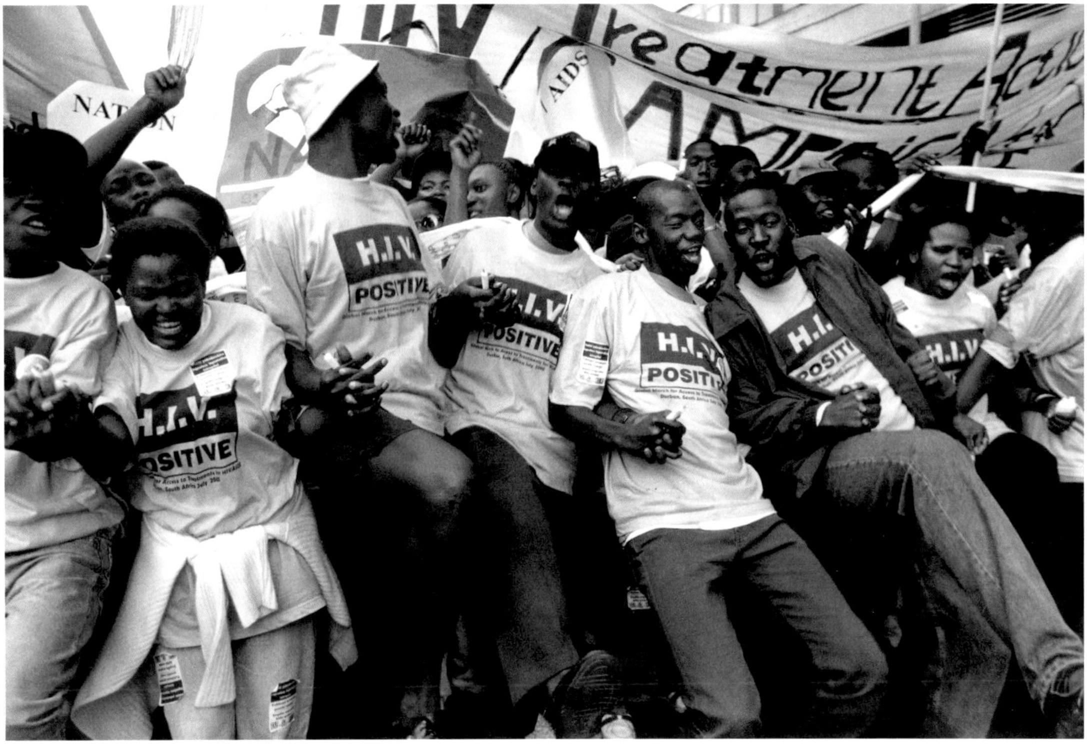
AIDS
Treatment
H.I.V.
POSITIVE

Conferencia «Romper el silencio». Durban, Sudáfrica

Los danzantes zulúes actuaron frente al Ayuntamiento de Durban durante la Conferencia Internacional sobre el Sida del año 2000.

Debe hacerse algo con la mayor urgencia. Y tras casi dos décadas de convivencia con la epidemia, tenemos experiencia sobre lo que realmente es efectivo.

Las experiencias de Uganda, Senegal y Tailandia han mostrado que las inversiones serias y las movilizaciones en torno a estas acciones constituyen la diferencia. Los tabúes y la discriminación pueden detenerse, pueden evitarse las nuevas infecciones, y la capacidad de las familias y las comunidades para cuidar a las personas con VIH y sida puede potenciarse.

El reto radica en pasar de la retórica a la acción, y a una acción con una intensidad y a una escala sin precedentes. Para ello debemos estar bien preparados, disponer de una buena estrategia y utilizar todos nuestros recursos y alianzas, así como perseverar en nuestros esfuerzos hasta ganar esta guerra.

NELSON MANDELA (EXTRACTO DE SU DISCURSO PARA LA CONFERENCIA)

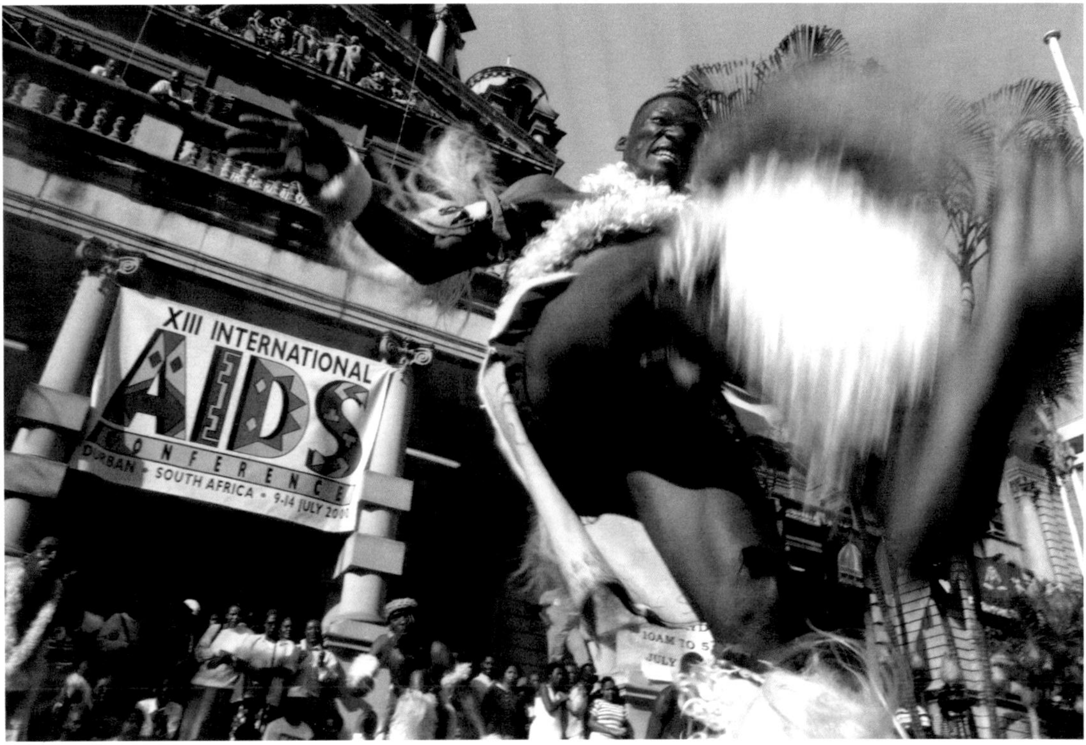
XIII INTERNATIONAL
AIDS
SOUTH AFRICA

Reverendo Gideon Byamugisha. Epílogo

Al hablar sobre el sida en África, la gente no debería considerar que el problema está «allá afuera». Es una falacia. Estamos en una aldea global y el sida se halla en todas partes. Y no sólo en África, pues no es únicamente un problema de África. Si hay guerra en Kosovo, no hay que creer que el sida no se extenderá porque Kosovo sea parte de Europa y no de África. Si la gente no sabe leer o escribir, si les falta autoestima y no es capaz de participar en la sociedad de la información, el sida se extenderá entre esa gente con idéntica facilidad ya se encuentre en Estados Unidos o en Uganda. A alguien que no disponga de suficiente comida, le será difícil protegerse contra el VIH y el sida, y si no tiene acceso a los preservativos cuando desea mantener relaciones sexuales, no estará protegido. Si la gente no gana suficiente dinero para poder tomar decisiones sobre su propia vida, o si una mujer está indefensa, tendrá un problema cuando deba decir «no» al sexo sin protección. El sida florece allí donde se dan estas condiciones.

El sida no es sólo una enfermedad. Es un síntoma de que algo más profundo no funciona bien en la familia global. Es la prueba de que se han roto las relaciones entre los individuos, las comunidades y las naciones. Muestra cómo nos tratamos y nos ayudamos, y en qué circunstancias nos quedamos callados. Muestra nuestras debilidades en la forma en que nos educamos unos a otros y en la manera en que nos vemos como comunidades, razas, naciones, clases, sexos y edades. El sida plantea la necesidad de sentarnos y abordar todos aquellos temas de los que hasta ahora nos hemos mantenido al margen, como la sexualidad, la pobreza o la forma en que afrontamos nuestras relaciones tanto a nivel familiar como social.

Este libro habla sobre el sida en África, pero la respuesta al sida debe comenzar a nivel individual, donde sea que vivamos. Antes de pensar en otras personas, preguntémonos: «¿Y yo? ¿Corro algún riesgo? ¿Cómo es mi vida sexual? ¿Cómo ha sido en el pasado y cómo es en este momento? ¿Podría ser seropositivo y no saberlo? ¿Sé todo lo que debo saber sobre la transmisión del VIH?». Las respuestas a estas preguntas nos darán una pauta sobre nuestra propia situación. Si no soy seropositivo, ¿cómo puedo protegerme para mantenerme así?

Y si lo soy, ¿cómo me protegeré contra la enfermedad y obtendré los servicios que necesito?

El siguiente cuestionario que debemos hacernos está relacionado con la familia. ¿Estoy poniendo en riesgo a mi familia de alguna manera? Si soy un padre de familia, ¿he informado lo suficiente a mis hijos sobre el VIH y el sida? ¿Saben cómo prevenirlo? ¿He reforzado su autoestima para que puedan decir «no» a un sexo sin protección y para que tengan la suficiente confianza para no involucrarse en situaciones que les hagan vulnerables a la enfermedad? Si no se abstienen del sexo, ¿les proporciono suficiente ayuda y apoyo?

En lo que respecta a la comunidad, hay muchas preguntas por las que los ciudadanos responsables deberían preocuparse. Si alguien en la comunidad es seropositivo, ¿dónde debe ir? ¿Una persona puede averiguar fácilmente si tiene o no el VIH? ¿Nuestra comunidad tiene un grupo de apoyo para las personas con el virus y un servicio de asistencia doméstica para las personas enfermas? ¿Existen mecanismos para despertar la conciencia de la comunidad sobre el VIH y el sida? ¿Animamos a los miembros de la comunidad a ser sinceros sobre su situación, y los líderes de la comunidad combaten los tabúes y los prejuicios? Quizá pueda hacerse mucho más. He visto anuncios que dicen: «Cuidado con el sida, el sida mata» al lado de la carretera, pero nunca los he visto en los hoteles en los que se practica el sexo. Hay una gran negación en la comunidad. ¿Cómo puedo ayudar a erradicarla?

Y llegamos al ámbito gubernamental, donde los problemas son los mismos pero a mayor escala. Hay 22 millones de personas en mi país, Uganda. ¿Cuántos están infectados y cuántos no lo están? ¿Qué puedo hacer por los que ya están infectados? ¿Qué hago para ayudar a aquellos que no están infectados para que continúen así? ¿Qué políticas se necesitan para asegurar que las infecciones no aumenten? ¿Qué puedo hacer para reducir los factores que alimentan la vulnerabilidad de mi pueblo, como el analfabetismo y la pobreza, los problemas culturales o las discriminaciones sexuales? ¿Qué papel puedo desempeñar para ayudar al gobierno a solucionar estos problemas?

Al empezar a pensar en estos problemas como miembros de la aldea global, nos planteamos la pregunta de por qué el sida se concentra en algunas naciones y sociedades, y no en otras. ¿Qué es lo que hace aumentar las infecciones de sida en las comunidades más afectadas? ¿Cuáles son las causas de las desigualdades?

Pronto descubrimos que lo que estas sociedades tienen en común es la pobreza. Están sujetas a políticas comerciales injustas y se enfrentan a enormes pagos por sus deudas. Dentro de los continentes y los países, el sida prospera en las comunidades marginadas y donde hay grupos de personas consideradas distintas por el resto de la sociedad. Les faltan información y servicios. Así pues, ¿ahora qué hago? ¿Puedo informar a aquellas personas que el mundo considera diferentes?

Arreglar nuestras relaciones rotas nos hará pensar no sólo como individuos o miembros de nacionalidades o tribus concretas, sino que nos hará reconocer que lo que hacemos en un país afecta a las personas de otros países. Lo que gasto aquí afecta a la manera de vivir de otra persona en el mundo. Lo que uno hace en el comercio y la industria, las leyes que rigen las patentes y los derechos, la forma en que el dinero se mueve entre naciones, todo esto afecta directamente a la vida de las personas.

Puesto que el sida se puede prevenir, y es una enfermedad manejable, el reto principal es de liderazgo. La historia querrá saber quiénes eran los líderes mundiales en el momento en que una enfermedad prevenible mató a millones de personas, debilitó a otros tantos millones y destruyó unas economías prometedoras en África, Asia y Latinoamérica. Hemos sufrido la indiferencia a nivel mundial, pero desafortunadamente también es cierto que este problema existe a nivel local. Los líderes de muchas comunidades o países infectados por esta epidemia aún no han explorado todas las posibilidades de la prevención del VIH, ni el cuidado y ayuda necesarios para los afectados de sida, que podrían ayudar a acabar con la epidemia. Conozco muchas comunidades cuyos dirigentes nunca han organizado una reunión, ni se han preocupado por saber cómo ayudar a los jóvenes a evitar la infección o cómo los individuos afectados pueden obtener un tratamiento básico.

En Uganda hemos tenido bastante suerte. El gobierno ha decidido correr el riesgo, y decir abiertamente: «Aquí tenemos un problema», mientras que otros gobiernos decían: «Sí, pero si lo declaramos nuestro turismo disminuirá» o «Quizá no sea tan malo como dicen». Uganda agarró el problema por los cuernos y está obteniendo resultados: ahora el nivel de infección ha bajado del 15 % al 8 %. El presidente Museveni tuvo la información en las primeras etapas, porque cuando salimos de la guerra y empezó a entrenar a su ejército, la mayoría de sus hombres estaban infectados. Reconoció que el sida sería un problema serio. Todos los miembros de la sociedad civil de Uganda —las instituciones gubernamentales, religiosas, comunitarias y sanitarias— trabajan codo a codo. Al romper el silencio, también se ha roto la cadena del contagio.

Muchas personas en Uganda han contribuido a la lucha contra el sida, y esto ha supuesto la diferencia. Se anima a la gente a que comparta su situación y la comunidad está dispuesta a aceptar y cuidar a las personas enfermas, a promocionar la abstinencia, a aceptar el empleo de preservativos. Incluso si las personas sólo disponen de un mínimo para sobrevivir, utilizan lo que tienen para ayudar a su hermano o hermana. No será suficiente, pero está allí. Nadie permanece pasivo. Si alguien no está de acuerdo con alguna política, no la combaten, pero ayudan a promover otras. Y ahora en Senegal hemos visto que el sida ha sido atacado antes de convertirse en un problema serio, de manera que ha sido combatido desde sus inicios.

Así que sabemos que, si hay dedicación, sinceridad y un deseo de innovar y pensar, esta enfermedad puede detenerse. El sufrimiento no es inevitable. Pueden evitarse las nuevas infecciones por VIH y la gente que sufre VIH o sida puede recibir cuidados y tratamientos. Cuando miramos a nuestro alrededor, vemos un gran deseo, a nivel mundial, de intentar rectificar lo que hemos hecho mal. La infinidad de conferencias y reuniones que se llevan a cabo apuntan a la idea de trabajar juntos y de aceptar nuestras responsabilidades como miembros de una aldea global. La historia nos pide más, pero ya hemos comenzado.

Hay una tendencia a considerar el sida como un problema demasiado grande y sobre el que no podemos hacer nada. Pero no somos débiles y la situación no es desesperanzadora: 36 millones de personas en el mundo son seropositivas, pero hay un total de seis mil millones en el mundo. Por supuesto que es posible que el sida empeore, pero está en nuestras manos el que no lo haga: tenemos la información necesaria; tenemos los medios y las capacidades para hacer algo; tenemos el dinero. Nuestro trabajo consiste en transformar nuestras relaciones rotas para que se recuperen y se refuercen.

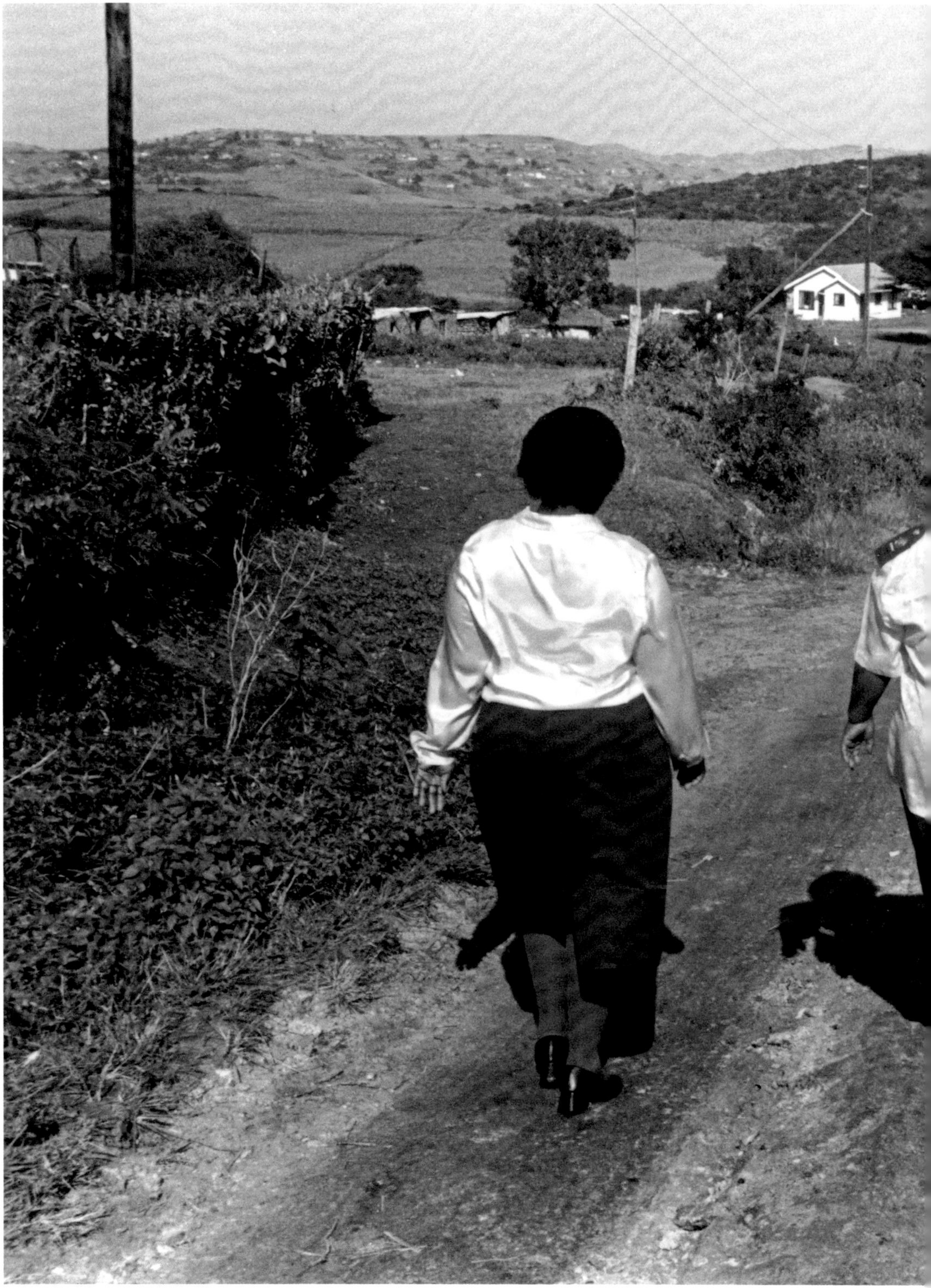

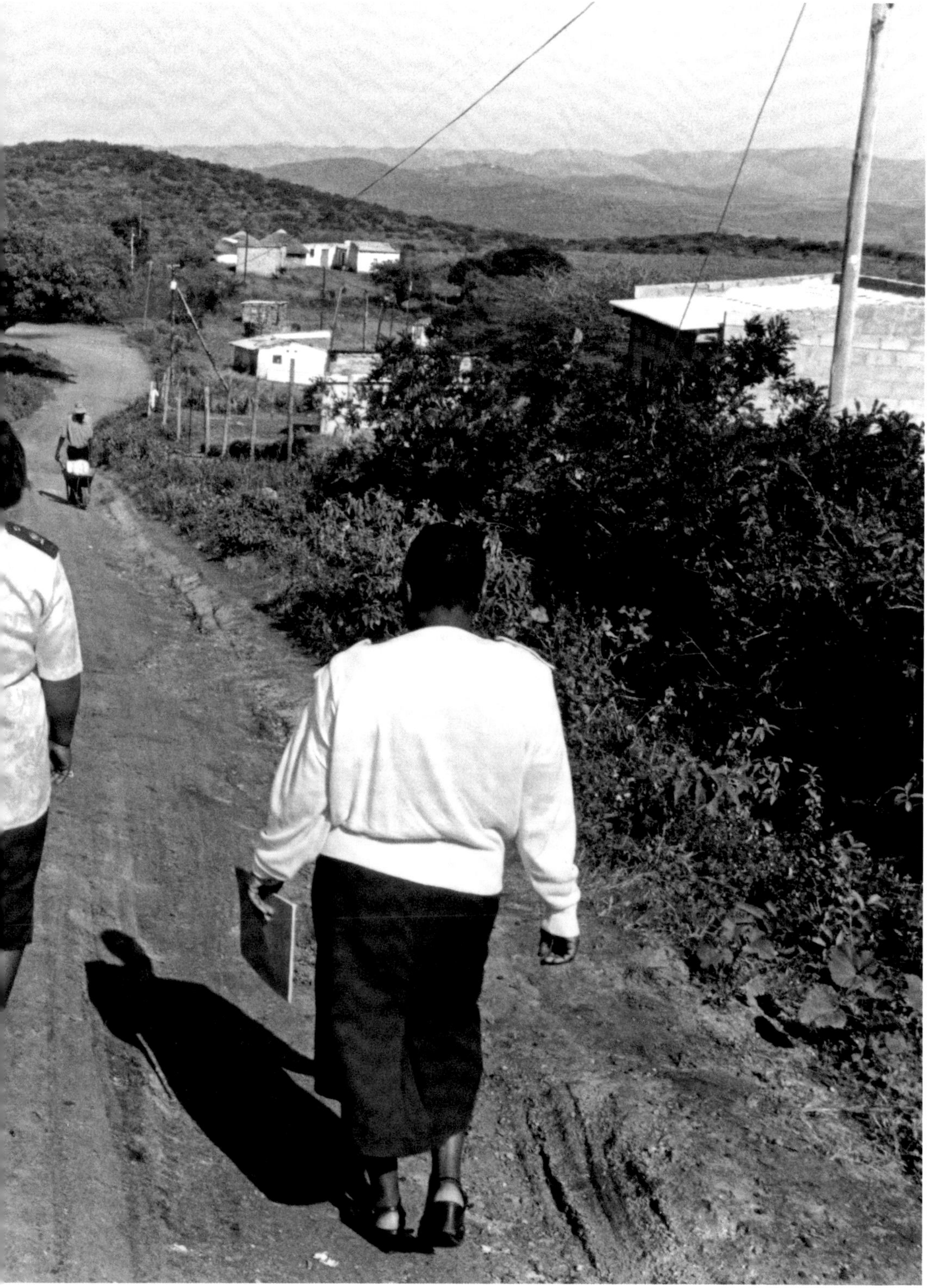

Fuentes

www.fase.es Apoyo a personas seropositivas, asesoramiento jurídico, escuelas de sida, salud y convivencia, prevención y línea gratuita de información.
www.aidsinfonyc.org La red de información del sida. Reseñas de tratamientos, glosarios de términos y otras publicaciones.
www.fipse.com Fundación para la investigación y la prevención del sida en España.
www.gesidaseimc.com Gesida: grupo de trabajo integrado por profesionales sanitarios para promover, fomentar y difundir el estudio e investigación de la infección por el VIH y sus complicaciones.
www.apoyopositivo.org Intento de satisfacer las necesidades críticas para el bienestar de los pacientes que sufren el sida.
www.omefa.org Organización educativa médico familiar acerca del sida y otras enfermedades.
www.adusalud.org.ar/vihsida.htm Asociación por los derechos de los usuarios de salud (Argentina).
www.clsida.org.co/ Corporación de lucha contra el sida (Colombia).
www.aids-sida.org Amigos contra el sida (México).
www.ssa.gob.mx/conasida Centro nacional para la prevención y control del VIH/sida (México).
www.internet.ve/accsi/ Acción ciudadana contra el sida (Venezuela).
www.unaids.org UNAIDS es el abogado de Naciones Unidas para una acción mundial contra el VIH y el sida.
www.undp.org El primer «Informe sobre el desarrollo humano» fue lanzado por el Programa de Desarrollo de Naciones Unidas en 1990.
www.worldbank.org/wdr El «Informe sobre el desarrollo mundial» emitido anualmente por el Banco Mundial es una guía sobre la situación económica, social y medioambiental del mundo hoy.

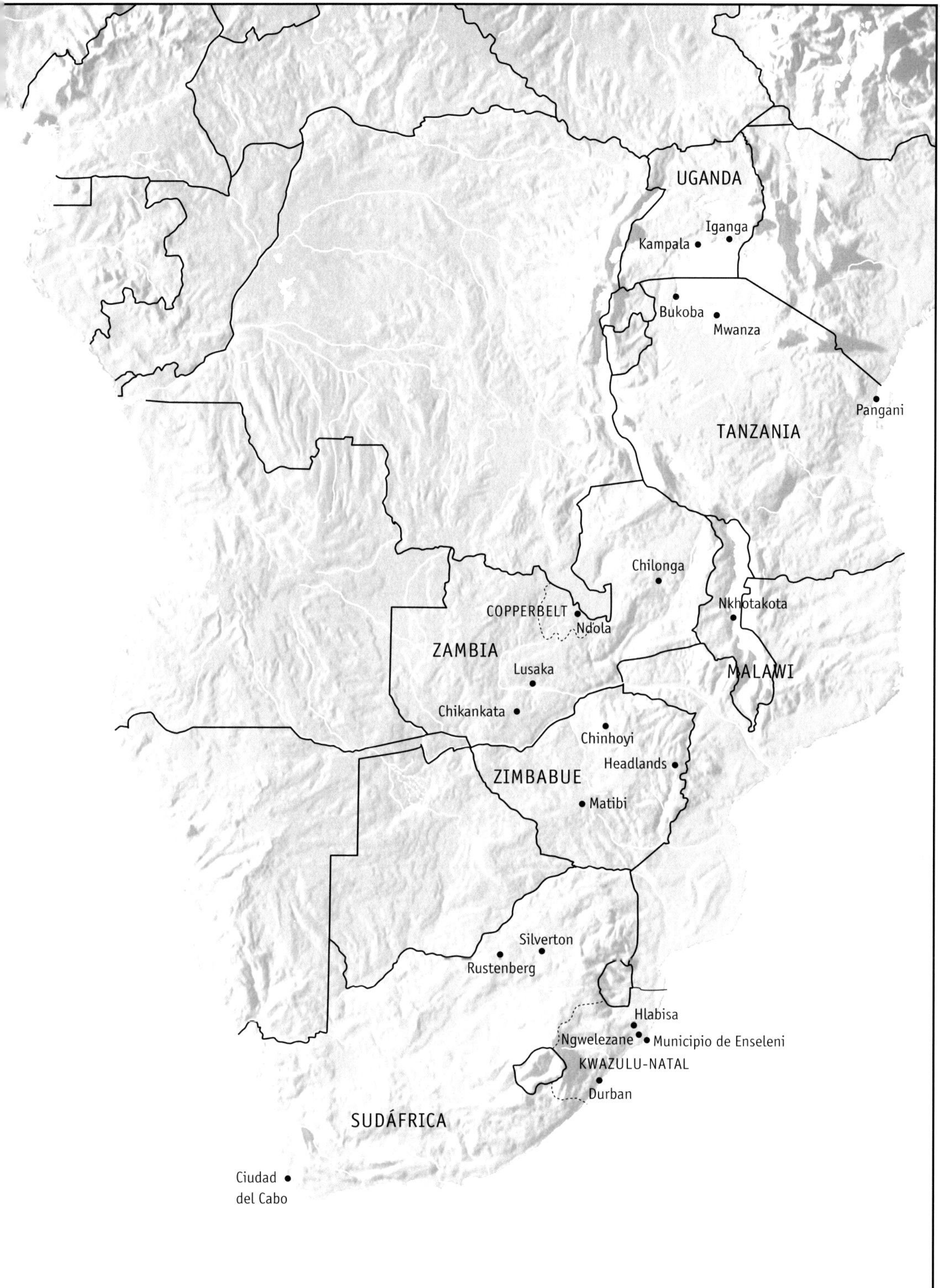
UGANDA
Iganga
Kampala
Bukoba
Mwanza
Pangani
TANZANIA
Chilonga
Nkhotakota
COPPERBELT
Ndola
ZAMBIA
Lusaka
MALAWI
Chikankata
Chinhoyi
Headlands
ZIMBABUE
Matibi
Silverton
Rustenberg
Hlabisa
Ngwelezane
Municipio de Enseleni
KWAZULU-NATAL
Durban
SUDÁFRICA
Ciudad
del Cabo

Mi trabajo sobre el VIH y el sida en África no habría sido posible sin la ayuda y la creatividad de Positive Lives, sus voluntarios y socios, el Terence Higgins Trust, Network y la Fundación Levi Strauss. Estoy en deuda con el W. Eugene Smith Memorial Fund, ActionAid (Ayuda en Acción) y el personal y los fotógrafos de Network.

Lyndall Stein ha sido un colaborador, consejero y amigo muy valioso durante todos estos años. También quisiera agradecer a Chris Boot, Stu Smith, Carolyn Watts, Colin Jacobson, Neil Burgess, Mitchell Warren, Alice Wynne Wilson, Nancy Khuswago, Tim Manchester, Thoko Luthuli, Nick Partridge, doctor Peter Haselau, Joseph Cabon, Stephen Mayes, Mike Kemp, Jaqueline Batoringaya, hermana Gloria, Lesley Lawson, capitanes Dean Pallant y Lena Jwili, doctor Aschwanden, David Goldblatt, doctor Ros Coleman, doctor Mara Rossi, Amanda Stone, Kevin Ryan, doctor Nathan Nshakira, David Nyirongo, Laurence Watts, Paul Kwangere, Sarah Stewart, Adam Hooper, Alick Mukuba, Mike Spry, Daniel Kalinaki, Joe Hanley, Glen Brent, Kevin Toolis, Rod Morris, Nicole Nogrette, Gary Wilson, doctor Ade Fukoya, Mia Diehl, Michelle McNally, Mark Porter, Peter Bussey, Vivien Hamley, Busi Chamane, Christian Aid, CAFOD, PSI, *Fortune Magazine*, *Guardian Weekend*, *L'Express* y Kodak Professional. Mi compañera Sarah Macauley ha sido mi fuente de inspiración y crítico durante todo el proceso.

Detrás de cada fotografía hay individuos, familias, cuidadores, organizaciones e instituciones, demasiado numerosas para nombrarlas a todas, cuya bondad y ayuda práctica no se olvidarán. Finalmente debo rendir homenaje a las personas que he fotografiado, la mayoría de las cuales conviven con el VIH o el sida, que me dieron la bienvenida a sus vidas con tanta generosidad y que han sido suficientemente valientes para dejarse fotografiar, creyendo que estas imágenes lograrán el cambio.

Gideon Mendel, julio de 2001

BLUME

Título original:
A Broken Landscape. HIV & AIDS in Africa

Producción y edición:
Chris Boot

Diseño:
SMITH

Traducción:
Maite Rodríguez Fischer

Coordinación de la edición en lengua española:
Cristina Rodríguez Fischer

Primera edición en lengua española 2002

Av. Mare de Déu de Lorda, 20
08034 Barcelona
Tel. 93 205 40 00 Fax 93 205 14 41
E-mail: info@blume.net

I.S.B.N.: 84-95939-00-2

Impreso en Italia

CONSULTE EL CATÁLOGO DE PUBLICACIONES *ON LINE*
INTERNET: HTTP://WWW.BLUME.NET